Antimicrobial Natural Products

Antimicrobial Natural Products

Editors

Fuhang Song
Yunjiang Feng

MDPI

Basel • Beijing • Wuhan • Barcelona • Belgrade • Novi Sad • Cluj • Manchester

Editors
Fuhang Song
School of Light Industry
Beijing Technology and
Business University
Beijing
China

Yunjiang Feng
Griffith Institute for Drug
Discovery (GRIDD)
Griffith University
Nathan
Australia

Editorial Office
MDPI
St. Alban-Anlage 66
4052 Basel, Switzerland

This is a reprint of articles from the Special Issue published online in the open access journal *Antibiotics* (ISSN 2079-6382) (available at: www.mdpi.com/journal/antibiotics/special_issues/Antimicrobial_Products).

For citation purposes, cite each article independently as indicated on the article page online and as indicated below:

Lastname, A.A.; Lastname, B.B. Article Title. *Journal Name* **Year**, *Volume Number*, Page Range.

ISBN 978-3-0365-9107-0 (Hbk)
ISBN 978-3-0365-9106-3 (PDF)
doi.org/10.3390/books978-3-0365-9106-3

Contents

antibiotics

Editorial

Antimicrobial Natural Products

Fuhang Song

School of Light Industry, Beijing Technology and Business University, Beijing 100048, China; songfuhang@btbu.edu.cn

Infectious diseases, resulting from microbial pathogens, are one of the major causes of morbidity and mortality worldwide. Natural products derived from foods, microbes, and plants played an important role in combating infections before the discovery of antibiotics. The start of the antibiotics era was marked by the clinical introduction of penicillin, which was the first antibiotic, discovered in 1928 by Alexander Fleming in the green mold *Penicillium notatum*. During the golden age of natural-product drug discovery (from the 1950s to the 1960s), the Nobel Prize in Physiology or Medicine was awarded to Alexander Fleming, Ernst B. Chain, and Sir Howard Florey in 1945 for their discovery of penicillin, and to Selman A. Waksman in 1952 for the discovery of streptomycin. The 2015 Nobel Prize in Physiology or Medicine was awarded to William C. Campbell and Satoshi Omura for their discovery of avermectins, a natural microbial product, and to Youyou Tu for her discovery of the natural plant product artemisinin; this heralded a new golden age of natural-product drug discovery [1].

This Special Issue features papers by experts working in natural product research, focusing on the control of microbial pathogens using natural products.

Propolis is a traditional, potentially medicinal product with several health benefits. Marta Peixoto and co-workers investigated the antimicrobial potential of ethanol extracts of propolis collected in Pereiro over a 5-year period (2011–2015) and two distinct apiaries/regions—Pereiro and Gerês—in selected years [2]. The results showed that a mixture of propolis and ethanol extracts had particularly interesting effects on *Bacillus subtilis*, *Propionibacterium acnes*, and *Staphylococcus aureus*. For *Saccharomyces cerevisiae* fungus, the mixtures mostly displayed MIC values similar to those of the most active single extract, except for mP (P11.EE + P13.EE), which was more active than the single extracts. This paper was a first attempt to evaluate the chemical profiles and bioactivity of mixed propolis samples from different years and regions. The findings of the study suggest great medicinal potential for propolis and are an important contribution to its valorization and standardization.

Bryophytes are important resources with the potential to produce unique natural compounds with antimicrobial properties. Valeeva and co-workers evaluated and characterized the antibacterial activity of intracellular and extracellular metabolites produced by the model mosses *Physcomitrium patens* and *Sphagnum fallax* [3]. Both polar (methanol-based) and non-polar (hexane-based) extracts of the two mosses inhibited the growth of Gram-negative *P. syringae* DC3000 bacteria. However, the bioactive metabolites were not stable during the extraction process; moreover, bioactivity completely ceased after lyophilization, while the secreted exudate fractions remained stable throughout the experiments. Exudates from both *P. patens* strains displayed selective high antimicrobial activity against Gram-positive *S. aureus*, *Enterococcus faecium*, and *Streptococcus pyogenes*. No antibacterial activity was observed with any exudates against Gram-negative bacteria *Salmonella*, *S. marcescens*, or *Escherichia coli*.

Natural products from fungi play a prominent role in the development of new drugs. Han and co-workers reported that the full molecular network of crude extracts of *Aspergillus westerdijkiae* L1295 showed several independent families of molecules; moreover, detailed

check for updates

Citation: Song, F. Antimicrobial Natural Products. *Antibiotics* **2022**, *11*, 1765. https://doi.org/10.3390/antibiotics11121765

Received: 21 November 2022
Accepted: 28 November 2022
Published: 7 December 2022

analysis of the molecular network revealed a cluster with 19 nodes representing a peptide family, showing MS/MS patterns containing the dipeptide [Ala-Phe] fragment. Guided by MS/MS and molecular networking, two new cyclic tetrapeptides (CTPs) (westertides A and B) with eight known compounds (ochratoxin A, ochratoxin A methyl ester, circumdatin F, circumdatin G, stachyline B, westerdijkin A, mellein, and 3-hydroxymellein) were identified from the fungus *A. westerdijkiae*, guided by OSMAC (one strain–many compounds) strategies [4]. Westertides A and B showed strong synergistic antifungal activity against *Candida albicans* with rapamycin. Furthermore, westertide A showed weak (histone deacetylase, HDAC) inhibitory activity.

Halogen substituents significantly impact the bioactivity and reactivity of organic compounds, so halogenated compounds play a profound role in the pharmaceutical industry. In the investigations by Luo and co-worders, *GedL*, a free-standing phenol flavin-dependent halogenase (FDH) from *A. terreus* NIH2624, is involved in the biosynthesis of geodin, and halogenates the substrate in late-stage biosynthesis. A flavin-dependent halogenase gene, *ptaK*, was identified from a cryptic BGC of endolichenic *Pestalotiopsis rhododendri* LF-19-12 via genome mining. A group of potential halogenated compounds with characteristic isotope patterns of two chloride atoms were detected in the crude extract of *P. rhododendri* LF-19-12 cultured in M2 medium using LC-MS and OSMAC strategies [5]. Then, two pairs of atropisomers (pestalachlorides A1a/A1b and A2a/A2b), together with two known compounds (pestalachloride A and SB87-H), were identified from *P. rhododendri* LF-19-12. Pestalachlorides exhibited antibacterial activity against drug-sensitive and drug-resistant *S. aureus* and *E. faecium*, with MIC values ranging from 4 mg/mL to 32 mg/mL.

Fungi from marine-derived environments are promising resources for the discovery of new chemical entries. Song and co-workers investigated the chemical constituents of the marine-derived *Talaromyces* sp. Fungus, isolated from a mud sample collected from the intertidal zones of the Yellow Sea in Qingdao, China. Three new compounds, including two new polyketide-derived oligophenalenone dimers (bacillisporins K and L) and one xanthoradone dimer (rugulosin D), along with four known compounds (bacillisporin B, macrosporusone D, rugulosin A, and penicillide), characterized this marine-derived fungus [6]. Bacillisporins K and L, bacillisporin B, macrosporusone D, rugulosin A, and penicillide exhibited antibacterial activity against *S. aureus*, with MIC values of 12.5, 25, 12.5, 6.25, 0.195, and 100 μg/mL, respectively.

Microorganism interactions offer each of the strains specific metabolic potential. Wang and co-workers studied the effect of co-culture on the secondary metabolites and bioactivity of two marine strains, *A. terreus* C23-3 and *A. unguis* DLEP2008001 [7]. Both of the strains grew well and produced metabolites when inoculated simultaneously, and *A. terreus* seemed to be more strongly induced by live *A. unguis*. Under some conditions, the extracts of co-culture showed higher antimicrobial activity than the axenic cultures. Different yields in the co-cultures vs. the corresponding axenic culture of fifteen MS-detectable and/or UV-active peaks were detected via LC-PDA-MS/MS analysis. Both strains produced chemical 'weapons' for antagonism. This study revealed the different responses of two *Aspergillus* strains in co-culture, which highlights new opportunities for antibiotic discovery.

Kartsev and co-workers synthesized a series of heteroaryl(aryl) thiazole derivatives based on a molecular hybridization approach [8]. Three of the synthesized compounds displayed antibacterial activity against pathogenic strains, including methicillin-resistant *S. aureus*, *P. aeruginosa*, and *E. coli*, with higher potential than the reference drug ampicillin. Some of the compounds exhibited better antifungal activity; the best compound was 4-butyl-1-hydroxy-*N*-(6-methylbenzo[d]thiazol-2-yl)-3-oxo-3,4-dihydronaphthalene-2-carboxamide with an MIC of 0.06–0.23 mg/mL and an MFC of 0.11–0.47 mg/mL. Docking studies revealed that the putative mechanisms are inhibition of the *E. coli* MurB enzyme and the inhibition of fungal 14a-lanosterol demethylase.

Drug delivery offers improvements to the therapeutic effects and systemic side effects of administered drugs. Khattak and co-workers studied the synergistic antibacterial activity of a chitosan-coated bacitracin cream under different in vitro characteristics such as rheology, pH, viscosity, and drug content. They revealed that the zones of inhibition in simple bacitracin-loaded cream were significantly smaller than those in chitosan-decorated bacitracin-loaded cream, indicating that chitosan synergistically improves the antimicrobial activity of bacitracin [9]. This contribution provided an effective method for the topical management of skin infections and wound healing.

Tran and co-workers focused on the antimicrobial compounds produced by *Bacillus strains*, their proposed mechanisms of action, and any research gaps in the mechanisms of these compounds. Omics approaches were also reviewed to clarify the mechanisms behind Bacillus probiotics [10].

Li and co-workers focused on the biocontrol of *C. albicans* by antagonistic microorganisms and bioactive compounds. In this review, the authors reported the bacteriostatic behavior of different antagonistic microorganisms (bacteria and fungi) against *C. albicans*. Moreover, they reviewed the natural products produced by microorganisms with unique structures and antifungal activity and their possible inhibitory mechanisms [11].

Maja Urošević and co-workers reviewed recent research on the biological and pharmaceutical aspects of curcumin, methods of sample preparation for its isolation, analytical methods for its identification and quantification in different matrices, and different techniques for developing formulations [12].

The emergence of multi-drug-resistant microbes is one of the most critical medical problems, and has prompted valuable contributions to new antibacterial drug development. This Special Issue presents multidisciplinary research focused on natural products with the aim of strengthening antimicrobial studies. The contributions collected herein provide valuable knowledge for researchers working in the field of natural-product chemistry.

Funding: This work was supported by grants from the National Natural Science Foundation of China (81973204), the Key Lab of Marine Bioactive Substance and Modern Analytical Technique, SOA (MBSMAT-2019-06), and Research Foundation for Advanced Talents of Beijing Technology and Business University (19008021176).

Conflicts of Interest: The author declares no conflict of interest.

References

1. Shen, B. A New Golden Age of Natural Products Drug Discovery. *Cell* **2015**, *163*, 1297–1300. [CrossRef] [PubMed]
2. Peixoto, M.; Freitas, A.S.; Cunha, A.; Oliveira, R.; Almeida-Aguiar, C. Mixing Propolis from Different Apiaries and Harvesting Years: Towards Propolis Standardization? *Antibiotics* **2022**, *11*, 1181. [CrossRef] [PubMed]
3. Valeeva, L.R.; Dague, A.L.; Hall, M.H.; Tikhonova, A.E.; Sharipova, M.R.; Valentovic, M.A.; Bogomolnaya, L.M.; Shakirov, E.V. Antimicrobial Activities of Secondary Metabolites from Model Mosses. *Antibiotics* **2022**, *11*, 1004. [CrossRef] [PubMed]
4. Han, J.; Wang, H.; Zhang, R.; Dai, H.; Chen, B.; Wang, T.; Sun, J.; Wang, W.; Song, F.; Li, E.; et al. Cyclic Tetrapeptides with Synergistic Antifungal Activity from the Fungus *Aspergillus westerdijkiae* Using LC-MS/MS-Based Molecular Networking. *Antibiotics* **2022**, *11*, 166. [CrossRef] [PubMed]
5. Luo, M.; Wang, M.; Chang, S.; He, N.; Shan, G.; Xie, Y. Halogenase-Targeted Genome Mining Leads to the Discovery of (+/−) Pestalachlorides A1a, A2a, and Their Atropisomers. *Antibiotics* **2022**, *11*, 1304. [CrossRef] [PubMed]
6. Song, F.; Dong, Y.; Wei, S.; Zhang, X.; Zhang, K.; Xu, X. New Antibacterial Secondary Metabolites from a Marine-Derived *Talaromyces* sp. Strain BTBU20213036. *Antibiotics* **2022**, *11*, 222. [CrossRef] [PubMed]
7. Wang, Y.; Glukhov, E.; He, Y.; Liu, Y.; Zhou, L.; Ma, X.; Hu, X.; Hong, P.; Gerwick, W.H.; Zhang, Y. Secondary Metabolite Variation and Bioactivities of Two Marine *Aspergillus* Strains in Static Co-Culture Investigated by Molecular Network Analysis and Multiple Database Mining Based on LC-PDA-MS/MS. *Antibiotics* **2022**, *11*, 513. [CrossRef] [PubMed]
8. Kartsev, V.; Geronikaki, A.; Zubenko, A.; Petrou, A.; Ivanov, M.; Glamoclija, J.; Sokovic, M.; Divaeva, L.; Morkovnik, A.; Klimenko, A. Synthesis and Antimicrobial Activity of New Heteroaryl(aryl) Thiazole Derivatives Molecular Docking Studies. *Antibiotics* **2022**, *11*, 1337. [CrossRef] [PubMed]
9. Khattak, R.Z.; Nawaz, A.; Alnuwaiser, M.A.; Latif, M.S.; Rashid, S.A.; Khan, A.A.; Alamoudi, S.A. Formulation, In Vitro Characterization and Antibacterial Activity of Chitosan-Decorated Cream Containing Bacitracin for Topical Delivery. *Antibiotics* **2022**, *11*, 1151. [CrossRef] [PubMed]
10. Tran, C.; Cock, I.E.; Chen, X.; Feng, Y. Antimicrobial *Bacillus*: Metabolites and Their Mode of Action. *Antibiotics* **2022**, *11*, 88. [CrossRef] [PubMed]

11. Li, H.; Yang, J.; Zhang, X.; Xu, X.; Song, F.; Li, H. Biocontrol of *Candida albicans* by Antagonistic Microorganisms and Bioactive Compounds. *Antibiotics* **2022**, *11*, 1238. [CrossRef] [PubMed]
12. Urošević, M.; Nikolić, L.; Gajić, I.; Nikolić, V.; Dinic, A.; Miljković, V. Curcumin: Biological Activities and Modern Pharmaceutical Forms. *Antibiotics* **2022**, *11*, 135. [CrossRef] [PubMed]

antibiotics

MDPI

Article

Synthesis and Antimicrobial Activity of New Heteroaryl(aryl) Thiazole Derivatives Molecular Docking Studies

Victor Kartsev [1], Athina Geronikaki [2,*], Alexander Zubenko [3], Anthi Petrou [2], Marija Ivanov [4], Jasmina Glamočlija [4], Marina Sokovic [4], Lyudmila Divaeva [5], Anatolii Morkovnik [5] and Alexander Klimenko [3]

1 InterBioScreen, 119019 Moscow, Russia
2 School of Health, Faculty of Pharmacy, Aristotle University, 54124 Thessaloniki, Greece
3 North-Caucasian Zonal Research Veterinary Institute, 346406 Novocherkassk, Russia
4 Mycological Laboratory, Department of Plant Physiology, Institute for Biological Research "Siniša Stanković", National Institute of Republic of Serbia, University of Belgrade, 11060 Belgrade, Serbia
5 Institute of Physical and Organic Chemistry, Southern Federal University, 344090 Rostov-on-Don, Russia
* Correspondence: geronik@pharm.auth.gr

Abstract: Herein, we report the design, synthesis, and evaluation of the antimicrobial activity of new heteroaryl (aryl) thiazole derivatives. The design was based on a molecular hybridization approach. The in vitro evaluation revealed that these compounds demonstrated moderate antibacterial activity. The best activity was achieved for compound **3**, with MIC and MBC in the range of 0.23–0.7 and 0.47–0.94 mg/mL, respectively. Three compounds (**2**, **3**, and **4**) were tested against three resistant strains, namely methicillin resistant *Staphylococcus aureus*, *P. aeruginosa*, and *E. coli*, which showed higher potential than the reference drug ampicillin. Antifungal activity of the compounds was better with MIC and MFC in the range of 0.06–0.47 and 0.11–0.94 mg/mL, respectively. The best activity was observed for compound **9**, with MIC at 0.06–0.23 mg/mL and MFC at 0.11–0.47 mg/mL. According to docking studies, the predicted inhibition of the *E. coli* MurB enzyme is a putative mechanism of the antibacterial activity of the compounds, while inhibition of 14a-lanosterol demethylase is probably the mechanism of their antifungal activity.

Keywords: antimicrobial; antibacterial; antifungal; heteroaryl (aryl) thiazole derivatives; docking

Citation: Kartsev, V.; Geronikaki, A.; Zubenko, A.; Petrou, A.; Ivanov, M.; Glamočlija, J.; Sokovic, M.; Divaeva, L.; Morkovnik, A.; Klimenko, A. Synthesis and Antimicrobial Activity of New Heteroaryl(aryl) Thiazole Derivatives Molecular Docking Studies. *Antibiotics* **2022**, *11*, 1337. https://doi.org/10.3390/antibiotics11101337

Academic Editors: Fuhang Song and Yunjiang Feng

Received: 2 August 2022
Accepted: 27 September 2022
Published: 30 September 2022

1. Introduction

There is an increasing demand for the development of new antibacterial agents, due to global emerging resistance to conventional antibiotics. During the last several decades, a plethora of different thiazolidine based compounds have been studied to evaluate their pharmacological potential [1,2]. The synthesis of thiazole derivatives has attracted widespread attention due to their diverse biological activities, including antimicrobial [3–9], anti-inflammatory [10–12], analgesic [13,14], antitumor [15–17], antidiabetic [18], anti-HIV [19,20], COX/LOX inhibitory [21,22], antioxidant [23,24], antileishmanial [25,26], and many others [27–30]. There are many drugs with this scaffold such as antitumor (dasatinib, tiazofurin,); antiviral (brecanavir, ritonavir); anti-infectious (nitazoxanide) [31]; antibacterial agents, including sulfathiazole [32] and penicillins [33]; and antifungal agents, such as ravuconazole [34], myxothiazol [35], abafungin [36], and ethaboxam [37] (Figure 1).

Phtalazino derivatives are also mentioned as antimicrobial agents [38–40]. On the other hand, sulfonamides have attracted the interest of researchers due to their wide spectrum of biological activities, including dihydrofolate reductase (DHFR) inhibitors [41], antitumor [42,43], carbonic anhydrase inhibitors [44,45], anti-inflammatory [46], antiretroviral activity [47,48], antimicrobial [49,50], and others [51–53].

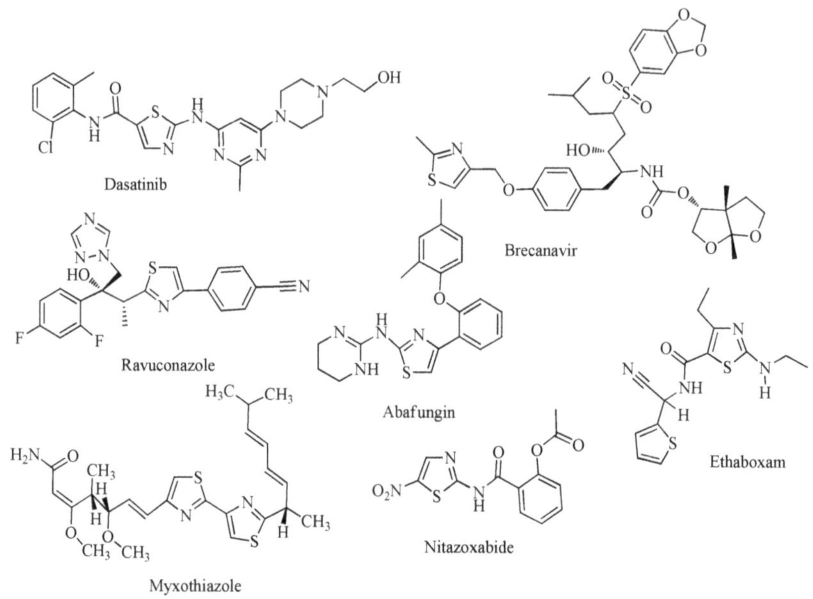

Figure 1. Thiazole-based approved drugs.

Sulfonamides are known as broad spectrum antimicrobial agents against *Gram-positive* and *Gram-negative* bacterial strains. These agents have low cost, low toxicity, and special activity against bacterial diseases. Sulfisoxazole, sulfamethizole, sulfamethoxazole, and sulfaphenazole are only some examples for the treatment of a diversity of bacterial infections.

Another example is the diuretic drug, chlorthalidone, which is used to treat hypertension or edema caused by heart failure, renal failure, hepatic cirrhosis, and estrogen therapy, as well as silver sulfadiazine, a topical sulfonamide antibiotic approved for the treatment of burns. Therefore, the design and development of hybrid molecules containing thiazolidinone phtalazine and thiazole cores, as well as sulfonamide groups, is a promising route in the search for novel antimicrobial agents. Molecular hybridization based on the amalgamation of two or more substitutions [54] is one of the new approaches in drug design. Hybridization is mainly aimed to improve the activity profile and to reduce undesired side effects [55].

Taking all of this information into account and based on our previous results [56–58], we designed and synthesized new derivatives incorporating thiazole, phtalazine moieties, and the sulfonamide group in the frame of one molecule.

2. Results and Discussion

2.1. Chemistry

In this work, we have described the synthesis of three structural series of new thiazole derivatives and presented the results of a study of their antibacterial and antifungal activity against a number of bacterial and fungal pathogens. All the target compounds were synthesized via four routes, as outlined in Scheme 1.

Starting acids for the synthesis of compounds **1**, **2**, **8**, and **9**; 1-(chloromethyl)-6,7-dimethoxy-3,4-dihydroisoquinoline for the synthesis of thiazole **3**; and 1-chlorophthalazine derivatives for the preparation of compounds **4–7** were provided by InterBioScreen Ltd (Moscow, Russia).

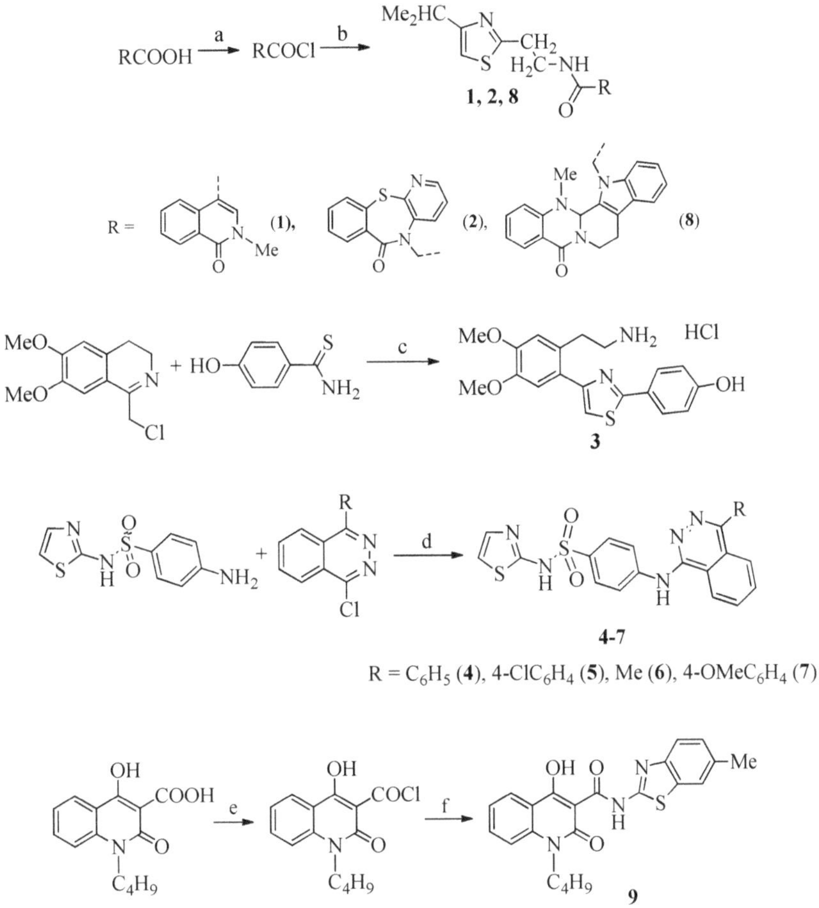

Scheme 1. Synthesis of target compounds. Reagents and conditions: (a) RCOOH, $SOCl_2$, $CHCl_3$, DMF, reflux; (b) 2-(4-isopropylthiazol-2-yl)ethan-1-amine, NEt_3, $CHCl_3$, 0–2 °C, saturated aqueous solution of $NaHCO_3$; (c) 1-(chloromethyl)-6,7-dimethoxy-3,4-dihydroisoquinoline, 4-hydroxybenzothioamide, i-PrOH, reflux, 2 h; (d) 1-chloro-4-R-phthalazine, 4-amino-N-(thiazol-2-yl)benzenesulfonamide, methyl cellosolve, reflux, 0.5 h, 95–100 °C, 1 h, 5% NH4OH; (e) 1-butyl-4-hydroxy-2-oxo-1,2-dihydroquinoline-3-carboxylic acid, $SOCl_2$, $CHCl_3$,DMF, reflux; (f) 6-methylbenzo[d]thiazol-2-amine, pyridine, DMF, 0 °C, 0.5 h, saturated aqueous solution of $NaHCO_3$.

Most of the compounds were synthesized according to the usual scheme, by acylation of the corresponding amines with acid chlorides. Compounds **1**, **2**, **8**, and **9** were obtained in these ways: acid chlorides 2-methyl-1-oxo-1,2-dihydroisoquinoline-4-carboxylic acid was used to synthesize compound **1**, 2-(6-oxobenzo[f]pyrido[2,3-b][1,4]thiazepin-5(6*H*)-yl)acetic acid for **2**, and 2-(14-methyl-5-oxo-7,8,13b,14-tetrahydroindolo[2′,3′:3,4]pyrido[2,1-b] quinazolin-13(5*H*)-yl)acetic acid for **8**, 1-butyl-4-hydroxy-2-oxo-1,2-dihydroquinoline-3-carboxylic acid for **9**; for these compounds, 2-(4-isopropylthiazol-2-yl)ethane-1-amine was used.

Compounds **4**–**7** were obtained by the reaction of nucleophilic substitution of the chlorine atom in 1-chloro-4-R-phthalazines. The best solvent for this reaction is methyl cellosolve, both in terms of yields and the purity of the reaction products. We especially note that the addition of ammonia for the conversion of product salts into bases should be carried out when the temperature of the reaction mixture is about 100 °C, since neutralization at ordinary temperature takes a very long time and does not guarantee the complete conversion of salts into bases.

Compound **3** was synthesized using our rather unusual recyclization reaction of 1-chloromethyl-3,4-dihydroisoquinolines under the action of thioamides and thioureas. In this article, the possibility of such recycling was confirmed by us, including using X-ray diffraction analysis and NMR spectroscopy (a one-proton singlet of the thiazole ring (H-5′) of all the compounds described in the article is observed in the region of 6.85–6.98 ppm).

In the present work, the spectra of the studied compounds are also characterized by the presence in their aromatic region of a one-proton singlet of the thiazole ring (H-5′) in the region of 6.01–7.04 ppm; in the case of compound **3**, this signal was detected at 7.44 ppm.

In the ^{1}H NMR spectra of compounds **1**, **2**, and **8** in deuterochloroform, the signals of the methyl groups of the isopropyl group are located in the upfield part of the spectra (1.21–1.28 ppm).

The amide proton NHCO in compounds **1** and **2** appears as a multiplet in the region 8.21–8.31, and in compound **8** it appears as a singlet at 8.08.

Compounds **4–7** contain two singlets: at 12.46–12.48 ppm of the sulfamide group $NHSO_2$ and at 9.42–9.49 ppm of the NH group.

2.2. *Biological Evaluation*

2.2.1. Antibacterial Activity

Synthesized compounds were tested for their antibacterial activity against a panel of six bacteria, using a microdilution method for the determination of minimal inhibitory and minimal bactericidal concentrations (MIC and MBC, respectively). The antibacterial activity of tested compounds was moderate to good, with MIC ranging from 0.17 to >3.75 mg/mL and MBC at 0.23–>3.75 mg/mL, as presented in Table 1. The order of activity can be presented as follows: **3** > **2** > **9** > **4** > **5** > **7** > **8** > **1** > **6**. The best activity was achieved for compound **3** with MIC and MBC at 0.23–0.70 mg/mL and 0.47–0.94 mg/mL, respectively. The most sensitive bacterium appeared to be *B. cereus*, whereas *E. coli* was the most resistant one.

Compound 4 exhibited the best activity among the compounds tested against *E. coli*, with MIC/MBC at 0.17/0.23 mg/mL, while compound **9** showed the same good activity against *B. cereus* and *S. Typhimurium*. Compounds **1** and **8** exhibited in vitro activity with MIC and MBC at 0.23/0.47 mg/mL against *E. cloacae*, compounds **3** and **5** against *E. coli*, while compound 3 also displayed good activity against *S. Typhimurium*. In general, these compounds showed moderate to low activity.

The study of structure–activity relationships revealed that the presence of 2-(3,4-dimethoxyphenyl)ethanamine as substituent at position 4 and phenol at position 2 of the thiazole ring (**3**) are beneficial for antibacterial activity. Among the group of compounds **1**, **2**, and **8**, the more favorable effect was observed in the case of 2-methylisoquinolin-1(2H)-one substituent connected via N-propylpropionamide with the thiazole ring (**2**). The presence of phenylphthalazine (**4**) as the substituent was positive in the case of compounds **4–7**. Introduction of 4-Cl substituent to phenylphthalazine (**5**) decreased a little activity, while the presence of the 4-OMe group decreased more activity. Finally, replacement of phenylphtalazine by 1-methylphtalazine (**6**) was detrimental for this group of compounds and in general for all tested compounds.

Table 1. Antibacterial activity of the title compounds (MIC/MBC in mg/mL).

№	Compounds		*S.a.*	*B.c.*	*L.m.*	*E.c.*	*S.T.*	*En.cl.*
1	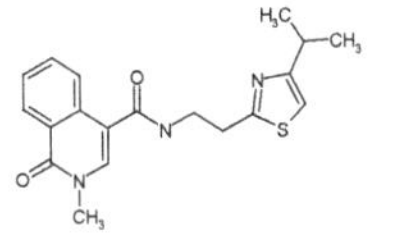	MIC	0.70 ± 0.19	0.35 ± 0.09	0.35 ± 0.09	>3.75	>3.75	0.23 ± 0.00
		MBC	0.94 ± 0.00	0.47 ± 0.00	0.47 ± 0.00	>3.75	>3.75	0.47 ± 0.00
2	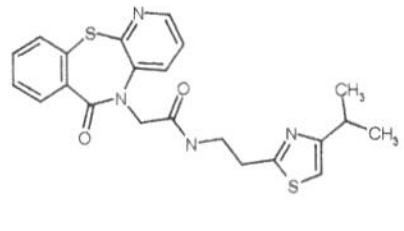	MIC	0.94 ± 0.00	0.47 ± 0.00	0.70 ± 0.19	0.35 ± 0.09	0.35 ± 0.09	0.35 ± 0.09
		MBC	1.88 ± 0.00	0.94 ± 0.00	0.94 ± 0.00	0.47 ± 0.00	0.47 ± 0.00	0.47 ± 0.00
3	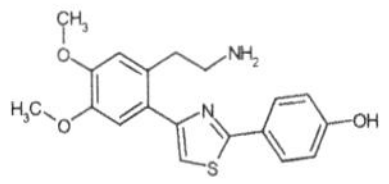	MIC	0.70 ± 0.19	0.23 ± 0.00	0.70 ± 0.19	0.23 ± 0.00	0.23 ± 0.00	0.70 ± 0.19
		MBC	0.94 ± 0.00	0.47 ± 0.00	0.94 ± 0.00	0.47 ± 0.00	0.47 ± 0.00	0.94 ± 0.00
4	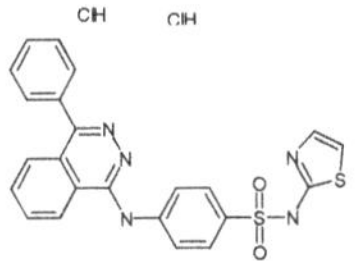	MIC	1.41 ± 0.38	0.70 ± 0.19	0.70 ± 0.19	0.17 ± 0.00	0.70 ± 0.19	0.70 ± 0.19
		MBC	1.88 ± 0.00	0.94 ± 0.00	0.94 ± 0.00	0.23 ± 0.00	0.94 ± 0.00	0.94 ± 0.00
5	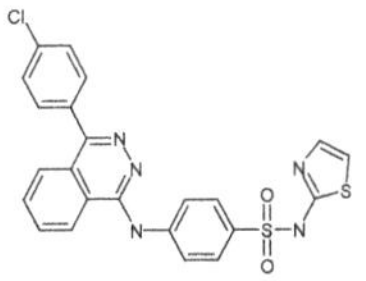	MIC	1.41 ± 0.38	0.70 ± 0.19	0.70 ± 0.19	0.23 ± 0.00	0.47 ± 0.00	0.70 ± 0.19
		MBC	1.88 ± 0.00	0.94 ± 0.00	0.94 ± 0.00	0.47 ± 0.00	0.94 ± 0.00	0.94 ± 0.00
6	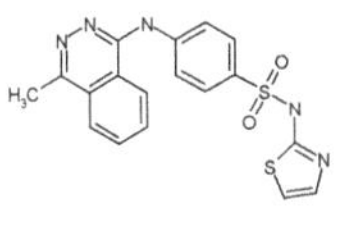	MIC	2.31 ± 0.76	1.41 ± 0.38	0.70 ± 0.19	>3.75	0.70 ± 0.19	0.70 ± 0.19
		MBC	3.75 ± 0.00	1.88 ± 0.00	0.94 ± 0.00	>3.75	0.94 ± 0.00	0.94 ± 0.00
7	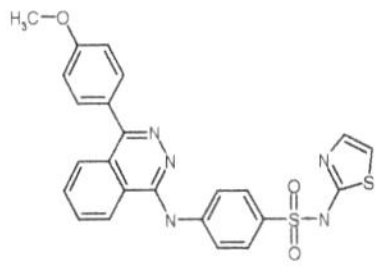	MIC	1.41 ± 0.00	0.70 ± 0.19	0.70 ± 0.19	0.35 ± 0.09	0.70 ± 0.19	0.70 ± 0.19
		MBC	1.88 ± 0.00	0.94 ± 0.00	0.94 ± 0.00	0.47 ± 0.00	0.94 ± 0.00	0.94 ± 0.00
8	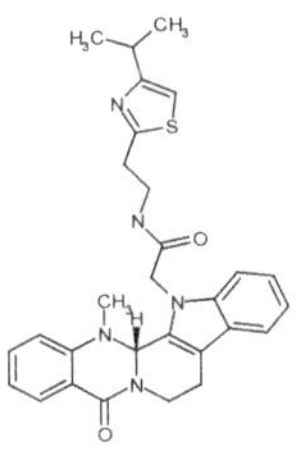	MIC	1.41 ± 0.38	0.47 ± 0.00	0.35 ± 0.09	>3.75	0.35 ± 0.09	0.23 ± 0.00
		MBC	1.88 ± 0.00	0.94 ± 0.00	0.47 ± 0.00	>3.75	0.47 ± 0.00	0.47 ± 0.00
9	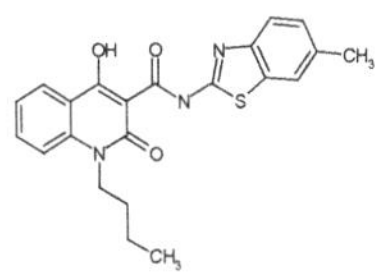	MIC	0.94 ± 0.00	0.17 ± 0.00	0.35 ± 0.09	0.70 ± 0.19	0.17 ± 0.19	0.70 ± 0.19
		MBC	1.88 ± 0.00	0.23 ± 0.00	0.47 ± 0.00	0.94 ± 0.00	0.23 ± 0.00	0.94 ± 0.00
	Streptomycin	MIC	0.10 ± 0.00	0.02 ± 0.00	0.15 ± 0.00	0.10 ± 0.00	0.10 ± 0.00	0.02 ± 0.00
		MBC	0.20 ± 0.01	0.05 ± 0.00	0.30 ± 0.01	0.20 ± 0.00	0.20 ± 0.01	0.05 ± 0.00
	Ampicillin	MIC	0.10 ± 0.00	0.10 ± 0.00	0.15 ± 0.00	0.15 ± 0.00	0.10 ± 0.00	0.10 ± 0.00
		MBC	0.15 ± 0.00	0.15 ± 0.00	0.30 ± 0.02	0.20 ± 0.01	0.20 ± 0.00	0.15 ± 0.01

The evaluation of three of the most active compounds (**2**, **3**, **4**) against three resistant strains, namely methicillin-resistant *Staphylococcus aureus* (MRSA), *P. aeruginosa*, and *E. coli*, revealed that all compounds were found to be more potent against *MRSA* than ampicillin and streptomycin, which did not show a bactericidal effect. Compound **4** also seems to be more active than ampicillin against *P. aeruginosa* strain, but no compound was more active than the reference drug against *E. coli* (Table 2). The compounds also were evaluated for their ability to inhibit the biofilm formation. Unfortunately, no compound showed good activity.

Table 2. Antibacterial activity and nhibition of biofil formation against resistant strains (MIC/MBC in mg/mL).

Compound		*MRSA*	*P.a.*	*E.c.*	MIC	0% MIC
2	MIC	0.94 ± 0.00	0.23 ± 0.00	0.94 ± 0.00	14.59	7.08
	MBC	1.88 ± 0.00	0.47 ± 0.00	1.88 ± 0.00		
3	MIC	0.47 ± 0.00	0.23 ± 0.00	0.47 ± 0.00	19.97	8.84
	MBC	0.94 ± 0.00	0.47 ± 0.00	0.94 ± 0.00		
4	MIC	0.94 ± 0.00	0.23 ± 0.00	0.94 ± 0.00	4.31	NE
	MBC	1.88 ± 0.00	0.47 ± 0.00	1.88 ± 0.00		
Streptomycin	MIC	0.10 ± 0.00	0.05 ± 0.00	0.10 ± 0.00	71.94	55.42
	MBC	/	0.10 ± 0.00	0.20 ± 0.00		
Ampicillin	MIC	/	0.20 ± 0.01	0.20 ± 0.00	67.36	30.35
	MBC	/	/	/		

2.2.2. Antifungal Activity

Synthesized thiazolyl derivatives (**1–9**) were evaluated for their antifungal activity. For the determination of minimal inhibitory/fungicidal activity, the microdilution method was used [59].

All compounds showed good antifungal activity, and the results are presented in Table 3. The antifungal potency of synthesized compounds can be presented as follows: **8** > **9** > **1** > **3** > **5** > **2** > **4** > **6** > **7**. The best antifungal activity is achieved for compound **8**, with MIC at 0.08–0.23 mg/mL and minimum fungicidal concentration (MFC) at 0.11–0.47 mg/mL, whereas the lowest activity was observed for compound **7**, with MIC at 0.23–0.47 mg/mL and MFC at 0.47–0.94 mg/mL.

Thus, the sensitivity of the most resistant strain, *Aspergillus fumigatus*, toward the compounds tested is **3** > **5** = **8** = **9** > **1** = **2** > **4** = **6** = **7**, while for the most susceptible one, which is *Trichoderma viride*, the susceptibility can be presented as **9** > **8** > **1** = **3** > **5** = **6** > **2** = **4** > **7**.

Ketoconazole showed antifungal potential at MIC 0.2–1.0 mg/mL and MFC 0.3–1.5 mg/mL, respectively, while bifonazole exhibited MIC at 0.1–0.2 and MFC at 0.2–0.25 mg/mL, respectively. Compounds **8** and **9** exhibited excellent activity, with MIC/MFC at 0.08/0.11 mg/mL, respectively, against *T. viride*, almost fourfold better than bifonazole and 29 times better than ketoconazole, as well as against *A. niger*, *A. versicolor*, *P. funiculosum*, and *P. cyclpoium var. verucosum*, with MIC/MFC at 0.11/0.23 mg/mL, respectievly. Good activity against *T. viride*, with MIC and MFC at 0.11 mg/mL and 0.23 mg/mL, respectively, was also displayed by compounds **1** and **3**, as well as by compounds **2**, **5**, and **6**, with MIC/MFC at 0.17/0.23 mg/mL, respectively. Compounds **3**, **5**, and **8** showed the same good activity against *A. niger*, with compound **8** also being potent against *P. cyclpoium var. verucosum*. It was observed that almost all compounds displayed better activity than ketoconazole against *T. viride*, with the exception of compounds **4** and **7**. In general, most of the compounds appeared to be more potent than ketoconazole against all fungi, except of *P.v.c.*, against which only three compounds (**1**, **8**, and **9**) were more active than ketoconazole.

Table 3. Antifungal activity of thiazole derivatives. (MIC and MBC in mg/mL).

Compounds		A.f.	A.n.	A.v.	P.f.	T.v.	P.v.c.
1	MIC	0.35 ± 0.08	0.08 ± 0.00	0.23 ± 0.00	0.17 ± 0.00	0.11 ± 0.00	0.17 ± 0.00
	MFC	0.47 ± 0.00	0.11 ± 0.00	0.47 ± 0.00	0.23 ± 0.00	0.23 ± 0.00	0.23 ± 0.00
2	MIC	0.35 ± 0.08	0.23 ± 0.00	0.35 ± 0.08	0.23 ± 0.00	0.17 ± 0.00	0.35 ± 0.08
	MFC	0.47 ± 0.00	0.47 ± 0.00	0.47 ± 0.00	0.47 ± 0.00	0.23 ± 0.00	0.47 ± 0.00
3	MIC	0.17 ± 0.00	0.11 ± 0.00	0.23 ± 0.00	0.23 ± 0.00	0.11 ± 0.00	0.35 ± 0.08
	MFC	0.23 ± 0.00	0.23 ± 0.00	0.47 ± 0.00	0.47 ± 0.00	0.23 ± 0.00	0.47 ± 0.00
4	MIC	0.47 ± 0.00	0.47 ± 0.00	0.23 ± 0.00	0.35 ± 0.08	0.23 ± 0.00	0.23 ± 0.00
	MFC	0.94 ± 0.00	0.94 ± 0.00	0.47 ± 0.00	0.47 ± 0.00	0.47 ± 0.00	0.47 ± 0.00
5	MIC	0.23 ± 0.00	0.06 ± 0.00	0.23 ± 0.00	0.23 ± 0.00	0.17 ± 0.00	0.17 ± 0.00
	MFC	0.47 ± 0.00	0.11 ± 0.00	0.47 ± 0.00	0.47 ± 0.00	0.23 ± 0.00	0.23 ± 0.00
6	MIC	0.47 ± 0.00	0.23 ± 0.00	0.47 ± 0.00	0.23 ± 0.00	0.17 ± 0.00	0.47 ± 0.00
	MFC	0.94 ± 0.00	0.47 ± 0.00	0.94 ± 0.00	0.47 ± 0.00	0.23 ± 0.00	0.94 ± 0.00
7	MIC	0.47 ± 0.00	0.23 ± 0.00	0.23 ± 0.00	0.47 ± 0.00	0.47 ± 0.00	0.23 ± 0.00
	MFC	0.94 ± 0.00	0.47 ± 0.00	0.47 ± 0.00	0.94 ± 0.00	0.94 ± 0.00	0.47 ± 0.00
8	MIC	0.23 ± 0.00	0.11 ± 0.00	0.11 ± 0.00	0.17 ± 0.00	0.08 ± 0.00	0.11 ± 0.00
	MFC	0.47 ± 0.00	0.23 ± 0.00	0.23 ± 0.00	0.23 ± 0.00	0.11 ± 0.00	0.23 ± 0.00
9	MIC	0.23 ± 0.00	0.17 ± 0.00	0.17 ± 0.00	0.17 ± 0.00	0.06 ± 0.00	0.17 ± 0.00
	MFC	0.47 ± 0.00	0.23 ± 0.00	0.23 ± 0.00	0.23 ± 0.00	0.11 ± 0.00	0.23 ± 0.00
Bifonazole	MIC	0.15 ± 0.00	0.15 ± 0.00	0.10 ± 0.00	0.20 ± 0.00	0.15 ± 0.00	0.10 ± 0.00
	MFC	0.20 ± 0.00	0.20 ± 0.00	0.20 ± 0.00	0.25 ± 0.00	0.20 ± 0.00	0.20 ± 0.00
Ketoconazole	MIC	0.20 ± 0.00	0.20 ± 0.00	0.20 ± 0.00	0.20 ± 0.00	1.00 ± 0.01	0.20 ± 0.00
	MFC	0.50 ± 0.00	0.50 ± 0.00	0.50 ± 0.00	0.50 ± 0.00	1.50 ± 0.00	0.30 ± 0.010

The study of the structure–activity relationship revealed that the presence of 14-methyl-7,8,13b,14-tetrahydroindolo[2′,3′:3,4]pyrido[2,1-b]quinazolin-5(13H)-one as the substituent for compound (**8**), connected to position 2 of the thiazole ring via N-propylpropionamide, is beneficial for antifungal activity. The replacement for this substituent with the presence of 2-methylisoquinolin-1(2H)-one led to compound (**1**), with decreased activity. For the series of compounds (**1**, **2**, and **8**), the presence of 2-methylisoquinolin-1(2H)-one (2) was negative for antifungal activity. In the case of the substituted phthalazine-1-yl)amino)-N-(thiazol-2-yl)benzenesulfonamides, the most favorable structure for antifungal activity is the presence of a 4-chlorophenyl substituent in position 4 of the phtalazin ring in compound (**5**). Removal of 4-Cl-pnenyl substituent led to compound (**4**) having reduced activity. The least potent among all the compounds tested appeared to be compounds **6** and **7**, with methylphtalazine and 4-OMe phenylphtalazine substituents decreasing activity even more. The presence of 4-Me as well as 4-OMe-Ph substituents appeared to be detrimental to antifungal activity.

As a conclusion, the antifungal activity depends upon the substituents on the thiazole ring and, in the case of phthalazin-1-yl)amino)-N-(thiazol-2-yl)benzenesulfonamides, upon the substituents on the phtalazine ring. It should be mentioned that the antifungal activity of the synthesized compounds is much better than that of the antibacterial compounds.

2.3. Docking Studies

2.3.1. Docking to Antibacterial Targets

In order to predict the possible mechanism of the activity of the tested compounds, docking studies were carried out on different targets. It is widely known that the most common mechanisms of activity of antibacterial agents are destroying the integrity of cell walls and cell membranes, inhibiting the expression of proteins, inhibiting the synthesis of nucleic acids, and affecting the energy metabolism of bacteria. In this direction, for the docking studies we used the enzymes responsible for these pathways, such as *E. coli* DNA gyrase, thymidylate kinase, *E. coli* primase, and *E. coli* MurA and *E. coli* MurB enzymes.

Analyzing the docking studies scores, a low Free Energy of Binding represents a strong binding of a ligand to the enzyme. Taking this into account, the docking studies revealed that the Free Energy of Binding of all tested compounds to *E. coli* DNA gyrase, thymidylate kinase, and *E. coli* primase and *E. coli* MurA enzymes was higher than that of *E. coli* MurB (−7.02–−9.96 kcal/mol); therefore, it may be suggested that inhibition of *E. coli* MurB is probably the most suitable mechanism of action of the compounds where binding scores were consistent with biological activity (Table 4).

Table 4. Molecular docking free binding energies (kcal/mol) to antibacterial targets.

Comp.	Est. Binding Energy (kcal/mol)					I-H	Residues
	E. coli Gyrase 1KZN	Thymidylate Kinase 4QGG	*E. coli* Primase 1DDE	*E. coli* MurA JV4T	*E. coli* MurB 2Q85	*E. coli* MurB	*E. coli* MurB
1	−3.46	-	-	−3.85	−7.02	1	Arg158
2	−4.52	−3.18	−2.94	−5.03	−9.16	1	Ser229
3	−4.88	−2.71	-	−5.14	−9.96	2	Gly47, Ser229
4	−3.82	−3.11	-	−4.69	−8.70	1	Ser229
5	−3.96	-	−2.91	−3.67	−7.53	1	Arg213
6	−4.62	−2.54	−3.47	−5.75	−8.65	2	Gly122, Arg213
7	−2.66	-	-	−4.12	−7.05	1	Arg213
8	−2.15	−3.54	−1.28	−4.63	−8.51	2	Arg213, Ser229
9	−2.74	-	-	−3.79	−8.52	1	Ser229

The docking pose of the most active compound **3** in *E. coli* MurB enzyme showed two favorable hydrogen bond interactions. The first one was between the oxygen atom of -OH group of the compound and the hydrogen of the side chain of Gly47 (distance 2.25 Å), and the other hydrogen bond interaction was between the oxygen atom of the –OCH_3 group of the compound and Ser229 residue (distance 2.73 Å). The NH_2 group interacts with positive ionizable interaction with the residue Glu325, stabilizing the complex compound-enzyme and playing a vital role proving the high inhibitory action of compound **3**. Moreover, the hydrogen bond formation with the residue Ser229 is crucial for the inhibitory action of compound **3** as well as for compounds **2**, **4**, **8**, and **9**, because this residue takes part in the proton transfer at the second stage of peptidoglycan synthesis [60] (Figures 2 and 3).

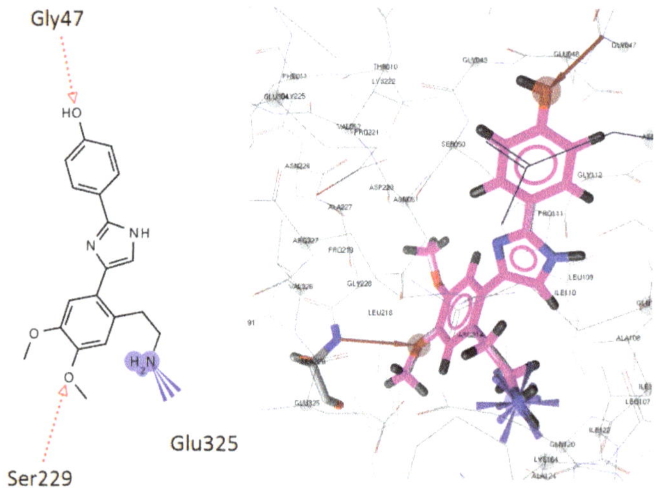

Figure 2. Docked conformation of the most active compound **3** in *E. coli* MurB. Red dotted arrows indicate H-bond, and blue lines indicate positive ionizable interactions.

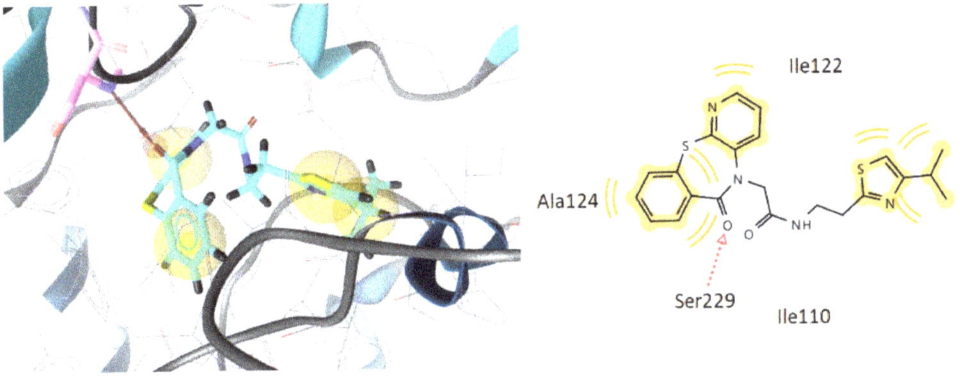

Figure 3. Docked conformation of the most active compound **2** in *E. coli* MurB. Red dotted arrows indicate H-bond, and yellow spheres indicate hydrophobic interactions.

The second-most-active compound, compound **2**, also forms this hydrogen bond interaction with the residue Ser229, which explains its high inhibitory action (Figure 3). Detailed analysis of the docking pose of the two most active compounds showed that they bind MurB in a similar way as FAD, and they fit into the binding center of the enzyme, forming a hydrogen bond with the residue Ser229 (Figure 4).

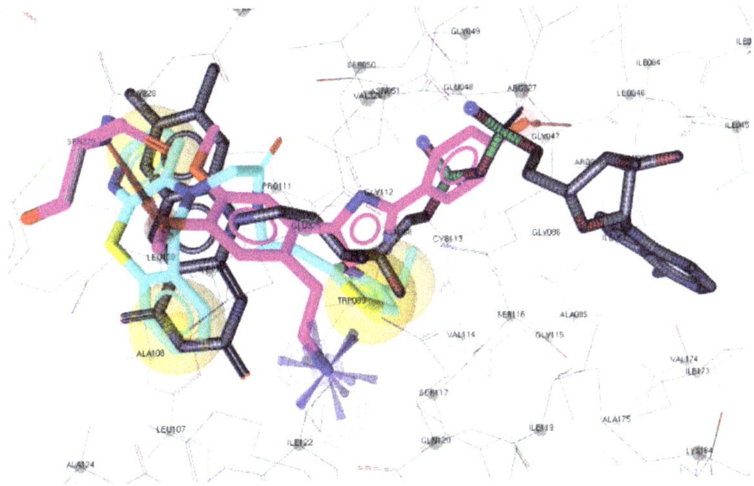

Figure 4. Superposition of compounds **3** (magenta), **2** (light blue), and FAD (gray) in *E. coli* MurB.

2.3.2. Docking to Antifungal Targets

All the synthesized compounds and the reference drug ketoconazole were docked to lanosterol 14α-demethylase of *C. albicans* and DNA topoisomerase IV (Table 5).

Table 5. Molecular docking free binding energies (kcal/mol) to antifungal targets.

N/N	Est. Binding Energy (kcal/mol) DNA TopoIV 1S16	Est. Binding Energy (kcal/mol) CYP51 of *C. albicans* 5V5Z	I-H	Residues CYP51 of *C. albicans*	Interactions with HEM601
1	−2.17	−9.15	1	Tyr132	Hydrophobic, Aromatic
2	−3.10	−7.95	1	Tyr132	Hydrophobic
3	-	−8.11	1	Tyr118	Hydrophobic
4	-	−7.52	1	Tyr118	Hydrophobic
5	−1.42	−7.50	1	Tyr118	Hydrophobic
6	−1.56	−8.64	1	Tyr64	Hydrophobic
7	−2.25	−7.12	-	-	Hydrophobic, Aromatic
8	-	−7.03	-	-	Hydrophobic
9	−2.71	−9.21	1	Tyr132	Hydrophobic, Aromatic
Ketoconazole	-	-8.23	1	Tyr64	Hydrophobic, Aromatic

Docking results showed that the most active compound **9** binds the enzyme alongside the heme group, interacting with heme throughout its benzene ring, which forms aromatic and hydrophobic interactions. In addition, a hydrogen bond with Tyr132 residue and an -OH group of the compound are formed. Moreover, hydrophobic interactions between Ile304, ile131, Ile379, Ty188, Phe233, Phe380, Leu376, and Met508 and the compound were detected. Interaction with the heme group was also observed with the benzene ring of ketoconazole, which also forms aromatic interactions (Figures 5 and 6). This property may account for why compound **9** has good antifungal activity. Superposition of compounds **9** and **1** and ketoconazole in the lanosterol 14α-demethylase of C. albicans ($CYP51_{ca}$) shows this interaction with the heme group proving this hypothesis (Figure 7).

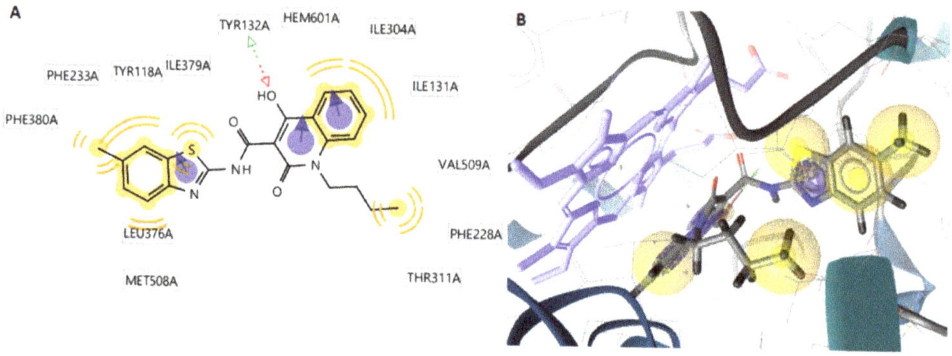

Figure 5. Docked conformation of the most active compound **9** in lanosterol 14α-demethylase of *C. albicans* ($CYP51_{ca}$) (**A**,**B**). Red and green dotted arrows indicate H-bond, blue arrows indicate aromatic interactions, and yellow spheres indicate hydrophobic interactions.

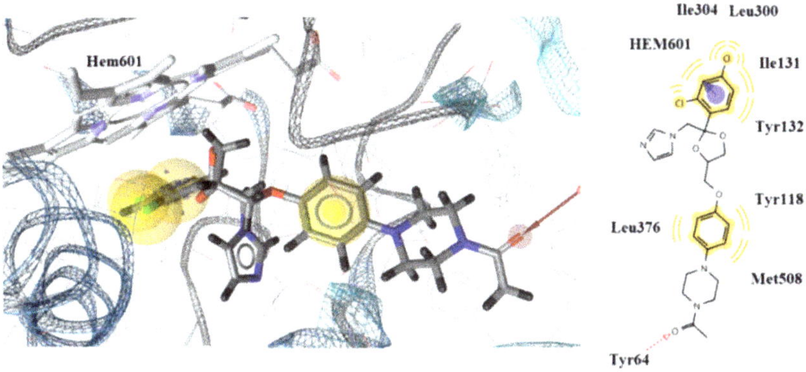

Figure 6. Docked conformation of ketoconazole in lanosterol 14α-demethylase of *C. albicans* ($CYP51_{ca}$).

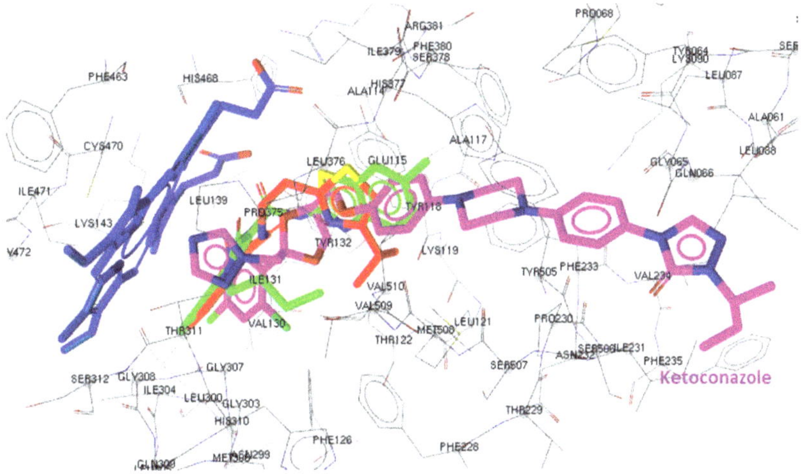

Figure 7. Superposition of compounds **9** (green), **1** (red), and ketoconazole (magenta) in lanosterol 14α-demethylase of C. albicans ($CYP51_{ca}$). Heme group is illustrated in blue.

2.4. Drug-Likeness

The bioavailability and drug-likeness scores of all compounds are shown in Table 6; according to prediction results, the bioavailability score of all compounds was about 0.55. Moreover, all compounds displayed good to excellent drug-likeness scores (−0.13–1.09). Figure 8 presents the bioavailability radar of some of the compounds. The best in the in silico predictions results was achieved for compounds **1** and **2**, with drug-likeness scores of 1.03 and 1.09, respectively, and with no violation of any rule.

Table 6. Drug-likeness predictions of tested compounds.

No	MW	Number of HBA [a]	Number of HBD [b]	Log $P_{o/w}$ (iLOGP) [c]	Log S [d]	TPSA [e]	BBB Permeant [f]	Lipinski, Ghose, Veber, Egan, and Muegge Violations	Bioavailability Score	Drug-Likeness Model Score
1	355.45	3	1	2.8	Poorly soluble	92.23	No	0	0.55	1.03
2	438.12	4	1	3.37	Poorly soluble	128.73	No	0	0.55	1.09
3	339.16	5	3	2.33	Poorly soluble	93.39	No	0	0.55	0.01
4	459.54	5	2	3.00	Poorly soluble	133.49	No	0	0.55	−0.12
5	493.99	5	2	3.20	Poorly soluble	133.49	No	3 *	0.55	−0.13
6	396.49	5	1	2.26	Poorly soluble	121.46	No	0	0.55	0.27
7	489.57	6	2	3.07	Poorly soluble	142.72	No	3 *	0.55	0.27
8	513.65	3	1	3.73	Poorly soluble	98.71	No	1 **	0.55	1.37
9	407.49	4	2	2.86	Poorly soluble	112.46	No	0	0.55	0.43

[a] Number of hydrogen bond acceptors; [b] number of hydrogen bond donors; [c] lipophilicity; [d] water solubility (SILICOS-IT (S = Soluble)); [e] topological polar surface area (Å^2); [f] blood–brain barrier permeant; * Ghose: 3 violations: MW > 480, WLOGP > 5.6, MR > 130; ** Lipinski: 1 violation: MW > 500; and Ghose: 2 violations: MW > 480, MR > 130.

Figure 8. Bioavailability radar of compounds **1** and **9**. The pink area represents the optimal range for each property for oral bioavailability. Lipophilicity (LIPO): XLOGP3 between −0.7 and +5.0; molecular weight (SIZE): MW between 150 and 500 g/mol; polarity (POLAR) TPSA between 20 and 130 Å^2; solubility (INSOLU): log S not higher than 6; saturation (INSATU): fraction of carbons in the sp3 hybridization not less than 0.25; flexibility (FLEX): no more than 9 rotatable bonds.

3. Materials and Methods

3.1. General Information

^{1}H NMR spectra of newly synthesized compounds were recorded on a spectrometer Bruker 400 (400 MHz); compound **6**—on spectrometer Bruker Fourier 300 (300 MHz) in DMSO-d_6. Chemical shifts of nuclei ^{1}H were measured relatively the residual signals of deuteron solvent (δ = 2.50 ppm; see Ref. (http://chem.ch.huji.ac.il/nmr/software/solvent.htmL (accessed on 1 August 2022)) and the literature cited therein). Coupling constants (*J*) are reported in Hz. Melting points were determined by using Fisher-Johns Melting Point Apparatus (Fisher Scientific) and are uncorrected. Elemental analysis was performed by the classical method of microanalysis. The reaction and purity of the obtained compounds were monitored by TLC (plates with Al_2O_3 III activity grade, eluent $CHCl_3$, and development of TLC plates by exposition to iodine vapors in "iodine chamber"). The solvents were purified according to standard procedures. The starting compounds and compound **9** were provided by InterBioscreen Ltd. (Russia).

3.1.1. General Procedure for the Synthesis of Compounds **1**, **2**, and **8**

A mixture of a corresponding acid (0.01 mol), $SOCl_2$ (1.43 g, 0.87 mL, 0.012 mol), $CHCl_3$ (20 mL), and DMF (0.05 mL) was refluxed until gas evolution stops and cooled, and the resulting solution of acid chloride was added dropwise at 0–2 °C to a solution of 2-(4-isopropylthiazol-2-yl)ethan-1-amine (1.7 g, 0.01 mol) and Et_3N (2.02 g, 2.78 mL, 0.02 mol) in $CHCl_3$ (15 mL). Then, $NaHCO_3$ (9.5 g) and water (100 mL) were added and stirred, the organic layer was separated and dried with Na_2SO_4, and the solvent was distilled off in vacuum at 30–40 °C. The residue was purified by recrystallization from a suitable solvent.

N-[2-(4-Isopropylthiazol-2-yl)ethyl]-2-methyl-1-oxo-1,2-dihydroisoquinoline-4-carboxamide (**1**). The starting compounds were 2-methyl-1-oxo-1,2-dihydroisoquinoline-4-carboxylic acid and 2-(4-isopropylthiazol-2-yl)ethan-1-amine. Yield 2.63 g (74%), colorless crystals, m.p. 105–107 °C (CCl_4). ^{1}H NMR (400 MHz, DMSO-d_6, δ, ppm): 1.28 (2s, 6H, 2Me), 3.24 (t, *J* 7.1, 2H), 3.56–3.67 (m, 6H, NMe, <u>CH</u>Me_2), 6.93 (s, 1H, H-5′), 7.50 (d, *J* 7.4, 1H, H-5), 7.68 (d, *J* 7.5, 1H, H-8), 7.77 (s, 1H, H-3), 8.21–8.31 (m, 3H, H-6, H-7, NH). ^{13}C NMR (126 MHz, DMSO-d_6, δ, ppm): 22.68(4C), 33.03(4C), 11.83(2C), 125.22(1C), 127.42(5C), 156.13 (5C). Found (%): C, 64.51; H, 6.15; N, 11.56, S, 9.34. Calc. for $C_{19}H_{21}N_3O_2S$ (%): C, 64.20; H, 5.95; N, 11.82, S, 9.02.

N-[2-(4-Isopropylthiazol-2-yl)ethyl]-2-(6-oxobenzo[f]pyrido[2,3-b][1,4]thiazepin-5(6H)-yl) acetamide (**2**). The starting compounds were 2-(6-oxobenzo[f]pyrido[2,3-b][1,4]thiazepin-5(6*H*)-yl)acetic acid and 2-(4-isopropylthiazol-2-yl)ethan-1-amine. Yield 3.60 g (82%), colorless crystals, m.p. 141–143 °C (EtOAc). ^{1}H NMR (400 MHz, DMSO-d_6, δ, ppm): 1.26 (s, 3H, Me), 1.28 (s, 3H, Me), 2.96–3.07 (m, 5H, <u>CH</u>Me_2), 3.13–3.18 (m, 2H, CH_2CO), 6.93 (s, 1H, H-7), 7.10 (s, 1H, H-5′), 7.40–7.43 (m, 2H, H-4, H-8), 7.51–7.54 (m, 1H, H-3), 7.62–7.66 (m, 1H, H-2), 7.99 (dd, *J* 8.2, 1.6, 1H, H-9), 8.26–8.28 (m, 2H, H-5, NH). ^{13}C NMR (126 MHz, DMSO-d_6, δ, ppm): 168.07 (NH<u>C</u>=O), 167.86 (N<u>C</u>=O), 166.85, 162.86, 159.71, 146.65, 140.74, 137.25, 136.18, 133.92, 132.12, 132.10, 131.84, 129.46, 125.11, 115.02, 111.83, 54.67, 39.26, 33.14, 31.00, 30.67, 25.69, 22.70. Found (%): C, 60.44; H, 5.31; N, 12.59; S, 14.78. Calc. for $C_{22}H_{22}N_4O_2S$ (%): C, 60.25; H, 5.06; N, 12.78; S, 14.62.

3.1.2. Synthesis of Compound **3** [61]

4-{4-[2-(2-Aminoethyl)-4,5-dimethoxyphenyl]thiazol-2-yl}phenol hydrochloride (**3**). The mixture of 1-(chloromethyl)-6,7-dimethoxy-3,4-dihydroisoquinoline (2.39 g, 0.01 моль), 4-hydroxybenzothioamide (1.53 g, 0.01 моль) and PrOH (15 mL) was boiled with stirring for 2 h. Then, it was cooled, and hydrochloride **3** was filtered off. Yield 2.65 г (67%), colorless crystals, m.p. 276–277 °C (PrOH). ^{1}H NMR (400 MHz, DMSO-d_6, δ, ppm): 3.15 (t, *J* 6.5, 2H, 2H-), 3.87 (s, 3H, OMe), 3.91 (s, 3H, OMe), 4.59 (t, *J* 6.5, 2H, 2H-), 7.04 (s, 1H, H-3″), 7.15 (d, *J* 8.4, 2H, H-3, H-5), 7.27 (s, 1H, H-6″), 7.44 (s, 1H, H-5′), 7.60–7.62 (m, 2H, NH_2), 7.68 (d, *J* 8.6, 2H, H-2, H-6), 8.76 (s, 1H, OH), 10.94 (s, 1H, $^+$NH). ^{13}C NMR (126 MHz, DMSO-d_6, δ, ppm): 166.52, 158.49 (<u>C</u>-OH), 153.82, 152.46 (<u>C</u>-OMe), 151.30 (<u>C</u>-OMe), 135.15 (2C), 134.82, 134.65, 133.58 (2C), 117.55 (2C), 117.16, 114.29 (2C), 48.19 (2C, <u>C</u>H_3), 43.62 (<u>C</u>H_2NH), 32.15. Found (%): C, 58.26; H, 5.60; Cl, 9.32; N, 7.00; S, 8.34. Calc. for $C_{19}H_{21}ClN_2O_3S$ (%): C, 58.08; H, 5.39; Cl, 9.02; N, 7.13; S, 8.16.

3.1.3. General Procedure for the Synthesis of Compounds **4**–**7**

A mixture of 1-chloro-4-R-phthalazine (0.01 mol), 4-amino-N-(thiazol-2-yl) benzenesulfonamide (2.55 g, 0.01 mol) in methyl cellosolve (20 mL) was boiled for 30 min, cooled to 95–100 °C, and poured into 5% NH_4OH (60 mL). Then, it was stirred for 1 h, filtered off, and washed with water (4–15 mL).

4-[(4-Phenylphthalazin-1-yl)amino]-N-(thiazol-2-yl)benzenesulfonamide (**4**). The starting compound was 1-chloro-4-phenylphthalazin. Yield 3.58 г (78%), colorless crystals, m.p. 297–299 °C (DMF). ^{1}H NMR (400 MHz, DMSO-d_6, δ, ppm): 6.70 (d, *J* 4.6, 1H, H-5′), 7.10 (d,

J 4.6, 1H, H-4′), 7.52–7.60 (m, 3H, H-3. H-5, H-4″), 7.64–7. 67 (m, 2H, H-3″, H-5″), 7.76–7.81 (m, 2H, H-2, H-6), 7.86–8.01 (m, 3H, H-2″, H-6″, H-6‴), 8.05–8.20 (m, 2H, H-5‴, H-7‴), 8.65 (d, *J* 8.2, 1H, H-8‴), 9.47 (s, 1H, NH), 12.48 (s, 1H, $NHSO_2$). ^{13}C NMR (126 MHz, DMSO-d_6, δ, ppm): 135.11 (2C), 132.97 (2C), 132.32, 130.15(3C), 129.19 (2C), 128.90 (4C), 127.17 (3C), 126.37, 119.89 (3C), 108.41. Found (%): C, 60.00; H, 3.51; N, 15.11; S, 14.21. Calc. for $C_{23}H_{17}N_5O_2S_2$ (%): C, 60.11; H, 3.73; N, 15.24; S, 13.96.

4-{[4-(4-Chlorophenyl)phthalazin-1-yl]amino}-N-(thiazol-2-yl)benzenesulfonamide (**5**). The starting compound was 1-chloro-4-(4-chlorophenyl)phthalazine. Yield 4.30 г (87%), colorless crystals, m.p. 280–281 °C (DMF). ^1H NMR (400 MHz, DMSO-d_6, δ, ppm): 6.70 (d, *J* 4.6, 1H, H-5′), 7.10 (d, *J* 4.7, 1H, H-4′), 7.56–7.60 (m, 2H, H-3, H-5), 7.65–7.70 (m, 2H, H-3″, H-5″), 7.76–7.81 (m, 2H, H-2, H-6), 7.90–7.91 (m, 2H, H-6‴, H-7‴), 7.96–8.02 (m, 1H, H-5‴,), 8.12–8.17 (m, 2H, H-2″, H-6″), 8.66 (d, *J* 8.2, 1H, H-8‴), 9.49 (s, 1H, NH), 12.48 (s, 1H, $NHSO_2$). ^{13}C NMR (126 MHz, DMSO-d_6, δ, ppm): 144.50 (NH-C̲), 135.80 (N=C̲), 134.08, 133.08, 132.43 (2C), 131.96 (3C), 128.98 (3C), 127.16 (3C), 126.18, 123.25, 119.99 (3C), 119.11, 108.43 (C̲=S). Found (%): C, 55.69; H, 3.05; Cl+S, 20.40; N, 14.02. Calc. for $C_{23}H_{16}ClN_5O_2S_2$ (%): C, 55.92; H, 3.26; Cl, 7.18; N, 14.18; S, 12.98.

4-[(4-Methylphthalazin-1-yl)amino]-N-(thiazol-2-yl)benzenesulfonamide (**6**). The starting compound was 1-chloro-4-methylphthalazine. Yield 3.70 г (93%), colorless crystals, m.p. 284–286 °C (DMF). ^1H NMR (400 MHz, DMSO-d_6, δ, ppm): 2.91 (s, 3H, Me), 6.69–6.71 (m, 1H, H-5′), 7.08–7.10 (m, 1H, H-4′), 7.75–7.82 (m, 2H, H-3, H-5), 7.98–8.12 (m, 4H, H-2, H-6, H-6″, H-7″), 8.21–8.28 (m, 1H, H-5‴), 8.79–8.85 (m, 1H, H-8‴). ^{13}C NMR (126 MHz, DMSO-d_6, δ, ppm): 169.06 (N=C̲S), 152.73 (2C), 136.10, 134.24 (2C), 133.71, 127.55, 127.16 (3C), 126.87, 123.94, 120.75 (2C), 120.14, 108.50, 18.23 ($\underline{C}H_3$). Found (%): C, 54.16; H, 3.62; N, 17.44; S, 16.45. Calc. for $C_{18}H_{15}N_5O_2S_2$ (%): C, 54.39; H, 3.80; N, 17.62; S, 16.13.

4-{[4-(4-Methoxyphenyl)phthalazin-1-yl]amino}-N-(thiazol-2-yl)benzenesulfonamide (**7**). The starting compound was 1-chloro-4-(4-methoxyphenyl)phthalazine. Yield 4.11 г (84%), colorless crystals, m.p. 142–143 °C (DMF). ^1H NMR (400 MHz, DMSO-d_6, δ, ppm): 3.89 (s, 3H, OMe), 6.69 (d, *J* 4.6, 1H, H-5′), 7.07–7.12 (m, 3H, H-4′, H-3″, H-5″), 7.56–7.63 (m, 2H, H-3, H-5), 7.76–7.79 (m, 2H, H-2″, H-6″), 7.85–7.90 (m, 1H, H-5‴), 7.92–8.00 (m, 2H, H-6‴, H-7‴), 8.12–8.17 (m, 2H, H-2, H-6), 8.63 (d, *J* 8.2, 1H, H-8‴), 9.42 (s, 1H, NH), 12.46 (s, 1H, $NHSO_2$). ^{13}C NMR (126 MHz, DMSO-d_6, δ, ppm): 160.11 (4C), 132.85, 132.20, 131.47 (3C), 127.17 (2C), 126.49, 119.76 (3C), 114.34, 108.40, 55.69 ($\underline{C}H_3$). Found (%): C, 58.65; H, 3.70; N, 14.12; S, 13.37. Calc. for $C_{24}H_{19}N_5O_3S_2$ (%): C, 58.88; H, 3.91; N, 14.31; S, 13.10.

N-[2-(4-Isopropylthiazol-2-yl)ethyl]-5-methyl-14-oxo-5a,6,12,14-tetrahydroindolo[2′,3′:4,5] pyrido[2,1-b]quinazoline-7(5H)-carboxamide (**8**). The starting compounds were 5-methyl-14-oxo-5a,6,12,14-tetrahydroindolo[2′,3′:4,5]pyrido[2,1-b]quinazoline-7(5H)-carboxylic acid and 2-(4-isopropylthiazol-2-yl)ethan-1-amine. Yield 3.54 g (71%), colorless crystals, m.p. 182–184 °C (EtOAc). ^1H NMR (400 MHz, DMSO-d_6, δ, ppm): 1.21 (s, 3H, Me), 1.24 (s, 3H, Me), 3.01–3.08 (m, 7H, NMe, 2H-α, 2H-7), 3.48 (d, J 6.8, 2H, 2H-α), 4.69–4.75 (m, 1H, C̲H̲Me_2), 4.92 (d, J 2.6, 2H, 2H-13), 6.01 (s, 1H, H-6), 6.78 (s, 1H, H-5′), 7.08 (d, J 7.2, 1H, H-4), 7.15–7.24 (m, 4H, H-2, H-3, H-10, H-11), 7.35 (d, J 8.1, 1H, H-12), 7.48–7.57 (m, 2H, H-2, H-3), 7.94 (d, J 7.8, 1H, H-9), 8.08 (s, 1H, NH). ^{13}C NMR (126 MHz, DMSO-d_6, δ, ppm): 167.97 ($CH_2\underline{C}$=O), 166.83 (2C), 162.82 (N-C̲=O), 128.47 (3C), 122.72, 119.84 (2C), 112.85, 111.61 (3C), 110.54 (4C), 46.51 (2C), 39.12 (2C), 32.97 (3C), 30.62, 22.65 (2C, $2\underline{C}H_3$), 20.16. Found (%): C, 67.11; H, 5.64; N, 14.00; S, 6.34. Calc. for $C_{28}H_{29}N_5O_2S$ (%): C, 67.31; H, 5.85; N, 14.02; S, 6.42.

4-Butyl-1-hydroxy-N-(6-methylbenzo[d]thiazol-2-yl)-3-oxo-3,4-dihydronaphthalene-2-carboxamide (**9**). The starting compounds were 1-butyl-4-hydroxy-2-oxo-1,2-dihydroquinoline-3-carboxylic acid and 6-methylbenzo[d]thiazol-2-amine. Yield 3.67 g (69%), colorless crystals, m.p. 97–99 °C (CCl_4). ^1H NMR (400 MHz, DMSO-d_6, δ, ppm): 1.05 (t, *J* 7.3, 3H, M̲e̲CH_2), 1.46–1.59 (m, 2H, 2H-β), 1.69–1.79 (m, 2H, 2H-γ), 2.48–2.51 (m, 5H, Me-6′, DMSO), 2.82 (s, 1H, H-4), 4.33–4.38 (m, 2H, H-α), 7.22 (dd, *J* 8.3, 1.7, 1H, H-8), 7.37 (t, *J* 7.6, 1H, H-6), 7.56–7.70 (m, 3H, H-5, H-7, H-5′), 7.80–7.83 (m, 1H, H-7′), 8.21–8.24 (m, 1H, H-4′), 13.81 (s, 1H, NH), 15.19 (s, 1H, OH). ^{13}C NMR (126 MHz, DMSO-d_6, δ, ppm): 170.85 (C̲=O),

166.82 (2C), 160.01, 151.12, 140.11, 135.47, 131.56, 129.78, 123.47 (2C), 120.05 (2C), 109.13 (2C), 108.87, 99.95, 52.61 ($\underline{C}H_2N$), 29.84, 20.15, 19.23, 11.45 ($\underline{C}H_3$).

3.2. *Biological Evaluation*

3.2.1. Antibacterial Action

The following Gram-negative bacteria, *Escherichia coli* (ATCC 35210), *Enterobacter cloacae (clinical isolate), Salmonella Typhimurium* (ATCC 13311), as well as Gram-positive bacteria, *Listeria monocytogenes* (NCTC 7973), *Bacillus cereus* (clinical isolate), and *Staphylococcus aureus* (ATCC 6538), were used. The bacterial strains are deposited at Mycological Laboratory, Department of Plant Physiology, Institute for Biological Research "Siniša Stankovic"—National Institute of Republic of Serbia, Belgrade, Serbia.

The minimum inhibitory and bactericidal (MIC/MBC) concentrations were defined as described previously [62,63]. Resistant strains used were isolates of *S. aureus*, *E. coli*, and *P.aeruginosa*, obtained as reported in Kartsev et al. [63].

3.2.2. Antifungal Activity

The examined strains were: Aspergillus niger (ATCC 6275), Aspergillus fumigatus (ATCC 1022), Aspergillus versicolor (ATCC 11730), Penicillium funiculosum (ATCC 36839), Trichoderma viride (IAM 5061), and Penicillium verrucosum var. cyclopium (food isolate). All experiments were performed in triplicate [64,65].

3.2.3. Inhibition of Biofilm Formation

The assays were performed as described before [66,67]. Briefly, *P. aeruginosa* resistant strain was incubated with MIC and subMIC of the tested compounds in tryptic soy broth enriched with 2% glucose at 37 °C for 24 h. Afterwards, each well was washed twice with sterile Phosphate buffered saline, pH 7.4 (PBS), and fixed with methanol for 10 min. Methanol was then removed, and the plate was air-dried. The biofilm was stained with 0.1% crystal violet (Bio-Merieux, France) for 30 min. The wells were washed with water, air-dried, and color dissolved in 96% ethanol (Zorka, Serbia). The absorbance was measured at 620 nm on a Multiskan FC Microplate Photometer, Thermo Scientific. The percentage of inhibition of biofilm formation was calculated by the formula:

$$[(A620control - A620sample)/A620control] \times 100. \quad (1)$$

3.3. *Molecular Modeling Studies*

The ligand preparation done by using chemdraw12.0, and geometries were optimized using LigandScout 4.4.5. The "Build/check/repair model" for the session "Prepare PDB file for docking programs" module was used for proteins preparation. For the final preparation of both ligands and protein preparation, Wizard of AutoDock tools 1.5.6 is used. Autodock 4 (ver. 4.2.6) was employed for docking simulations and Autogrid4 for affinity grid maps preparation. The resulting poses and potential interactions were visualized using LigandScout 4.4.5.

X-ray crystal structures of *E. coli* DNA GyrB, thymidylate kinase, *E. coli* MurA, *E. coli* primase, *E. coli* MurB, DNA topoisomerase IV, and CYP51 of *C. albicans* (PDB ID: 1KZN, AQGG, 1DDE, JV4T, 2Q85, 1S16, and 5V5Z, respectively) with bound inhibitors were retrieved from Brookhaven Protein Data Bank (PDB). The pdb files of proteins were submitted to "Build/check/repair model" for the session "Prepare PDB file for docking programs"; missing side chains were modeled in, water positions and symmetry were corrected, and hydrogen atoms were added. Only chain A of each enzyme of the repaired pdb file was evaluated and passed to AutodockTools (ADT ver.1.5.6) for PDBQT file preparation. ADT assigned polar hydrogens, water molecules and nonstandard residues were removed, so only polar hydrogen was maintained, and Gasteiger charges were computed for protein atoms. AutoDock saved the prepared file in PDBQT format.

All molecules were sketched in Chemdraw12.0 program. The geometry of built compounds was optimized using the molecular mechanical force fields 94 (MMFF94) energy via LigandScout [68], partial charges were also calculated, comformers of each ligand were generated, and the one with the best conformation was maintained and saved as mol2 file that was passed, as usual, to ADT for PDBQT file preparation. There, polar hydrogen was added to each structure, followed by computing Gasteiger and Kollman charges and the torsions.

Autodock 4 (ver. 4.2.6) was employed for docking simulations. A computationally (relatively) 'hybrid' force field that contains terms based on molecular, mechanics, and empirical terms is used by AutoDock. The evaluation step includes: First, calculation of the energy of protein and ligand in the unbound state. Second, calculation of the energy of the ligand–protein complex. Third, taking the difference between first and second steps.

$$\Delta G = \left(V_{\text{bound}}^{L-L} - V_{\text{unbound}}^{L-L}\right) + \left(V_{\text{bound}}^{P-P} - V_{\text{unbound}}^{P-P}\right) + \left(V_{\text{bound}}^{P-L} - V_{\text{unbound}}^{P-L} + \Delta S_{\text{conf}}\right)$$

where P refers to the protein, L refers to the ligand, V are the pair-wise evaluations mentioned above, and ΔS_{conf} denotes the loss of conformational entropy upon binding [69]. The ligand molecule is in an arbitrary conformation, orientation, and position, and this molecular docking program finds favorable poses in a protein-binding site using Lamarckian genetic algorithms implemented therein to search for the best conformers.

A Lamarckian genetic algorithm was used as the search engine, with a total of 100 runs. The region of interest, used by Autodock4 for docking runs and by Autogrid4 for affinity grid maps preparation, was defined in such a way to comprise the whole catalytic binding site using a grid of 50 × 50 × 50 points with a grid space of 0.375 Å. All parameters used in docking were default. The translation, quaternion, and torsions steps were taken from default values in AutoDock. The Lamarckian genetic algorithm and the pseudo-Solis and Wets methods were applied for minimization using default parameters. The number of docking runs was 100. After docking, the 100 solutions were clustered into groups, with RMS lower than 1.0 E. The clusters were ranked by the lowest energy representative of each cluster. Upon completion of docking, the best poses were screened by examination of binding energy ($\Delta G_{\text{binding}}$, kcal/mol) and number in cluster. In order to describe the ligand-binding pocket interactions, the top-ranked binding mode found by AutoDock in complex with the binding pocket of enzyme was selected. The resulting poses and potential interactions were visualized using LigandScout.

4. Conclusions

In this work, three structural series of new thiazole derivatives were synthesized and evaluated for their antibacterial and antifungal activity against a series of bacterial and fungal pathogens. The antibacterial activity of the tested compounds is moderate to good, with MIC at 0.23–>3.75 mg/mL and MBC at 0.35–>3.75 mg/mL. Compounds **4** and **9** demonstrated the best activity among the tested compounds against *E. coli* and *B. cereus* and *S.Typhimurium*, respectively, with MIC/MBC at 0.17/0.23 mg/mL, respectively.

Three of the most active compounds (**2**, **3**, and **4**) were also evaluated against three resistant strains, *MRSA*, *E. coli*, and *P. artuginosa*, demonstrating better activity than the reference drugs against MRSA, while compound **4** also was active against *P. aeruginosa*.

According to the results on antifungal activity, all compounds are active, but the best activity was observed for compound **8**, with MIC and MFC in the range of 0.08–0.23 and 0.11–0.47 mg/mL, respectively.

Docking analysis indicated a probable involvement of MurB inhibition in the antibacterial mechanism of the compounds tested, while the docking analysis to 14α-lanosterol demethylase (CYP51) of *Candida albicans* indicated a probable implication of CYP51 reductase at the antifungal activity of the compounds. Finally, compound **8** showed the best drug-likeness model score.

Author Contributions: Conceptualization, V.K. and A.G.; methodology, A.Z.; software, A.P.; validation, A.P.; formal analysis, L.D., A.M. and A.K.; investigation, M.I., J.G. and M.S.; data curation, A.G., A.Z. and M.I.; writing—original draft preparation, A.Z., A.G. and M.I.; writing—review and editing, A.G. and M.I.; visualization, V.K.; supervision, A.G. and A.Z.; funding acquisition, A.Z. and M.S. All authors have read and agreed to the published version of the manuscript.

Funding: This work has been supported by the Ministry of Science and Higher Education of the Russian Federation (Southern Federal University, 2020, project FENW-2020-0031), by Fundamental Scientific Research of the State Academies of Sciences for 2014–2021 (grant no. 0710-2019-0044), and by financial support from the Serbian Ministry of Education, Science and Technological Development (project no. 451-03-68/2020-14/200007).

Institutional Review Board Statement: Not applicable.

Informed Consent Statement: Not applicable.

Data Availability Statement: Not applicable.

Conflicts of Interest: The authors declare no conflict of interest.

References

1. Tripathi, A.C.; Gupta, S.J.; Fatima, G.N.; Sonar, P.K.; Verma, A.; Saraf, S.K. 4-Thiazolidinones: The advances continue .... *Eur. J. Med. Chem.* **2014**, *72*, 52–77. [CrossRef] [PubMed]
2. Kaminskyy, D.; Kryshchyshyn, A.; Lesyk, R. 5-Ene-4-thiazolidinones—An efficient tool in medicinal chemistry. *Eur. J. Med. Chem.* **2017**, *140*, 542–594. [CrossRef] [PubMed]
3. Bikobo, D.S.N.; Vodnar, D.C.; Stana, A.; Tiperciuc, B.; Nastasă, C.; Douchet, M.; Oniga, O. Synthesis of 2-phenylamino-thiazole derivatives as antimicrobial agents. *J. Saudi Chem. Soc.* **2017**, *21*, 861–868. [CrossRef]
4. AlThagafi, I.; El-Metwaly, N.; Farghaly, T.A. New Series of Thiazole Derivatives: Synthesis, Structural Elucidation, Antimicrobial Activity, Molecular Modeling and MOE Docking. *Molecules* **2019**, *24*, 1741. [CrossRef] [PubMed]
5. Biernasiuk, A.; Kawczyńska, M.; Berecka-Rycerz, A.; Rosada, B.; Gumieniczek, A.; Malm, A.; Dzitko, K.; Łączkowski, K.Z. Synthesis, antimicrobial activity, and determination of the lipophilicity of ((cyclohex-3-enylmethylene)hydrazinyl)thiazole derivatives. *Med. Chem. Res.* **2019**, *28*, 2023–2036. [CrossRef]
6. Salem, M.A. Synthesis of New Thiazole, Bithiazolidinone and Pyrano[2,3-d]thiazole Derivatives as Potential Antimicrobial Agents. *Croat. Chem. Acta* **2017**, *90*, 7–15. [CrossRef]
7. Bondock, S.; Fouda, A.M. Synthesis and evaluation of some new 5-(hetaryl)thiazoles as potential antimicrobial agents. *Synth. Commun.* **2018**, *48*, 561–573. [CrossRef]
8. Borcea, A.-M.; Ionuț, I.; Crișan, O.; Oniga, O. An Overview of the Synthesis and Antimicrobial, Antiprotozoal, and Antitumor Activity of Thiazole and Bisthiazole Derivatives. *Molecules* **2021**, *26*, 624. [CrossRef]
9. Carbone, A.; Cascioferro, S.; Parrino, B.; Carbone, D.; Pecoraro, C.; Schillaci, D.; Cusimano, M.G.; Cirrincione, G.; Diana, P. Thiazole Analogues of the Marine Alkaloid Nortopsentin as Inhibitors of Bacterial Biofilm Formation. *Molecules* **2021**, *26*, 81. [CrossRef]
10. Kamble, R.D.; Meshram, R.J.; Hese, S.V.; More, R.A.; Kamble, S.S.; Gacche, R.N.; Dawane, B.S. Synthesis and in silico investigation of thiazoles bearing pyrazoles derivatives as anti-inflammatory agents. *Comput. Biol. Chem.* **2016**, *61*, 86–96. [CrossRef]
11. Mohareb, R.; Al-Omran, F.; Abdelaziz, M.; Ibrahim, R. Anti-inflammatory and Anti-ulcer Activities of New Fused Thiazole Derivatives Derived from 2-(2-Oxo-2H-chromen-3-yl)thiazol-4(5H)-one. *Acta Chim. Slov.* **2017**, *64*, 349–364. [CrossRef] [PubMed]
12. Gümüş, M.; Yakan, M.; Koca, İ. Recent advances of thiazole hybrids in biological applications. *Future Med. Chem.* **2019**, *11*, 1979–1998. [CrossRef] [PubMed]
13. Muhammad, Z.A.; Masaret, G.S.; Amin, M.; Abdallah, M.A.; Farghaly, T. Anti-inflammatory, Analgesic and Anti-ulcerogenic Activities of Novel bis-thiadiazoles, bis-thiazoles and bis-formazanes. *Med. Chem.* **2017**, *13*, 226–238. [CrossRef] [PubMed]
14. Kryshchyshyn, A.; Roman, O.; Lozynskyi, A.; Lesyk, R. Thiopyrano[2,3-d]Thiazoles as New Efficient Scaffolds in Medicinal Chemistry. *Sci. Pharm.* **2018**, *86*, 26. [CrossRef] [PubMed]
15. Ayati, A.; Emami, S.; Moghimi, S.; Foroumadi, A. Thiazole in the targeted anticancer drug discovery. *Future Med. Chem.* **2019**, *11*, 1929–1952. [CrossRef]
16. Carbone, D.; Vestuto, V.; Ferraro, M.R.; Ciaglia, T.; Pecoraro, C.; Sommella, E.; Cascioferro, S.; Salviati, E.; Novi, S.; Tecce, M.F.; et al. Metabolomics-assisted discovery of a new anticancer GLS-1 inhibitor chemotype from a nortopsentin-inspired library: From phenotype screening to target identification. *Eur. J. Med. Chem.* **2022**, *234*, 114233. [CrossRef]
17. Di Franco, S.; Parrino, B.; Gaggianesi, M.; Pantina, V.D.; Bianca, P.; Nicotra, A.; Mangiapane, L.R.; Iacono, M.L.; Ganduscio, G.; Veschi, V.; et al. CHK1 inhibitor sensitizes resistant colorectal cancer stem cells to nortopsentin. *iScience* **2021**, *24*, 102664. [CrossRef]
18. Khatik, G.; Datusalia, A.K.; Ahsan, W.; Kaur, P.; Vyas, M.; Mittal, A.; Nayak, S. A Retrospect Study on Thiazole Derivatives as the Potential Antidiabetic Agents in Drug Discovery and Developments. *Curr. Drug Discov. Technol.* **2018**, *15*, 163–177. [CrossRef]

19. Petrou, A.; Eleftheriou, P.; Geronikaki, A.; Akrivou, M.G.; Vizirianakis, I. Novel Thiazolidin-4-ones as Potential Non-nucleoside Inhibitors of HIV-1 Reverse Transcriptase. *Molecules* **2019**, *24*, 3821. [CrossRef]
20. Xu, Z.; Ba, M.; Zhou, H.; Cao, Y.; Tang, C.; Yang, Y.; He, R.; Liang, Y.; Zhang, X.; Li, Z.; et al. 2,4,5-Trisubstituted thiazole derivatives: A novel and potent class of non-nucleoside inhibitors of wild type and mutant HIV-1 reverse transcriptase. *Eur. J. Med. Chem.* **2014**, *85*, 27–42. [CrossRef]
21. Liaras, K.; Fesatidou, M.; Geronikaki, A. Thiazoles and Thiazolidinones as COX/LOX Inhibitors. *Molecules* **2018**, *23*, 685. [CrossRef] [PubMed]
22. Jacob, P.J.; Manju, S.L. Identification and development of thiazole leads as COX-2/5-LOX inhibitors through in-vitro and in-vivo biological evaluation for anti-inflammatory activity. *Bioorg. Chem.* **2020**, *100*, 103882. [CrossRef] [PubMed]
23. Grozav, A.; Porumb, I.-D.; Găină, L.I.; Filip, L.; Hanganu, D. Cytotoxicity and Antioxidant Potential of Novel 2-(2-((1H-indol-5yl)methylene)-hydrazinyl)-thiazole Derivatives. *Molecules* **2017**, *22*, 260. [CrossRef] [PubMed]
24. Djukic, M.; Fesatidou, M.; Xenikakis, I.; Geronikaki, A.; Angelova, V.T.; Savic, V.; Pasic, M.; Krilovic, B.; Djukic, D.; Gobeljic, B.; et al. In vitro antioxidant activity of thiazolidinone derivatives of 1,3-thiazole and 1,3,4-thiadiazole. *Chem. Interact.* **2018**, *286*, 119–131. [CrossRef]
25. Brito, C.C.B.; Silva, H.V.C.D.; Brondani, D.J.; Faria, A.R.; Ximenes, R.M.; Silva, I.M.D.; Albuquerque, J.F.C.; Castilho, M.S. Synthesis and biological evaluation of thiazole derivatives as LbSOD inhibitors. *J. Enzym. Inhib. Med. Chem.* **2019**, *34*, 333–342. [CrossRef] [PubMed]
26. Rodrigues, C.A.; dos Santos, P.F.; da Costa, M.O.L.; Pavani, T.; Xander, P.; Geraldo, M.M.; Mengarda, A.C.A.; de Moraes, J.; Rando, D.G.G. 4-Phenyl-1,3-thiazole-2-amines as scaffolds for new antileishmanial agents. *J. Venom. Anim. Toxins Incl. Trop. Dis.* **2018**, *24*, 26. [CrossRef]
27. Sowjanya, C.H.; Swamy, S.S.; Gomathi, S.; Babu, A.K. Synthesis, Chemistry and Anti-Hypertensive Activity of Some New Thiazole-Thiadiazole Derivatives. *Int. J. Adv. Res. Med. Pharm. Sci.* **2016**, *1*, 6–10.
28. Salar, U.; Khan, K.M.; Chigurupati, S.; Taha, M.; Wadood, A.; Vijayabalan, S.; Ghufran, M.; Perveen, S. New Hybrid Hydrazinyl Thiazole Substituted Chromones: As Potential α-Amylase Inhibitors and Radical (DPPH & ABTS) Scavengers. *Sci. Rep.* **2017**, *7*, 16980. [CrossRef]
29. Kharb, R.; Sharma, P.C.; Yar, M.S. Pharmacological significance of triazole scaffold. *J. Enzym. Inhib. Med. Chem.* **2011**, *26*, 1–21. [CrossRef]
30. Kenchappa, R.; Bodke, Y.D.; Telkar, S.; Sindhe, M.A. Antifungal and anthelmintic activity of novel benzofuran derivatives containing thiazolo benzimidazole nucleus: An in vitro evaluation. *J. Chem. Biol.* **2016**, *10*, 11–23. [CrossRef]
31. White, C.A., Jr. Nitazoxanide: A new broad spectrum antiparasitic agent. *Expert Rev. Anti. Infect. Ther.* **2004**, *2*, 43–49. [CrossRef] [PubMed]
32. Rouf, A.; Tanyeli, C. Bioactive thiazole and benzothiazole derivatives. *Eur. J. Med. Chem.* **2015**, *97*, 911–927. [CrossRef] [PubMed]
33. Kardos, N.; Demain, A.L. Penicillin: The medicine with the greatest impact on therapeutic outcomes. *Appl. Microbiol. Biotechnol.* **2011**, *92*, 677–687. [CrossRef] [PubMed]
34. Pasqualotto, A.C.; O Thiele, K.; Goldani, L.Z. Novel triazole antifungal drugs: Focus on isavuconazole, ravuconazole and albaconazole. *Curr. Opin. Investig. drugs* **2010**, *11*, 165–174. [PubMed]
35. Thierbach, G.; Reichenbach, H. Myxothiazol, a new antibiotic interfering with respiration. *Antimicrob. Agents Chemother.* **1981**, *19*, 504–507. [CrossRef] [PubMed]
36. Borelli, C.; Schaller, M.; Niewerth, M.; Nocker, K.; Baasner, B.; Berg, D.; Tiemann, R.; Tietjen, K.; Fugmann, B.; Lang-Fugmann, S.; et al. Modes of Action of the New Arylguanidine Abafungin beyond Interference with Ergosterol Biosynthesis and in vitro Activity against Medically Important Fungi. *Chemotherapy* **2008**, *54*, 245–259. [CrossRef]
37. Li, X.-H.; Yang, X.-L.; Ling, Y.; Fan, Z.-J.; Liang, X.-M.; Wang, D.-Q.; Chen, A.F.-H.; Li, Z.-M. Synthesis and Fungicidal Activity of Novel 2-Oxocycloalkylsulfonylureas. *J. Agric. Food Chem.* **2005**, *53*, 2202–2206. [CrossRef]
38. Abbasi, M.; Nazifi, S.M.R.; Nazifi, Z.S.; Massah, A.R. Synthesis, characterization and in vitro antibacterial activity of novel phthalazine sulfonamide derivatives. *J. Chem. Sci.* **2017**, *129*, 1257–1266. [CrossRef]
39. Sławiński, J.; Pogorzelska, A.; Żołnowska, B.; Kedzia, A.; Ziółkowska-Klinkosz, M.; Kwapisz, E. Synthesis and Anti-Yeast Evaluation of Novel 2-Alkylthio-4-chloro-5-methyl-N-[imino-(1-oxo-(1H)-phthalazin-2-yl)methyl]benzenesulfonamide Derivatives. *Molecules* **2014**, *19*, 13704–13723. [CrossRef]
40. Mourad, A.K.; Makhlouf, A.A.; Soliman, A.Y.; Mohamed, S.A. Phthalazines and phthalazine hybrids as antimicrobial agents: Synthesis and biological evaluation. *J. Chem. Res.* **2020**, *44*, 31–41. [CrossRef]
41. Hussein, E.M.; Al-Rooqi, M.M.; Abd El-Galil, S.M.; Ahmed, S.A. Design, synthesis, and biological evaluation of novel N4-substituted sulfonamides: Acetamides derivatives as dihydrofolate reductase (DHFR) inhibitors. *BMC Chem.* **2019**, *13*, 91. [CrossRef] [PubMed]
42. Kachaeva, M.V.; Hodyna, D.M.; Semenyuta, I.V.; Pilyo, S.G.; Prokopenko, V.M.; Kovalishyn, V.V.; Metelytsia, L.O.; Brovarets, V.S. Design, synthesis and evaluation of novel sulfonamides as potential anticancer agents. *Comput. Biol. Chem.* **2018**, *74*, 294–303. [CrossRef] [PubMed]
43. Irfan, A.; Rubab, L.; Rehman, M.U.; Anjum, R.; Ullah, S.; Marjana, M.; Qadeer, S.; Sana, S. Coumarin sulfonamide derivatives: An emerging class of therapeutic agents. *Heterocycl. Commun.* **2020**, *26*, 46–59. [CrossRef]

44. Bonardi, A.; Nocentini, A.; Bua, S.; Combs, J.; Lomelino, C.; Andring, J.; Lucarini, L.; Sgambellone, S.; Masini, E.; McKenna, R.; et al. Sulfonamide Inhibitors of Human Carbonic Anhydrases Designed through a Three-Tails Approach: Improving Ligand/Isoform Matching and Selectivity of Action. *J. Med. Chem.* **2020**, *63*, 7422–7444. [CrossRef] [PubMed]
45. Gokcen, T.; Gulcin, I.; Ozturk, T.; Goren, A.C. A class of sulfonamides as carbonic anhydrase I and II inhibitors. *J. Enzym. Inhib. Med. Chem.* **2016**, *31* (Suppl. 2), 180–188. [CrossRef] [PubMed]
46. Esirden, I.; Tanç, M.; Supuran, C.T.; Kaya, M. Microwave assisted synthesis of novel tetrazole/sulfonamide derivatives based on octahydroacridine, xanthene and chromene skeletons as inhibitors of the carbonic anhydrases isoforms I, II, IV and VII. *Bioorg. Med. Chem. Lett.* **2016**, *27*, 86–89. [CrossRef]
47. Zhang, J.; Tan, Y.; Li, G.; Chen, L.; Nie, M.; Wang, Z.; Ji, H. Coumarin Sulfonamides and Amides Derivatives: Design, Synthesis, and Antitumor Activity In Vitro. *Molecules* **2021**, *26*, 786. [CrossRef]
48. Azzam, R.A.; Elboshi, H.A.; Elgemeie, G.H. Novel Synthesis and Antiviral Evaluation of New Benzothiazole-Bearing *N*-Sulfonamide 2-Pyridone Derivatives as USP7 Enzyme Inhibitors. *ACS Omega* **2020**, *5*, 30023–30036. [CrossRef]
49. Dash, R.N.; Moharana, A.K.; Subudhi, B.B. Sulfonamides: Antiviral Strategy for Neglected Tropical Disease Virus. *Curr. Org. Chem.* **2020**, *24*, 1018–1041. [CrossRef]
50. Qadir, M.A.; Ahmed, M.; Iqbal, M. Synthesis, Characterization, and Antibacterial Activities of Novel Sulfonamides Derived through Condensation of Amino Group Containing Drugs, Amino Acids, and Their Analogs. *BioMed Res. Int.* **2015**, *2015*, 938486. [CrossRef]
51. Beheshtimaal, K.; Khazaeili, T.; Asakere, N.; Mousavi, F.; Massah, A.R.; Assakere, N. Synthesis of Some Novel Sulfonamide-imines as Potential Antimicrobial Agents. *Lett. Org. Chem.* **2018**, *15*, 111–117. [CrossRef]
52. Gao, H.-D.; Liu, P.; Yang, Y.; Gao, F. Sulfonamide-1,3,5-triazine–thiazoles: Discovery of a novel class of antidiabetic agents via inhibition of DPP-4. *RSC Adv.* **2016**, *6*, 83438–83447. [CrossRef]
53. Zajdel, P.; Marciniec, K.; Maślankiewicz, A.; Grychowska, K.; Satała, G.; Duszyńska, B.; Lenda, T.; Siwek, A.; Nowak, G.; Partyka, A.; et al. Antidepressant and antipsychotic activity of new quinoline- and isoquinoline-sulfonamide analogs of aripiprazole targeting serotonin 5-HT_1A/5-HT_2A/5-HT_7 and dopamine D_2/D_3 receptors. *Eur. J. Med. Chem.* **2013**, *60*, 42–50. [CrossRef] [PubMed]
54. Gouda, M.; Hussein, B. Synthesis and Anti-Oxidant Evaluation of Some Novel Sulfa Drugs. *Lett. Drug Des. Dis.* **2017**, *14*, 1425–1432. [CrossRef]
55. Ivasiv, V.; Albertini, C.; Gonçalves, A.E.; Rossi, M.; Bolognesi, M.L. Molecular Hybridization as a Tool for Designing Multitarget Drug Candidates for Complex Diseases. *Curr. Top. Med. Chem.* **2019**, *19*, 1694–1711. [CrossRef] [PubMed]
56. Viegas-Junior, C.; Danuello, A.; da Silva Bolzani, V.; Barreiro, E.J.; Fraga, C.A.M. Molecular Hybridization: A Useful Tool in the Design of New Drug Prototypes. *Curr. Med. Chem.* **2007**, *14*, 1829–1852. [CrossRef]
57. Horishny, V.; Kartsev, V.; Matiychuk, V.; Geronikaki, A.; Anthi, P.; Pogodin, P.; Poroikov, V.; Ivanov, M.; Kostic, M.; Soković, M.D.; et al. 3-Amino-5-(indol-3-yl)methylene-4-oxo-2-thioxothiazolidine Derivatives as Antimicrobial Agents: Synthesis, Computational and Biological Evaluation. *Pharmaceuticals* **2020**, *13*, 229. [CrossRef]
58. Horishny, V.; Kartsev, V.; Geronikaki, A.; Matiychuk, V.; Petrou, A.; Glamoclija, J.; Ciric, A.; Sokovic, M. 5-(1H-Indol-3-ylmethylene)-4-oxo-2-thioxothiazolidin-3-yl)alkancarboxylic Acids as Antimicrobial Agents: Synthesis, Biological Evaluation, and Molecular Docking Studies. *Molecules* **2020**, *25*, 1964. [CrossRef]
59. Rotilie, C.A.; Fass, R.J.; Prior, R.B.; Perkins, R.L. Microdilution technique for antimicrobial susceptibility testing of anaerobic bacteria. *Antimicro. Agents Chemother.* **1975**, *7*, 311–315. [CrossRef]
60. Benson, T.E.; Walsh, C.T.; Massey, V. Kinetic Characterization of Wild-Type and S229A Mutant MurB: Evidence for the Role of Ser 229 as a General Acid. *Biochemistry* **1997**, *36*, 796–805. [CrossRef]
61. Zubenko, A.A.; Morkovnik, A.S.; Divaeva, L.N.; Demidov, O.P.; Kartsev, V.G.; Sochnev, V.S.; Klimenko, A.I.; Dobaeva, N.M.; Borodkin, G.S.; Bodryakov, A.N.; et al. Thiourea assisted recyclization of 1-(chloromethyl)dihydroisoquinolines: A convenient route to β-(o-thiazolylaryl)ethylamines. *Mendeleev Commun.* **2021**, *31*, 125–127. [CrossRef]
62. Simakov, S.; Kartsev, V.; Petrou, A.; Nicolaou, I.; Geronikaki, A.; Ivanov, M.; Kostic, M.; Glamočlija, J.; Soković, M.; Talea, D.; et al. 4-(Indol-3-yl)thiazole-2-amines and 4-ındol-3-yl)thiazole Acylamines as Novel Antimicrobial Agents: Synthesis, In Silico and In Vitro Evaluation. *Pharmaceuticals* **2021**, *14*, 1096. [CrossRef] [PubMed]
63. Victor, K.; Boris, L.; Athina, G.; Anthi, P.; Marija, S.; Marina, K.; Oliver, R.; Marina, S. Design, synthesis and antimicrobial activity of usnic acid derivatives. *MedChemComm* **2018**, *9*, 870–882. [CrossRef] [PubMed]
64. Kostić, M.; Smiljković, M.; Petrović, J.; Glamočlija, J.; Barros, L.; Ferreira, I.C.F.R.; Ćirić, A.; Soković, M. Chemical, nutritive composition and a wide range of bioactive properties of honey mushroom Armillaria mellea (Vahl: Fr.) Kummer. *Food Funct.* **2017**, *8*, 3239–3249. [CrossRef]
65. Kritsi, E.; Matsoukas, M.-T.; Potamitis, C.; Detsi, A.; Ivanov, M.; Sokovic, M.; Zoumpoulakis, P. Novel Hit Compounds as Putative Antifungals: The Case of Aspergillus fumigatus. *Molecules* **2019**, *24*, 3853. [CrossRef] [PubMed]
66. Aleksić, M.; Stanisavljević, D.; Smiljković, M.; Vasiljević, P.; Stevanović, M.; Soković, M.; Stojković, D. Pyrimethanil: Between efficient fungicide against Aspergillus rot on cherry tomato and cytotoxic agent on human cell lines. *Ann. Appl. Biol.* **2019**, *175*, 228–235. [CrossRef]

67. Cady, N.C.; McKean, K.A.; Behnke, J.; Kubec, R.; Mosier, A.P.; Kasper, S.H.; Burz, D.S.; Musah, R.A. Inhibition of Biofilm Formation, Quorum Sensing and Infection in Pseudomonas aeruginosa by Natural Products-Inspired Organosulfur Compounds. *PLoS ONE* **2012**, *7*, e38492. [CrossRef]
68. Wolber, G.; Lange, T. LigandScout: 3-D Pharmacophores Derived from Protein-Bound Ligands and Their Use as Virtual Screening Filters. *J. Chem. Inf. Model.* **2005**, *45*, 160–169. [CrossRef]
69. Huey, R.; Morris, G.; Olson, A.J.; Goodsell, D.S. A semiempirical free energy force field with charge-based desolvation. *J. Comput. Chem.* **2007**, *28*, 1145–1152. [CrossRef]

antibiotics

Article

Halogenase-Targeted Genome Mining Leads to the Discovery of (±) Pestalachlorides A1a, A2a, and Their Atropisomers

Mengna Luo [†], Mengyuan Wang [†], Shanshan Chang, Ning He, Guangzhi Shan * and Yunying Xie *

CAMS Key Laboratory of Synthetic Biology for Drug Innovation, Institute of Medicinal Biotechnology, Chinese Academy of Medical Sciences & Peking Union Medical College, Tiantan Xili No.1, Beijing 100050, China
* Correspondence: shanguangzhi@imb.pumc.edu.cn (G.S.); xieyy@imb.pumc.edu.cn (Y.X.)
† These authors contributed equally to this work.

Abstract: Genome mining has become an important tool for discovering new natural products and identifying the cryptic biosynthesis gene clusters. Here, we utilized the flavin-dependent halogenase GedL as the probe in combination with characteristic halogen isotope patterns to mine new halogenated secondary metabolites from our in-house fungal database. As a result, two pairs of atropisomers, pestalachlorides A1a (**1a**)/A1b (**1b**) and A2a (**2a**)/A2b (**2b**), along with known compounds pestalachloride A (**3**) and SB87-H (**4**), were identified from *Pestalotiopsis rhododendri* LF-19-12. A plausible biosynthetic assembly line for pestalachlorides involving a putative free-standing phenol flavin-dependent halogenase was proposed based on bioinformatics analysis. Pestalachlorides exhibited antibacterial activity against sensitive and drug-resistant *S. aureus* and *E. faecium* with MIC values ranging from 4 μg/mL to 32 μg/mL. This study indicates that halogenase-targeted genome mining is an efficient strategy for discovering halogenated compounds and their corresponding halogenases.

Keywords: pestalachlorides; halogenase; genome-mining; flavin-dependent halogenases (FDHs); atropisomers

check for updates

Citation: Luo, M.; Wang, M.; Chang, S.; He, N.; Shan, G.; Xie, Y. Halogenase-Targeted Genome Mining Leads to the Discovery of (±) Pestalachlorides A1a, A2a, and Their Atropisomers. *Antibiotics* **2022**, *11*, 1304. https://doi.org/10.3390/antibiotics11101304

Academic Editors: Fuhang Song, Yunjiang Feng and Michal Letek

Received: 29 July 2022
Accepted: 22 September 2022
Published: 25 September 2022

1. Introduction

Halogenated compounds play a profound role in the pharmaceutical industry as halogen substituents can significantly impact the bioactivity and reactivity of organic compounds [1–3]. According to an economic report, 88% of the 100 top-selling drugs employed chlorine in their final pharmaceutical products or the manufacturing process [4]. Nature is an important source of halogenated compounds. To date, over 5000 halogenated natural products have been discovered from fungi, bacteria, algae, cyanobacteria, plants, et al. [5]. Amongst, fungi as the third kingdom in nature, contributed nearly one-fifth (988) of halogenated metabolites [6] and are expected to harbor many more halogenated natural products to be identified [7].

Nature usually orchestrates halogen-carbon bond formation by a variety of halogenases. Several types of halogenases have been identified so far, including heme- or vanadium-dependent haloperoxidases, *S*-adenosyl-L-methionine-dependent halogenases, nonheme-iron α-ketoglutarate-dependent halogenases, and flavin-dependent halogenases (FDHs) [2,8–10]. Amongst halogenases, FDHs are widely distributed across all kingdoms of life [11] and are particularly notable for their strong regioselectivity and substrate diversity [2,12]. Almost all FDHs have the following two conserved motifs: A flavin-binding motif GxGxxG, for binding of the diffusible flavin adenine dinucleotide (FAD) [3], and a structural motif WxWxIP, thought to prevent a monooxygenation reaction by blocking direct contact between the substrate and hydroperoxy flavin [13,14]. These signature motifs can be used as probes for promptly identifying putative FDHs from genomic sequences. [2,3]

Fungi are a rich source of flavin-dependent halogenases (FDHs). Up to now, twenty-three halogenases have been reported from fungi, twenty of which are FDHs [7]. As for the substrates, FDHs prefer electron-rich precursors, such as phenols, indoles, or pyrroles. The phenol-containing structure is the most common substrate of the identified fungal FDHs, such as GedL from *Aspergillus terreus* NIH2624, which dichlorinated the phenol unit of sulochrin to produce dihydrogeodin, PtaM from *Pestalotiopsis fici*, which assembled one chloride atom to isosulochrin in the pestheic acid biosynthesis, and GsfI from *Penicillium aethiopicum*, which decorated griseophenone C with one chloride atom. Apart from the identified FDHs, thousands of putative FDHs were inferred in fungal genomes according to bioinformatics analysis [7], which hints that there are more halogenated metabolites or haloganases in fungi to be awaiting exploration. With the development of bioinformatics and the increasing decrease in genome sequencing costs, genome mining has become a powerful strategy for discovering new natural products or unearthing cryptic biosynthesis gene clusters [15–17]. An increasing number of chemical scaffolds, such as unusual post-translationally modified ribosomal peptide linaridins [18], PKS-NRPS hybrid aspyridones [19], and noncanonical polyketide burkholderic acid [20], have been discovered by genome mining. Although genome mining often involves genetic manipulation, including heterologous expression, in vitro reconstitution, and activation of the BGC in the native host, the non-genetic method sometimes shows high efficiency for mining the metabolites with characteristic features that can be easily detected using specific analysis methods. Halogenated compounds often exhibit characteristic isotope patterns in their mass spectra due to the presence of chlorine or bromine atoms, which makes them readily detectable from a complicated background. Additionally, according to the Natural Products Atlas database [6], more than 99% of halogenated microbial natural products are chlorinated or brominated ones, which consolidates the power of LC-MS in genome mining of halogenated natural products.

As a part of our efforts to investigate new natural products [21–25], we used the fungal FDH GedL [26] as a probe to explore the halogenase-containing BGCs from our in-house fungal genome database. A putative halogenase gene, *ptlK*, was mined from an endolichenic fungus, *Pestalotiopsis rhododendri* LF-19-12, and further bioinformatics analysis disclosed that *ptlK* was located in a cryptic BGC *ptl*. Subsequently, LC-MS was employed to interrogate the production of halogenated metabolites. As a result, a family of potential chlorinated compounds with characteristic chlorine isotope patterns were detected in the crude extract of *Pestalotiopsis rhododendri* LF-19-12 culture in the M2 medium. LC-UV-MS guided isolation led to obtaining two pairs of atropisomers, pestalachlorides A1a (**1a**)/A1b (**1b**) and A2b (**2a**)/A2b (**2b**), along with known compounds pestalachloride A (**3**) [27] and SB87-H (**4**) [28] (Figure 1). Here, we reported their discovery, isolation, structural elucidation, and biosynthesis.

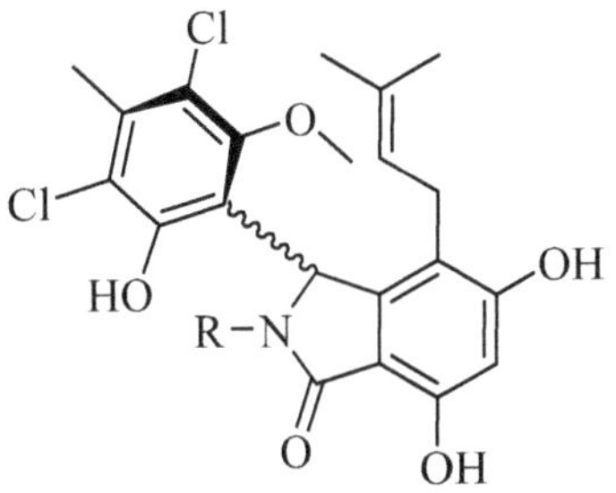

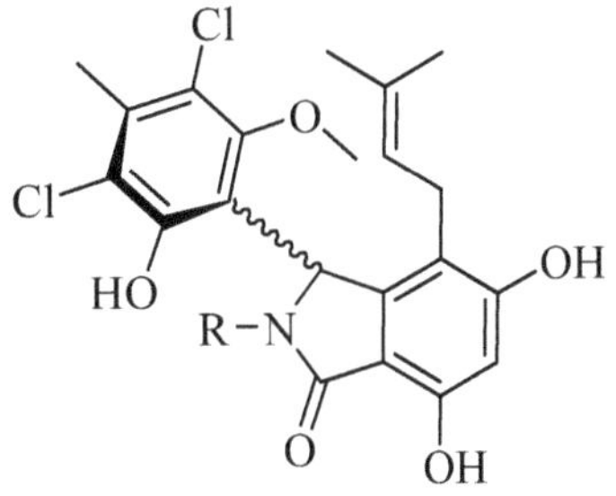

1a (rac) R= $-CH_2CH_2OH$

2a (rac) R= $-CH_2CH_2CH_2COOH$

1b (rac) R= $-CH_2CH_2OH$

2b (rac) R= $-CH_2CH_2CH_2COOH$

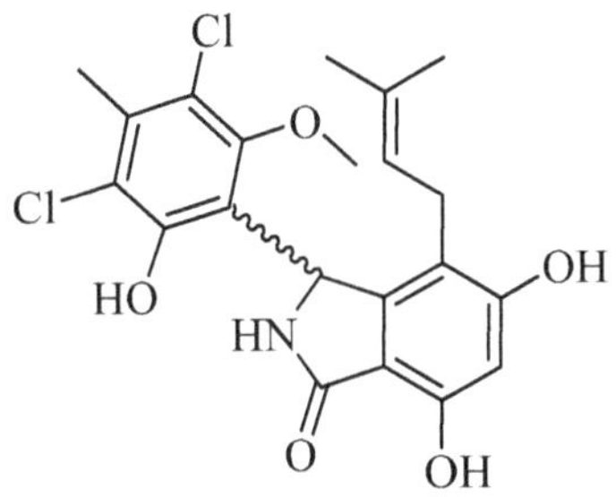

3 (rac, atropisomers)

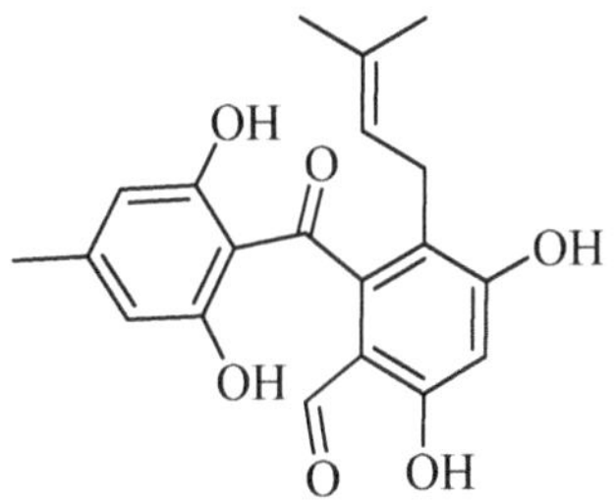

4

Figure 1. Structures of pestalachlorides A1a (**1a**), A1b (**1b**), A2a (**2a**), A2b (**2b**), and A (**3**) as well as SB87-H (**4**).

2. Results

2.1. Genome Mining of the Halogenase-Containing Biosynthesis Gene Cluster

Flavin-dependent halogenases (FDHs), the most characterized halogenases based on their substrates, can be categorized into the following five main classes: free-standing phenol, free-standing indole, carrier protein-dependent phenol, carrier protein-dependent pyrrole, and aliphatic FDHs [9]. GedL is a free-standing phenol FDH from *Aspergillus terreus* NIH2624 [26]. It is involved in the biosynthesis of geodin and halogenates the substrate at the late stage of biosynthesis [26]. Here, we used GedL as the probe to conduct tBlastp analysis on our in-house fungal genome sequences. An antiSMASH analysis was subsequently performed, and a gene *ptaK*, encoding a putative flavin-dependent halogenase with 51% amino acid sequence identity to GedL [26], was found to be contained in a cryptic BGC of endolichenic *Pestalotiopsis rhododendri* LF-19-12. Succeeding phylogenetic analysis showed that PtlK grouped with free-standing phenol FDHs (Figure 2), suggesting that its substrate might hold a phenol moiety.

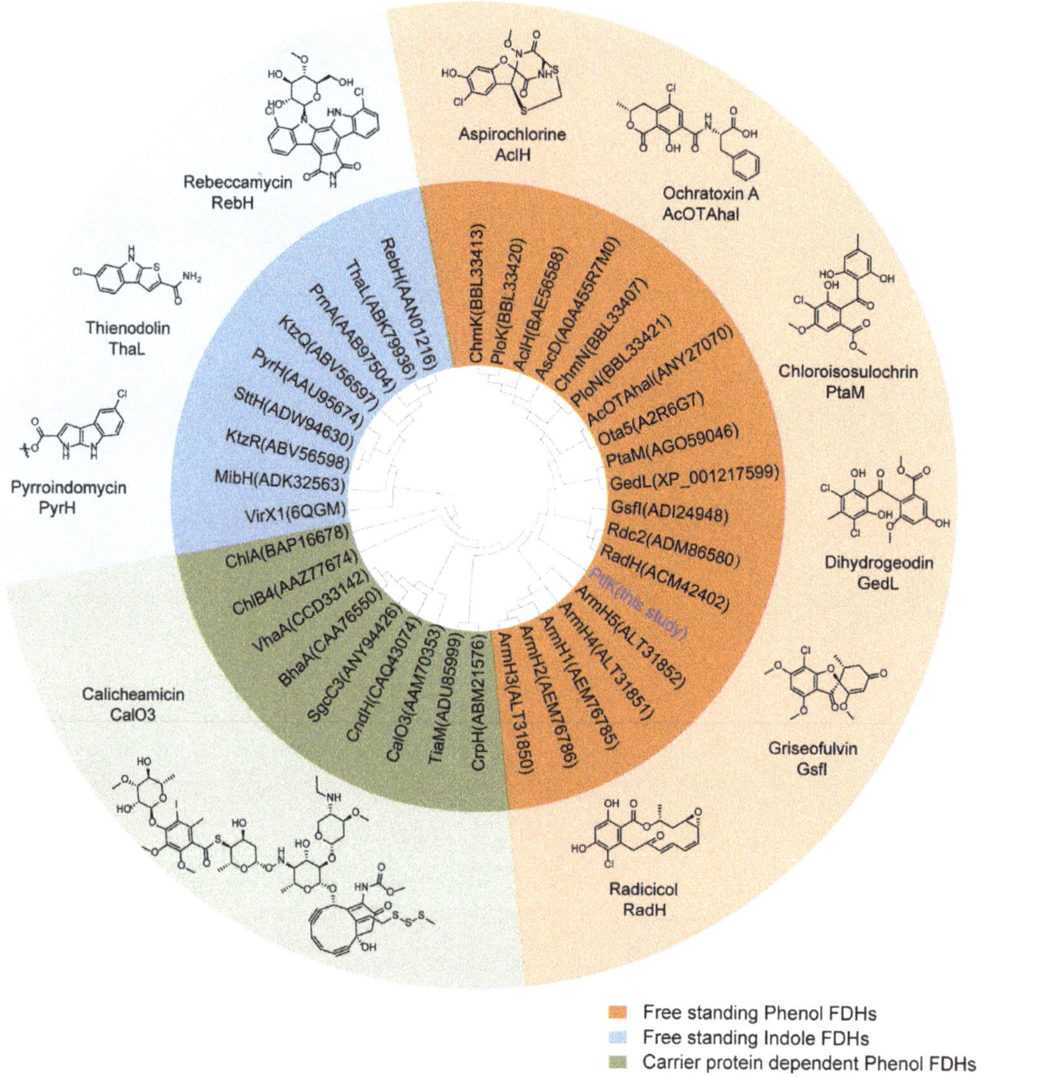

Figure 2. Phylogenetic tree based on amino acid sequences of PtlK and the selected flavin-dependent halogenases (FDHs). GenBank, UniProtKB, or PDB accession numbers are given in parentheses. The three dominant categories of FDHs: free-standing phenol, free-standing indole, and carrier protein-dependent phenol FDHs are highlighted in orange, blue, and green, respectively. Representative products are shown beside. The phylogenetic tree was constructed using the UPGMA method. Visualization was conducted with MEGA7.

Subsequently, LC-MS and OSMAC strategies were employed to exploit the production of halogenated secondary metabolites. *Pestalotiopsis rhododendri* LF-19-12 was cultured in four different media (M1, M2, PDB, and YES) and then extracted using MeOH. The obtained material was applied to LC-MS analysis. As a result, a group of potential halogenated compounds with characteristic isotope patterns of two chloride atoms were detected in the crude extract of the *Pestalotiopsis rhododendri* LF-19-12 culture in the M2 medium (Figure 3).

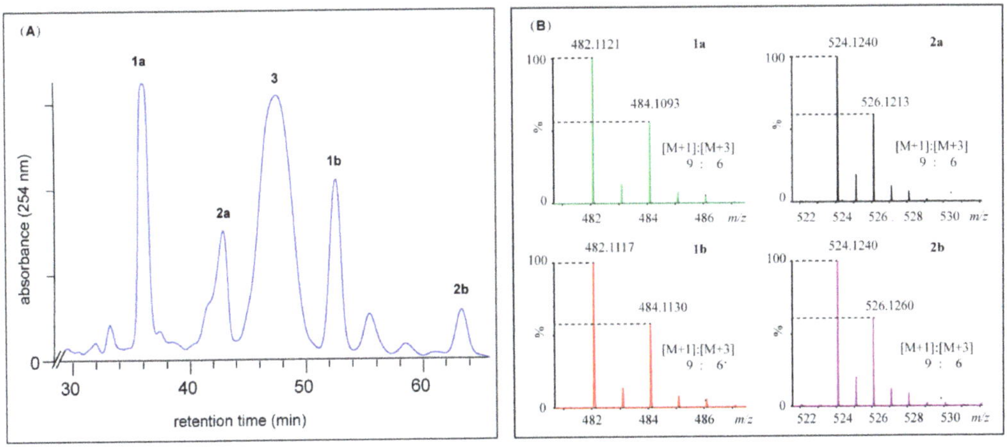

Figure 3. HPLC-UV chromatogram of the crude extract of *Pestalotiopsis rhododendri* LF-19-12 cultured in M2 medium (**A**) and the isotope patterns of peaks **1a**, **1b**, **2a**, and **2b** (**B**), indicating that two chloride atoms were contained.

2.2. Structural Elucidation for (±) Pestalachlorides A1a, A1b, A2a, and A2b

Pestalotiopsis rhododendri LF-19-12 was fermented in the M2 medium, and pestalachlorides A1a (**1a**), A1b (**1b**), A2a (**2a**), A2b (**2b**), and A (**3**), as well as SB87-H (**4**), were isolated and purified by an LC-UV-MS-guided method from a 9-day broth culture. Briefly, the culture of *Pestalotiopsis rhododendri* LF-19-12 was filtered, and the obtained mycelia were extracted with acetone. The yielded crude extract was fractionated and separated sequentially using silica gel, an ODS flash column, and further purified by semi-preparative chromatography to yield **1a** (23.2 mg), **1b** (2.2 mg), **2a** (1.9 mg), and **2b** (0.2 mg).

The HR-ESIMS spectrum of compound **1a** revealed a characteristic isotope pattern of double chlorides (Figure 3). Furthermore, analysis of HR-ESIMS and ^{13}C NMR data disclosed that **1a** has a molecular formula of $C_{23}H_{26}Cl_2NO_6$ ($[M+H]^+$ m/z 482.1121, calcd. 482.1137). Interpretation of the ^{1}H, ^{13}C NMR, and HSQC data for **1a** (Table 1, Figures S1–S3) disclosed a carbonyl group (δ_C 168.6), 12 aromatic carbons, one of which is protonated, a trisubstituted olefin, a methine, three methylene units, one of which is attached to an oxygen atom, four methyl moieties, one of which is methoxy, and three phenolic hydroxyl groups. All the above interpretations accounted for 8 degrees of unsaturation and required **1a** to incorporate three rings, two of which should be aryl rings.

^{1}H-^{1}H COSY correlations (Figures 4 and S3) revealed two isolated proton spin-systems attributed to -CH_2-CH_2-OH and -CH_2-CH= (Figure 4). Furtherly, an isoprenyl unit in **1a** was established by HMBC correlations (Figures 4 and S5) from H-4′ and H-5′ to vinylic carbons C-3′ and C-2′. HMBC correlations from H-1′ and H-2′ to C-6 suggested that the isoprenyl group was connected to the aromatic ring at C-6. Two phenolic hydroxyl groups at C-5 and C-3, respectively, can be inferred by the downfield chemical shifts of C-3 and C-5. Further correlations from H-4 to C-2, C-6, C-3, C-5, and C-1, from H-8 to C-6, C-2, and C-1, as well as from H-16 to C-8 and C-1, allowed construction of the substituted isoindole-1-one scaffold.

Table 1. ^{1}H and ^{13}C NMR data of pestalachlorides A1a (**1a**)/A1b (**1b**) and A2b (**2a**)/A2b (**2b**) in DMSO-d_6.

No.	1a		1b *		2a		2b *
	δ_C, Type	δ_H, Multi. (J in Hz)	δ_C, Type	δ_H, Multi. (J in Hz)	δ_C, Type	δ_H, Multi. (J in Hz)	δ_H, Multi. (J in Hz)
1	168.6, C		169.2, C		168.5, C		
2	109.2, C		109.7, C		109.0, C		
3	153.9, C		153.6, C		153.9, C		
4	101.3, CH	6.32, s	101.2, CH	6.28, s	101.5, CH	6.32, s	6.29, s
5	159.7, C		159.6, C		159.7, C		
6	114.1, C		114.3, C		114.2, C		
7	146.5, C		145.7, C		146.5, C		
8	55.1, CH	6.13, s	57.1, CH	6.02, s	54.5, CH	6.09, s	5.96, s
9	117.8, C		117.0, C		119.1, C		
10	151.1, C		151.6, C		151.2, C		
11	117.8, C		117.7, C		117.9, C		
12	134.9, C		134.8, C		134.9, C		
13	119.1, C		118.5, C		117.7, C		
14	154.7, C		154.3, C		154.7, C		
15	59.8, -OCH$_3$	3.05, s	61.5, -OCH$_3$	3.97, s	59.8, -OCH$_3$	3.04, s	3.97, s
16	41.8, CH$_2$	2.58, dt (13.3, 5.4) 3.70, dt (13.9, 6.9)	42.3, CH$_2$	2.85–2.78, m 3.40–3.35, m	38.4, CH$_2$	2.56, dt (13.9, 5.9) 3.63, dt (14.7, 7.6)	2.84, dt (13.6, 6.6) 3.54–3.46, m
17	58.3, CH$_2$	3.45, dt (11.5, 5.1)	58.8, CH$_2$	3.59–3.50, m	23.2, CH$_2$	1.69, m	1.69, m
18	-	-	-	-	31.1, CH$_2$	2.16, t (7.4)	2.18, t (6.6)
1′	23.8, CH$_2$	2.78, dd (15.1, 6.7) 2.93, dd (15.4, 5.2)	23.6, CH$_2$	2.73, dd, (15.4, 6.2) 2.96, dd, (15.5, 5.5)	23.9, CH$_2$	2.77, dd (15.1, 6.7) 2.94, dd (15.4, 5.0)	2.72, dd (15.7, 5.6) 2.94, dd, (15.4, 5.0)
2′	122.3, CH	4.37, t (5.4)	122.6, CH	4.16, t (5.8)	122.3, CH	4.36, m	4.16, m
3′	129.7, C		129.5, C		129.7, C		
4′	25.1, CH$_3$	1.31, s	25.1, CH$_3$	1.29, s	25.1, CH$_3$	1.31, s	1.29, s
5′	17.4, CH$_3$	1.39, s	17.5, CH$_3$	1.40, s	17.4, CH$_3$	1.38, s	1.39, s
1″	18.1, CH$_3$	2.39, s	18.1, CH$_3$	2.38, s	18.1, CH$_3$	2.39, s	2.38, s
OH-3		9.08, s		8.98, s		9.09, s	8.98, s
OH-5		9.86, s		9.25, s		9.88, s	9.25, s
OH-10		10.03, s		9.79, s		10.06, s	9.81, s
OH-17		3.36, s		4.78, s	-	-	-
COOH	-	-	-	-	174.0, C	11.99, s	11.99, s

*: For **1b**, the data was obtained from the mixture of **1a** and **1b** by comparison with those of **1a**. For **2b**, only ^{1}H NMR spectrum was obtained due to its trace amount and instability.

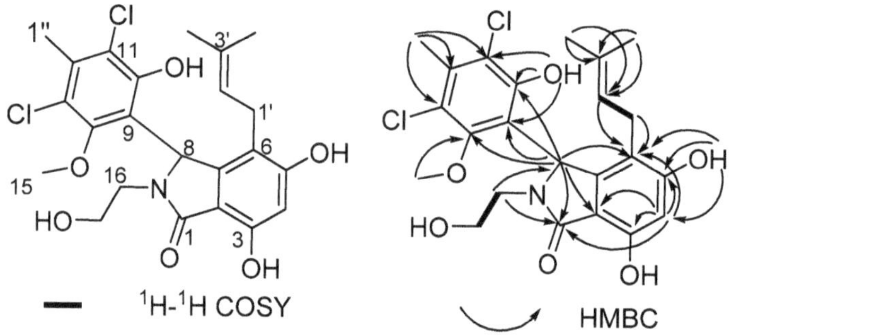

Figure 4. The planar structure and key 2D NMR correlations of **1a**.

HMBC correlations from H-1″ to C-11, C-12, and C-13, from the phenolic proton at δ 10.03 to C-9, C-10, and C-11, from the methoxy protons at δ 3.05 to C-14, from H-8 to C-9, C-10, and C-14 indicated that a hexasubstituted benzene ring was attached to C-8 via a C-C bond. As a result, the two chlorine atoms in **1a** could only be located at C-11 and C-13. Therefore, the planar structure of **1a** was assembled as shown in Figure 4.

The structure of **1a** was further confirmed by the single crystal X-ray analysis. The crystallographic data disclosed that **1a** featured a centrosymmetric space group *P*121/*c*1, suggestive of its being a racemate of 8*R* and 8*S* enantiomers (Figure 5).

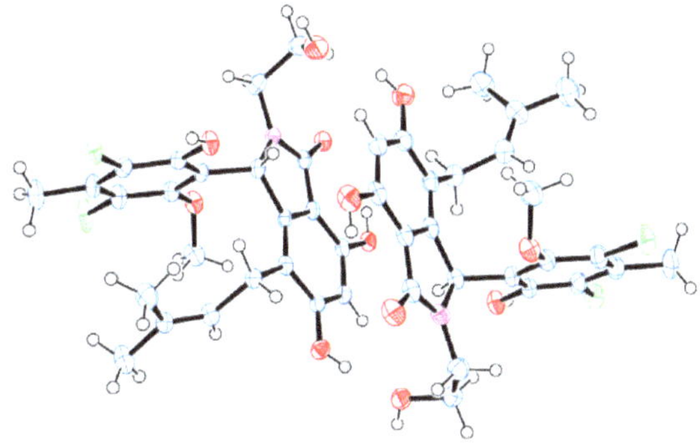

Figure 5. Perspective ORTEP drawing for (±) pestalachloride A1a (**1a**).

Compound **1b**, an isomer of **1a** by HR-ESIMS analysis, was quickly converted into **1a** in acetonitrile aqueous. Therefore, only a mixture of **1a** and **1b** was obtained. The ^{1}H NMR spectrum of the mixture displayed the following two sets of signals (Figure S6): one set of signals is identical to that of **1a**, and the others are nearly similar to those of **1a** except for the methoxy proton chemical shifts (δ 3.97, deshielded for **1b** vs. δ 3.05, shielded for **1a**), suggestive of **1b** as an atropisomer of **1a**. The ^{13}C NMR spectrum (Figure S7) further supported the above hypothesis. To our knowledge, the analog of **1a** and **1b**, pestalachloride A (**3**) from *Pestalotiopsis adusta*, also has atropisomer axial chirality due to the hindered rotation around the C8-C9 bond, but its two atropisomers could not be chromatographically separated [27].

The HR-ESIMS analysis of compound **2a** returned a molecular formula of $C_{25}H_{29}Cl_2NO_7$ ($[M+H]^+$ *m/z* 524.1240, calcd. 524.1243). The ^{1}H and ^{13}C NMR signals of **2a** are very closely related to those of **1a** except for the signals of the *N*-substituent as follows: $\delta_{C\text{-}16}$ 38.4/$\delta_{H\text{-}16}$ 2.56, 3.63, $\delta_{C\text{-}17}$ 23.2/$\delta_{H\text{-}17}$ 1.69, $\delta_{C\text{-}18}$ 31.1/$\delta_{H\text{-}18}$ 2.16, and $\delta_{C\text{-}19}$ 174.0/$\delta_{H\text{-}COOH}$ 11.99, indicative of a fragment of -CH_2-CH_2-CH_2-COOH (Figures S8–S10). The above proposed *N*-substituent was further confirmed by ^{1}H-^{1}H COSY correlations of H-16/H-17/H-18 and HMBC cross signals from H-17 and H-18 to C-19 as well as from H-16 to C-1 and C-8, respectively (Figures S11 and S12). The upfield methoxyl proton signals at δ 3.04 indicated that the methoxy was located in the shielded area of the isoindole-1-one residue. A careful examination of the NMR spectra of **2a** disclosed the presence of a minor component **2b,** which was subsequently proved to be an atropisomer of **2a**. Compound **2a** showed no optical activity, suggestive of it also being a racemate.

HR-ESIMS analysis revealed **2b** as an isomer of **2a.** By comparison with the ^{1}H NMR spectra of **2a**, that of **2b** exhibited nearly identical signals to those of the minor component in **2a** (Figure S13). The methoxyl proton signals of **2b** (δ 3.97), downfield relative to those of **2a** (δ 3.04), inferred that the methoxy in **2b** was located in the deshielded area of the isoindole-1-one residue. **2a** and **2b** can also be interconverted with each other at room temperature.

As proved above, axial chirality was present for pestalachlorides A, A1, and A2, which resulted in time-dependent atropisomerism. To interrogate the stability of pestalachlorides atropisomers, we calculated the relative Gibbs energy barriers for the atropisomers interconversions at the M062X/def2TZVP/SMD (H_2O)//B3LYP/6-31G(d)/PCM (H_2O) level. The results disclosed that the barriers of **1a** to **1b** and **1b** to **1a** were 24.6 kcal/mol and 24.4 kcal/mol, and the corresponding interconversion half-times were 34 h and 24 h at room temperature, respectively, in agreement with the fact that **1b** is a little more unstable than **1a**; the barriers between two atropisomers of pestalachloride A were 21.4 kcal/mol and 21.6 kcal/mol, and the corresponding interconversion half-times were 0.15 h and 0.23 h, respectively, supporting their inseparability; the barriers of **2a** and **2b** interconversion were

26.9 kcal/mol and 27.4 kcal/mol, respectively, indicating that they can also interconvert with each other [29].

2.3. Proposed Biosynthetic Pathway for Pestalachlorides

Based on an analysis of the functions of genes within the *ptl* cluster, as well as a comparison with the previous reported biosynthetic assembly lines of geodin [26], pestheic acid [30], and monodictyphenone [31] (Table 2), a plausible biosynthesis pathway for pestalachlorides was proposed (Scheme 1). The non-reducing polyketide synthase PtlB, showing 87%, 66%, and 63% of sequence identity with PtaB [30], mdpG [31], and GedC [26], respectively, was proposed to assemble and cyclize atrochrysone thioester (**5**). PltA, with high amino acid identity (92%) to PtaB [30], was reasoned to hydrolyze the thioester bond of **5** to release atrochrysone carboxylic acid (**6**) from PtlB. The following concerted decarboxylation and dehydration were proposed according to monodictyphenone biosynthetic logic. However, no gene encoding putative decarboxylase as MdpH was found within and near the *ptl* cluster, suggesting that the intermediate **6** might undergo spontaneous decarboxylation and dehydration to form emodin anthrone (**7**) as indicated by an earlier study [31,32]. **7** was subsequently oxidized to emodin (**8**) by a putative anthrone oxygenase PtlJ, which showed 44%, 41%, and 43% identity to GedH [26], PtaC [30], and MdpH2 [33], separately. According to the retrosynthesis analysis of pestalachlorides, **8** should be converted to alatinone (**9**); however, the mechanism of this conversion remains to be determined. Subsequently, as the previous study suggested [34], **9** might be cleaved to **10** via oxidation and thioesterification catalyzed by PtlC, a putative Baeyer-Villiger oxidase (47% identity with PtaJ [30]), and PtlF, a putative glutathione *S*-transferase (39% identity with MdpJ [31]), respectively. A succeeding reduction of **10** to the aldehyde **11** might be catalyzed by oxidoreductase PtlM, which displayed 51% identity to MdpK, and the latter was proposed to reduce thioester to benzaldehyde in arugosin F biosynthesis [34]. A putative prenyltransferase PtlH, the homolog of which is absent in Geodin, pestheic acid, and monodictyphenone biosynthetic assembly lines, showed 40% sequence identity with xanthone prenyltransferase A [35] and thus was postulated to C-prenylate **11** to give **4**. Subsequently, halogenation by FDH halogenase PtlK and methylation by O-methyltransferase PtlI occurred to give pestalone, which can be spontaneously reacted with primary amide to give compounds **1a/1b**, **2a/2b**, and **3** [36,37].

Table 2. Analysis of the pestalachloride biosynthetic gene cluster by comparison with geodin (*ged*), pestheic acid (*pta*), and monodictyphenone (*mdp*) BGCs.

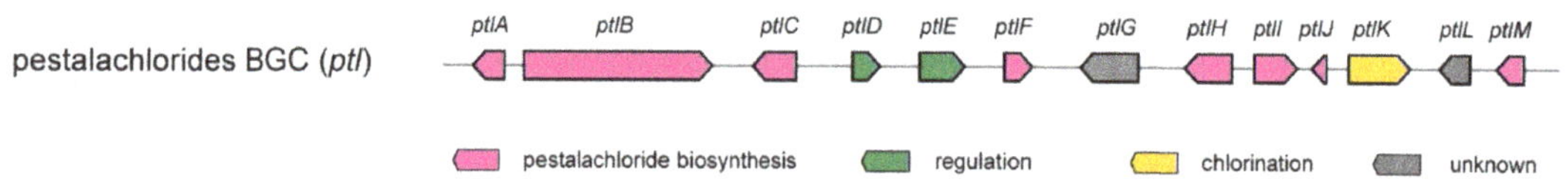

Genes	Putative Function	Ged Homolog (% id.)	Pta Homolog (% id.)	Mdp Homolog (% id.)
ptlA	Lactamase B	GedB (70)	PtaB (92)	MdpF (67)
ptlB	non-reducing PKS	GedC (63)	PtaA (87)	MdpG (66)
ptlC	Baeyer-Villiger oxidase	GedK (45)	PtaJ (47)	MdpL (43)
ptlD	transcriptional regulator	GedD (39)	PtaR1 (36)	MdpA (38)
ptlE	transcriptional regulator	GedR (61)	PtaR2 (29)	MdpE (29)
ptlF	Glutathione S-transferase	-	-	MdpJ (39)
ptlG	Pyranose dehydrogenase	-	-	-
ptlH	Xanthone prenyltransferase	-	-	-
ptlI	O-methyltransferase	-	-	-
ptlJ	Anthrone oxygenase	GedH (44)	PtaC (41)	MdpH2 (43)
ptlK	Flavine halogenase	GedL (51)	PtaM (47)	-
ptlL	Short-chain dehydrogenase	-	-	MdpC (25)
ptlM	Oxidoreductase	GedF (48)	PtaF (49)	MdpK (51)

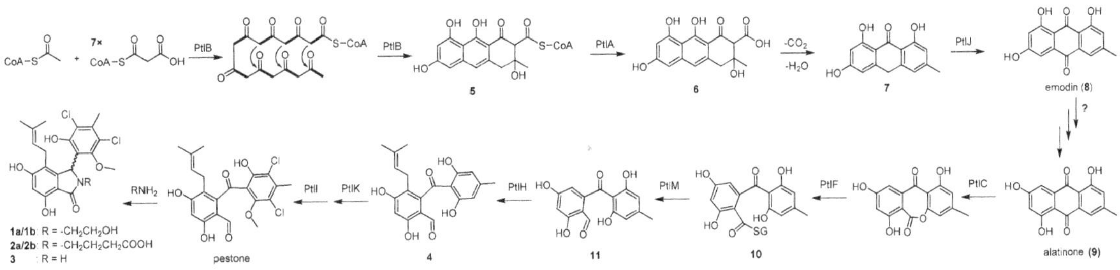

Scheme 1. A plausible biosynthesis pathway for pestalachlorides A1a (**1a**), A1b (**1b**), A2a (**2a**), A2b (**2b**), and A (**3**).

2.4. Antimicrobial Activities of Pestalachlorides

The analog of pestalachloride A1a and A2a, pestalachloride A, was previously reported to show antibacterial activity against the standard and methicillin-resistant Staphylococcus aureus (MIC = 10 μg/mL) and the plant pathogenic fungus Fusarium culmorum (MIC = 3.2 μg/mL). To preliminarily explore the bioactivity of new compounds and the structure-activity relationship of pestalachlorides, compounds A1a and A2a, together with pestalachloride A (3), were evaluated for their activity against standard *Staphylococcus aureus* ATCC 29213 and methicillin-resistant *Staphylococcus aureus* (MRSA), as well as other human pathogenic microbes, *Enterococcus faecium* ATCC 35667, Vancomycin-Resistant *Enterococcus faecium* (VRE), and *Candida albicans* ATCC 10231. The pestalachloride A showed moderate activity against *Staphylococcus aureus* ATCC 29213, MRSA, and VRE with minimum inhibitory concentrations (MIC) of 8 μg/mL, 4 μg/mL, and 8 μg/mL, respectively. Pestalachloride A1a showed weak antibacterial activity against four Gram-positive bacteria (MIC = 32 μg/mL), while pestalachloride A2a showed no antibacterial activity within the tested concentration range, indicating that the bulky *N*-substituents can reduce the antibacterial activity of these compounds (Table 3). On the other hand, all of the tested three compounds showed no activity against the fungus *Candida albicans*.

Table 3. Antibacterial activity of pestalachlorides A1a (1a), A2a (2a), and A (3).

	Antibacterial Activity (MIC, μg/mL)					
	Compounds			**Positive Controls**		
	1a	**2a**	**3**	**Fluconazole**	**Vancomycin**	**Meropenem**
S. Aureus[a]	32	> 32	8	ND	0.5	ND
MRSA [b]	32	> 32	4	ND	0.5	16
E. faecium[c]	32	> 32	16	ND	0.25	ND
VRE [d]	32	> 32	8	ND	16	ND
C. albicans[e]	> 32	> 32	> 32	0.5	ND	ND

[a] *Staphylococcus aureus* ATCC 29213, [b] methicillin-resistant *Staphylococcus Aureus*, [c] *Enterococcus faecium* ATCC 35667, [d] Vancomycin-Resistant *Enterococcus faecium*, [e] *Candida albicans* ATCC 10231; ND: not detected.

3. Discussion

With the development of sequencing and bioinformatics, genome mining has increasingly become an important strategy for identifying new compounds and cryptic enzymes and exploring new biosynthetic logics. Here we succeeded in discovering new pestalachloride analogs and thus unearthing their biosynthetic gene cluster by utilizing the strategy of halogenase-targeted genome mining combined with characteristic isotope patterns of halogen atoms. Pestalachlorides A1a, A2a, and their analog pestalachloride A share an isoinodin-1-one core structure that occurs in a number of bioactive compounds [37]. From the biosynthesis view, pestalachlides belong to the pestalone-type benzophenones [37]. This class of compounds features a prenyl group attached to a benzophone that is often clorinated. Although a total of 21 natural analogs of pestalone, including SB87-Cl and SB87-H from *Chrysosporium* sp. [38], pestalone from *Pestalotia* sp. CNL-365 [39], pestalachloride A-C from *Pestalotiopsis adusta* [27], (±)pestalachloride D from *Pestalotiopsis* sp. ZJ-2009-7-6 [40], pestalachlorides E and F from *Pestalotiopsis* sp. ZJ-2009-7-6 [41], pestalones B-H from *Pestalotiopsis neglecta* F9D003 [42], and pestalotinones A–D from *Pestalotiopsis trachicarpicola* SC-J551 [28] have been discovered, no biosynthesis gene clusters responsible for their assembly are reported. To our knowledge, this is the first report of the biosynthesis gene clusters of pestalachlorides and their analogs, pestalone-type benzophenones. So far, there are lots of known natural metabolites that are still not connected with their biosynthesis gene clusters, which hinders the further mining of natural products. Given a large part of them contain halogen atoms, halogenase-targeted genome mining reported here might be an efficient strategy to uncover their biosynthesis origin.

PtlK, assembling double chloride atoms to the phenol residue of pestalachlorides at the late stage of biosynthesis, was reasoned to be a free-standing phenol FDH. Free-standing FDHs, including indole and phenol FDHs, have gained broader interest because it is easier to use them in biotransformation. Amongst, free-standing indole FDHs have been deeply investigated and engineered [13,14,43–46]; however, the counterpart researches on free-standing phenol FDHs are still scarce. Although free-standing phenol FDHs are widely distributed in fungi, only a few are connected with their products, and none of their structures have been determined [7], which hinders the application of these enzymes. Further mining of fungal free-stand phenol FDHs and their products will benefit their structural determination and engineering for biocatalytic application.

4. Materials and Methods

4.1. General Experimental Details

UV measurements were recorded on a Shimadzu UV-2550 spectrophotometer. NMR spectra were acquired with Varian Mercury 600 spectrometers using DMSO-d_6 as solvent. HR-ESIMS and ESIMS/MS data were obtained on a Waters Xevo G2-XS QTof mass spectrometer (Waters, Manchester, UK) with an ACQUITY UPLC® CSH™ C_{18} column (Waters, 1.7 μm, 2.1 × 100 mm) or CORTECS® C_{18} (waters, 2.7 μm, 2.1 × 50 mm) HPLC analyses were performed on an Agilent 1200 or Shimadzu LC-20A instrument using an XBridge C_{18} column (3.5 μm, 4.6 × 150 mm) or Reprosil-Pur Basic-C_{18} column (5 μm, 250 × 10 mm). The genomic DNA was sequenced using the IlluminaHiSeq platform (Illumina, San Diego, CA, USA), assembled via SPAdes 3.13.0 software [47], and uploaded onto Genbank (JALYBT000000000).

4.2. Genome Mining of the Halogenase-Containing Biosynthesis Gene Clusters

TBlastp analysis was performed using fungal FDH GedL as the probe to explore new halogenated secondary metabolites from our in-house fungal genomic database. The hit-containing sequences were further analyzed by antiSMASH and the putative halogenase potentially involved in secondary metabolite biosynthesis were picked out for further phylogenetic analysis with characterized FDHs. The characterized FDHs are Rdc2 (ADM86580), KtzR (ABV56598), RebH (AAN01216), ThaL (ABK79936), PrnA (AAB97504), BhaA (CAA76550), SttH (ADW94630), MibH (ADK32563), TiaM (ADU85999), PyrH (AAU95674), VhaA (CCD33142), CndH (CAQ43074), CrpH (ABM21576), SgcC3 (ANY94426), GsfI (ADI24948), AcOTAhal (ANY27070), AclH (BAE56588), KtzQ (ABV56597), PtaM (AGO59046), CalO3 (AAM70353), ChlB4 (AAZ77674), ChmK (BBL33413), ChmN (BBL33407), PloN (BBL33421), PloK (BBL33420), AscD (A0A455R7M0), Ota5 (A2R6G7), GedL (XP_001217599), RadH (ACM42402), ArmH5 (ALT31852), ArmH4 (ALT31851), ArmH1 (AEM76785), ArmH2 (AEM76786), ArmH3 (ALT31850), ChlA (BAP16678), and VirX1 (6QGM). The amino acid sequence of the putative halogenase PtlK combined with the selected known halogenases was aligned by MUSCLE [48], and their phylogenetic tree was constructed based on the UPGMA [49] method and visualized with MEGA 7.0.26 [50].

4.3. Culture Condition Prioritization for the Production of Chlorinated Compounds

Pestalotiopsis rhododendri LF-19-12 was originally isolated from a lichen sample collected from Tibet, China, and identified based on phylogenetic NJ tree based on ITS sequences (Figure S14). To explore the production of chlorinated compounds, four culture media, M1 (peptone 2 g, yeast powder 4 g, starch 10 g, 1 L distilled water), M2 (mannitol 40 g, maltose 40 g, yeast powder 10 g, K_2HPO_4 2 g, $MgSO_4{\cdot}7H_2O$ 0.5 g, $FeSO_4{\cdot}7H_2O$ 0.01 g, 1 L distilled water), PDB (200 g potato, 20 g glucose, 1 L distilled water), and YES media (sucrose 150 g, yeast powder 20 g, $MgSO_4{\cdot}7H_2O$ 0.5 g, $ZnSO_4{\cdot}7H_2O$ 0.01 g, $CuSO_4{\cdot}5H_2O$ 0.005 g, 1 L distilled water) were selected for culturing *Pestalotiopsis rhododendri* LF-19-12. The fungus *Pestalotiopsis* LF-19-12 was first cultured in 250 mL Erlenmeyer flasks containing 50 mL of potato dextrose broth (PDB) medium and incubated on a rotary shaker at 220 rpm and 28 °C for 48 h to yield the seed culture. Then 50 mL of the seed culture was inoculated into a 500 mL Erlenmeyer flask containing 100 mL of fermentation medium and incubated at 220 rpm and 28 °C for 9 days. The fermentation for each culture medium was carried out in triplicate. Subsequently, 2 mL of culture was filtrated, and the obtained mycelia were extracted using methanol. The obtained crude extract was pretreated with ODS and then analyzed using HR-ESIMS/MS.

4.4. Fermentation and Isolation

The spores of *Pestalotiopsis* LF-19-12 were inoculated into 3 × 500 mL Erlenmeyer flasks each containing 100 mL of potato dextrose broth (PDB) medium to be precultured at 28 °C and 220 rpm for 48 h. Then, the obtained 3 × 100 mL of preculture were inoculated into 3 × 5 L Erlenmeyer flask each containing 1 L of M2 medium (mannitol 40 g, maltose 40 g, yeast powder 10 g, K_2HPO_4 2 g, $MgSO_4{\cdot}7H_2O$ 0.5 g, $FeSO_4{\cdot}7H_2O$ 0.01 g, 1 L distilled water), and incubated on a rotary shaker at 220 rpm and 28 °C. After 9 days, the mycelia were harvested and extracted six times with acetone, yielding 17.72 g of crude extract. The obtained extract was subjected to a silica gel (300–400 mesh, Yantai Chemical Industry Research Institute, Yantai, China) column eluted with a stepwise gradient of CH_2Cl_2−MeOH mixtures (1:0, 100:1, 100:2, 100:4, 100:6, 100:8, 10:1, 5:1, 4:1, 2:1, 0:1, *v*/*v*) to give Fr A-G. The eluents were analyzed by LC-MS, and the targeted compounds were mainly found in the Fr.E (576.95 mg) and Fr.F (172.64 mg).

Fr.E was further separated with an ODS flash column eluted with a gradient ACN−H_2O solution from 15% ACN to 70% ACN. **1a** (2.3 mg) was crystalized from the eluent of 45% ACN. The other eluents were combined into five fractions (E1: 72.5 mg; E2: 52.3 mg; E3: 57.1 mg; E4: 36.7 mg; E5: 145.7 mg) according to LC-MS analysis results. The Fr. E3 was further purified by semi-preparative RP HPLC (Reprosil-Pur Basic-C_{18}, 5 μm, 250 × 10 mm, 30% ACN−H_2O, 30 °C, 2.5 mL/min) to yield **1a** (t_R 40.54 min, 20.9 mg) and **1b** (t_R 54.50 min, 2.2 mg). The Fr. E4 was further purified by semi-preparative RP HPLC (Xbridge™ Prep C_{18}, 5 μm, 250 × 10 mm, 40% ACN−H_2O, 30 °C, 2.5 mL/min) to afford **3** (t_R 48.03 min, 2.6 mg). The Fr. E5 was purified using semi-preparative RP HPLC (Xbridge™ Prep C_{18}, 5 μm, 250 × 10 mm, 40% ACN−H_2O, 30 °C, 2.5 mL/min) to give **4** (t_R 17.25 min, 13.5 mg).

Fr.F was separated with an ODS flash column eluted with a stepwise gradient of ACN in water (20%, 30%, 50%, 70%, and 100%; *v*/*v*; each for 5 min). All eluents were analyzed by LC−MS, and those containing halogenated compounds were combined to yield fraction Fr.F1 (37.3 mg). The Fr.F1 was subsequently purified by semi-preparative chromatography (XSelect CSH C_{18} OBD ™ prep column, 5 μm, 250 × 10 mm, 45% ACN aqueous containing 0.1% TFA, 30 °C, 2.5 mL/min), to yield **A2a** (t_R 30.5 min, 1.9 mg) and **A2b** (t_R 48.8 min,0.2 mg).

Pestalachloride A1a (**1a**): white powder; 0 (*c* 0.1, MeOH); UV (MeOH) λ_{max} (log ε) 222.4 (4.78) 258.8 (32.43), 297.2 (26.26); 1D and 2D NMR data (DMSO-d_6) see Table 1 and Supplementary Material; HR-ESI(+)MS $[M+H]^+$ *m/z* 482.1121 (calcd. for $C_{23}H_{26}Cl_2NO_6$, 482.1137, 3.3 ppm).

Pestalachloride A1b (**1b**): white powder; ^{1}H and ^{13}C NMR data (DMSO-d_6) see Table 1 and Supplementary Material; HR-ESI(+)MS [M+H]$^+$ *m/z* 482.1117 (calcd. for $C_{23}H_{26}Cl_2NO_6$, 482.1137, 4.1 ppm).

Pestalachloride A2a (**2a**): white powder; 0 (*c* 0.1, MeOH); UV (MeOH) λ_{max} (log ε) 208.2 (3.6), 258.8 (38.7), 297.0 (30.4); 1D and 2D NMR data (DMSO-d_6) see Table 1 and Supplementary Material; HR-ESI(+)MS [M+H]$^+$ *m/z* 524.1240 (calcd. for $C_{25}H_{28}Cl_2NO_7$, 524.1243, 0.57 ppm).

Pestalachloride A2b (**2b**): white powder; ^{1}H NMR data (600 MHz, DMSO-d_6) see Table 1 and Supplementary Material; HR-ESI(+)MS [M+H]$^+$ *m/z* 524.1237 (calcd. for $C_{25}H_{28}Cl_2NO_7$, 524.1243, 1.1 ppm).

4.5. The Calculation of the Relative Gibbs Energy Barriers

S-configuration structures of **1a**, **1b**, **2a**, **2b**, and atropisomers of **3** were first optimized using Gaussian 16 at the B3LYP/6-31G (d)/PCM (H_2O) level. Then relaxed dihedral angle (rotation between C8–C9) scans were performed at the same level. The Gibbs energies for the calculation of barriers were calculated at the M062X/def2TZVP/SMD (H_2O)//B3LYP/6-31G (d)/PCM (H_2O) level.

4.6. Antibacterial Bioassay

The minimum inhibitory concentration (MIC) values of the obtained compounds against *Staphylococcus aureus* ATCC 29213, *Enterococcus faecium* ATCC 35667, Methicillin-Resistant *Staphylococcus aureus*, Vancomycin-Resistant *Enterococcus faecium*, and *Candida albicans* ATCC 10231 were determined using a broth microdilution protocol [51]. Briefly, 50 μL of bacterial or fungal suspension (5×10^5 CFU/mL) was added to each well of the 96-well plate. Subsequently, 50 μL of each work solution of pestalachlorides A1a, A2a, A and the corresponding positive drugs (64, 32, 16, 8, 4, 2, 1, 0.5, 0.25, 0.125, 0.0625, 0.03125 μg/mL) were added and incubated at 33 °C for 18 h. The lowest concentration that completely prevents the growth of the assayed organism was defined as the MIC.

Supplementary Materials: The following supporting information can be downloaded at: https://www.mdpi.com/article/10.3390/antibiotics11101304/s1. Figures S1–S5: 1D/2D NMR of pestalachloride A1a. Figures S6 and S7: 1D NMR of pestalachloride A1b. Figures S8–S12: 1D/2D NMR of pestalachloride A2a. Figure S13: 1H NMR of pestalachloride A2b by comparison with that of pestalachlorides A2a (**2a**). Figure S14: phylogenetic NJ tree of *Pestalotiopsis* sp. LF-19-12 and related type strains.

Author Contributions: Conceptualization, Y.X. and G.S.; methodology, M.L., M.W. and N.H.; software, Y.X.; validation, Y.X. and S.C.; formal analysis, M.L., M.W. and Y.X.; investigation, M.L. and M.W.; resources, M.L. and N.H.; data curation, M.L., M.W. and Y.X.; writing—original draft preparation, M.L. and M.W.; writing—review and editing, Y.X. and S.C.; visualization, Y.X.; supervision, Y.X. and G.S.; project administration, Y.X.; funding acquisition, Y.X. and S.C. All authors have read and agreed to the published version of the manuscript.

Funding: This research was co-funded by The National Natural Science Foundation of China (81973219 and 82104047) and CAMS Innovation Fund for Medical Sciences (CIFMS, 2021-I2M-1-028).

Data Availability Statement: The data presented in this study are available on request from the corresponding author.

Acknowledgments: We thank OE Biotech Co., Ltd. (Shanghai, China) for sequencing the genome DNA sequence.

Conflicts of Interest: The authors declare no conflict of interest. The funders had no role in the design of the study; in the collection, analyses, or interpretation of data; in the writing of the manuscript, or in the decision to publish the results.

References

1. Parisini, E.; Metrangolo, P.; Pilati, T.; Resnati, G.; Terraneo, G. Halogen bonding in halocarbon-protein complexes: A structural survey. *Chem. Soc. Rev.* **2011**, *40*, 2267–2278. [CrossRef]
2. Crowe, C.; Molyneux, S.; Sharma, S.V.; Zhang, Y.; Gkotsi, D.S.; Connaris, H.; Goss, R.J.M. Halogenases: A palette of emerging opportunities for synthetic biology–synthetic chemistry and C–H functionalisation. *Chem. Soc. Rev.* **2021**, *50*, 9443–9481. [CrossRef]
3. Latham, J.; Brandenburger, E.; Shepherd, S.A.; Menon, B.R.K.; Micklefield, J. Development of Halogenase Enzymes for Use in Synthesis. *Chem. Rev.* **2018**, *118*, 232–269. [CrossRef]
4. Whitfield, R.; Brown, F. The Benefits of Chlorine Chemistry in Pharmaceuticals in the United States and Canada. *IHS Econ.* **2016**.
5. Gribble, G.W. A recent survey of naturally occurring organohalogen compounds. *Environ. Chem.* **2015**, *12*, 396–405. [CrossRef]
6. van Santen, J.A.; Poynton, E.F.; Iskakova, D.; McMann, E.; Alsup, T.A.; Clark, T.N.; Fergusson, C.H.; Fewer, D.P.; Hughes, A.H.; McCadden, C.A.; et al. The Natural Products Atlas 2.0: A database of microbially-derived natural products. *Nucleic Acids Res.* **2022**, *50*, D1317–D1323. [CrossRef] [PubMed]
7. Cochereau, B.; Meslet-Cladière, L.; Pouchus, Y.F.; Grovel, O.; Roullier, C. Halogenation in Fungi: What Do We Know and What Remains to Be Discovered? *Molecules* **2022**, *27*, 3157. [CrossRef] [PubMed]
8. Ludewig, H.; Molyneux, S.; Ferrinho, S.; Guo, K.; Lynch, R.; Gkotsi, D.S.; Goss, R.J.M. Halogenases: Structures and functions. *Curr. Opin. Struct. Biol.* **2020**, *65*, 51–60. [CrossRef]
9. Menon, B.R.K.; Richmond, D.; Menon, N. Halogenases for biosynthetic pathway engineering: Toward new routes to naturals and non-naturals. *Catal. Rev.* **2022**, *64*, 533–591. [CrossRef]
10. Zeng, J.; Zhan, J. Chlorinated Natural Products and Related Halogenases. *Isr. J. Chem.* **2019**, *59*, 387–402. [CrossRef]
11. Agarwal, V.; Miles, Z.D.; Winter, J.M.; Eustáquio, A.S.; El Gamal, A.A.; Moore, B.S. Enzymatic Halogenation and Dehalogenation Reactions: Pervasive and Mechanistically Diverse. *Chem. Rev.* **2017**, *117*, 5619–5674. [CrossRef] [PubMed]
12. Phintha, A.; Prakinee, K.; Chaiyen, P. Chapter Eleven—Structures, mechanisms and applications of flavin-dependent halogenases. In *The Enzymes*; Chaiyen, P., Tamanoi, F., Eds.; Academic Press: Cambridge, MA, USA, 2020; Volume 47, pp. 327–364.
13. Dong, C.; Flecks, S.; Unversucht, S.; Haupt, C.; van Pée, K.H.; Naismith, J.H. Tryptophan 7-halogenase (PrnA) structure suggests a mechanism for regioselective chlorination. *Science* **2005**, *309*, 2216–2219. [CrossRef] [PubMed]
14. Zhu, X.; De Laurentis, W.; Leang, K.; Herrmann, J.; Ihlefeld, K.; van Pée, K.-H.; Naismith, J.H. Structural Insights into Regioselectivity in the Enzymatic Chlorination of Tryptophan. *J. Mol. Biol.* **2009**, *391*, 74–85. [CrossRef] [PubMed]
15. Scherlach, K.; Hertweck, C. Mining and unearthing hidden biosynthetic potential. *Nat. Commun.* **2021**, *12*, 3864. [CrossRef] [PubMed]
16. Rajwani, R.; Ohlemacher, S.I.; Zhao, G.; Liu, H.B.; Bewley, C.A. Genome-Guided Discovery of Natural Products through Multiplexed Low-Coverage Whole-Genome Sequencing of Soil Actinomycetes on Oxford Nanopore Flongle. *mSystems* **2021**, *6*, e0102021. [CrossRef]
17. Miethke, M.; Pieroni, M.; Weber, T.; Bronstrup, M.; Hammann, P.; Halby, L.; Arimondo, P.B.; Glaser, P.; Aigle, B.; Bode, H.B.; et al. Towards the sustainable discovery and development of new antibiotics. *Nat. Rev. Chem.* **2021**, 726–749. [CrossRef]
18. Claesen, J.; Bibb, M. Genome mining and genetic analysis of cypemycin biosynthesis reveal an unusual class of posttranslationally modified peptides. *Proc. Natl. Acad. Sci. USA* **2010**, *107*, 16297–16302. [CrossRef]
19. Bergmann, S.; Schümann, J.; Scherlach, K.; Lange, C.; Brakhage, A.A.; Hertweck, C. Genomics-driven discovery of PKS-NRPS hybrid metabolites from *Aspergillus nidulans*. *Nat. Chem. Biol.* **2007**, *3*, 213–217. [CrossRef]
20. Franke, J.; Ishida, K.; Hertweck, C. Genomics-Driven Discovery of Burkholderic Acid, a Noncanonical, Cryptic Polyketide from Human Pathogenic *Burkholderia* Species. *Angew. Chem. Int. Ed.* **2012**, *51*, 11611–11615. [CrossRef]
21. Luo, M.; Chang, S.; Li, Y.; Xi, X.; Chen, M.; He, N.; Wang, M.; Zhao, W.; Xie, Y. Molecular Networking-Based Screening Led to the Discovery of a Cyclic Heptadepsipeptide from an Endolichenic *Xylaria* sp. *J. Nat. Prod.* **2022**, *85*, 972–979. [CrossRef]
22. Li, Y.; Liu, L.; Zhang, G.; He, N.; Guo, W.; Hong, B.; Xie, Y. Potashchelins, a Suite of Lipid Siderophores Bearing Both L-threo and L-erythro Beta-Hydroxyaspartic Acids, Acquired From the Potash-Salt-Ore-Derived Extremophile *Halomonas* sp. MG34. *Front. Chem.* **2020**, *8*, 197. [CrossRef] [PubMed]
23. Li, Y.; He, N.; Luo, M.; Hong, B.; Xie, Y. Application of untargeted tandem mass spectrometry with molecular networking for detection of enniatins and beauvericins from complex samples. *J. Chromatogr. A* **2020**, *1634*, 461626. [CrossRef] [PubMed]
24. Shi, Y.; Wang, X.; He, N.; Xie, Y.; Hong, B. Rescrutiny of the sansanmycin biosynthetic gene cluster leads to the discovery of a novel sansanmycin analogue with more potency against *Mycobacterium tuberculosis*. *J. Antibiot.* **2019**, *72*, 769–774. [CrossRef]
25. Shi, Y.; Gu, R.; Li, Y.; Wang, X.; Ren, W.; Li, X.; Wang, L.; Xie, Y.; Hong, B. Exploring novel herbicidin analogues by transcriptional regulator overexpression and MS/MS molecular networking. *Microb. Cell Factories* **2019**, *18*, 175. [CrossRef] [PubMed]
26. Nielsen, M.T.; Nielsen, J.B.; Anyaogu, D.C.; Holm, D.K.; Nielsen, K.F.; Larsen, T.O.; Mortensen, U.H. Heterologous reconstitution of the intact geodin gene cluster in *Aspergillus nidulans* through a simple and versatile PCR based approach. *PLoS ONE* **2013**, *8*, e72871. [CrossRef]
27. Li, E.; Jiang, L.; Guo, L.; Zhang, H.; Che, Y. Pestalachlorides A–C, antifungal metabolites from the plant endophytic fungus *Pestalotiopsis adusta*. *Bioorg. Med. Chem.* **2008**, *16*, 7894–7899. [CrossRef] [PubMed]
28. Jiang, Z.; Wu, P.; Li, H.; Xue, J.; Wei, X. Pestalotinones A-D, new benzophenone antibiotics from endophytic fungus *Pestalotiopsis trachicarpicola* SC-J551. *J. Antibiot.* **2022**, *75*, 207–212. [CrossRef] [PubMed]

29. LaPlante, S.R.; Edwards, P.J.; Fader, L.D.; Jakalian, A.; Hucke, O. Revealing atropisomer axial chirality in drug discovery. *ChemMedChem* **2011**, *6*, 505–513. [CrossRef]
30. Xu, X.; Liu, L.; Zhang, F.; Wang, W.; Li, J.; Guo, L.; Che, Y.; Liu, G. Identification of the first diphenyl ether gene cluster for pestheic acid biosynthesis in plant endophyte *Pestalotiopsis* fici. *ChemBioChem* **2014**, *15*, 284–292. [CrossRef]
31. Chiang, Y.M.; Szewczyk, E.; Davidson, A.D.; Entwistle, R.; Keller, N.P.; Wang, C.C.; Oakley, B.R. Characterization of the *Aspergillus nidulans* monodictyphenone gene cluster. *Appl. Environ. Microbiol.* **2010**, *76*, 2067–2074. [CrossRef] [PubMed]
32. Szewczyk, E.; Chiang, Y.M.; Oakley, C.E.; Davidson, A.D.; Wang, C.C.; Oakley, B.R. Identification and characterization of the asperthecin gene cluster of *Aspergillus nidulans*. *Appl. Environ. Microbiol.* **2008**, *74*, 7607–7612. [CrossRef] [PubMed]
33. Szwalbe, A.J.; Williams, K.; Song, Z.; de Mattos-Shipley, K.; Vincent, J.L.; Bailey, A.M.; Willis, C.L.; Cox, R.J.; Simpson, T.J. Characterisation of the biosynthetic pathway to agnestins A and B reveals the reductive route to chrysophanol in fungi. *Chem. Sci.* **2019**, *10*, 233–238. [CrossRef] [PubMed]
34. Simpson, T.J. Genetic and Biosynthetic Studies of the Fungal Prenylated Xanthone Shamixanthone and Related Metabolites in *Aspergillus* spp. Revisited. *ChemBioChem* **2012**, *13*, 1680–1688. [CrossRef]
35. Sanchez, J.F.; Entwistle, R.; Hung, J.H.; Yaegashi, J.; Jain, S.; Chiang, Y.M.; Wang, C.C.; Oakley, B.R. Genome-based deletion analysis reveals the prenyl xanthone biosynthesis pathway in *Aspergillus nidulans*. *J. Am. Chem. Soc.* **2011**, *133*, 4010–4017. [CrossRef]
36. Slavov, N.; Cvengros, J.; Neudorfl, J.M.; Schmalz, H.G. Total synthesis of the marine antibiotic pestalone and its surprisingly facile conversion into pestalalactone and pestalachloride A. *Angew. Chem. Int. Ed.* **2010**, *49*, 7588–7591. [CrossRef] [PubMed]
37. Augner, D.; Gerbino, D.C.; Slavov, N.; Neudörfl, J.M.; Schmalz, H.G. N-Capping of primary amines with 2-acyl-benzaldehydes to give isoindolinones. *Org. Lett.* **2011**, *13*, 5374–5377. [CrossRef] [PubMed]
38. Wachi, Y.; Yamashita, T.; Komatsu, K.; Yoshida, S. JP Patent JKXXAF JP 07061950 A2 19950307, 1995.
39. Cueto, M.; Jensen, P.R.; Kauffman, C.; Fenical, W.; Lobkovsky, E.; Clardy, J. Pestalone, a New Antibiotic Produced by a Marine Fungus in Response to Bacterial Challenge. *J. Nat. Prod.* **2001**, *64*, 1444–1446. [CrossRef]
40. Wei, M.-Y.; Li, D.; Shao, C.-L.; Deng, D.-S.; Wang, C.-Y. (±)−Pestalachloride D, an Antibacterial Racemate of Chlorinated Benzophenone Derivative from a Soft Coral-Derived Fungus *Pestalotiopsis* sp. *Mar. Drugs* **2013**, *11*, 1050–1060. [CrossRef]
41. Xing, Q.; Gan, L.-S.; Mou, X.-F.; Wang, W.; Wang, C.-Y.; Wei, M.-Y.; Shao, C.-L. Isolation, resolution and biological evaluation of pestalachlorides E and F containing both point and axial chirality. *RSC Adv.* **2016**, *6*, 22653–22658. [CrossRef]
42. Wang, W.; Park, C.; Oh, E.; Sung, Y.; Lee, J.; Park, K.-H.; Kang, H. Benzophenone Compounds, from a Marine-Derived Strain of the Fungus *Pestalotiopsis neglecta*, Inhibit Proliferation of Pancreatic Cancer Cells by Targeting the MEK/ERK Pathway. *J. Nat. Prod.* **2019**, *82*, 3357–3365. [CrossRef]
43. Payne, J.T.; Andorfer, M.C.; Lewis, J.C. Regioselective arene halogenation using the FAD-dependent halogenase RebH. *Angew. Chem. Int. Ed.* **2013**, *52*, 5271–5274. [CrossRef] [PubMed]
44. Bitto, E.; Huang, Y.; Bingman, C.A.; Singh, S.; Thorson, J.S.; Phillips, G.N., Jr. The structure of flavin-dependent tryptophan 7-halogenase RebH. *Proteins* **2008**, *70*, 289–293. [CrossRef] [PubMed]
45. Moritzer, A.C.; Minges, H.; Prior, T.; Frese, M.; Sewald, N.; Niemann, H.H. Structure-based switch of regioselectivity in the flavin-dependent tryptophan 6-halogenase Thal. *J. Biol. Chem.* **2019**, *294*, 2529–2542. [CrossRef]
46. Frese, M.; Sewald, N. Enzymatic halogenation of tryptophan on a gram scale. *Angew. Chem. Int. Ed.* **2015**, *54*, 298–301. [CrossRef]
47. Bankevich, A.; Nurk, S.; Antipov, D.; Gurevich, A.A.; Dvorkin, M.; Kulikov, A.S.; Lesin, V.M.; Nikolenko, S.I.; Pham, S.; Prjibelski, A.D.; et al. SPAdes: A new genome assembly algorithm and its applications to single-cell sequencing. *J Comput Biol* **2012**, *19*, 455–477. [CrossRef] [PubMed]
48. Edgar, R.C. MUSCLE: Multiple sequence alignment with high accuracy and high throughput. *Nucleic Acids Res.* **2004**, *32*, 1792–1797. [CrossRef] [PubMed]
49. Sneath, P.H.; Sokal, R.R. Numerical taxonomy. *Nature* **1962**, *193*, 855–860. [CrossRef]
50. Kumar, S.; Stecher, G.; Tamura, K. MEGA7: Molecular Evolutionary Genetics Analysis Version 7.0 for Bigger Datasets. *Mol. Biol. Evol.* **2016**, *33*, 1870–1874. [CrossRef]
51. Clinical and Laboratory Standards Institute. *Methods for Dilution Antimicrobial Susceptibility Tests for Bacteria that Grow Aerobically—Tenth Edition, M7-A10*; Clinical and Laboratory Standards Institute: Wayne, PA, USA, 2015.

antibiotics

MDPI

Review

Biocontrol of *Candida albicans* by Antagonistic Microorganisms and Bioactive Compounds

Honghua Li [1], Jinpeng Yang [2], Xinwan Zhang [2], Xiuli Xu [2], Fuhang Song [1,*] and Hehe Li [1,3,*]

1 School of Light Industry, Beijing Technology and Business University, Beijing 100048, China
2 School of Ocean Sciences, China University of Geosciences, Beijing 100083, China
3 Key Laboratory of Brewing Molecular Engineering of China Light Industry, Beijing Technology and Business University, Beijing 100048, China
* Correspondence: songfuhang@btbu.edu.cn (F.S.); lihehe@btbu.edu.cn (H.L.)

Abstract: *Candida albicans* is an endogenous opportunistic pathogenic fungus that is harmless when the host system remains stable. However, *C. albicans* could seriously threaten human life and health when the body's immune function declines or the normal flora is out of balance. Due to the increasing resistance of candidiasis to existing drugs, it is important to find new strategies to help treat this type of systemic fungal disease. Biological control is considered as a promising strategy which is more friendly and safer. In this review, we compare the bacteriostatic behavior of different antagonistic microorganisms (bacteria and fungi) against *C. albicans*. In addition, natural products with unique structures have attracted researchers' attention. Therefore, the bioactive nature products produced by different microorganisms and their possible inhibitory mechanisms are also reviewed. The application of biological control strategies and the discovery of new compounds with antifungal activity will reduce the resistance of *C. albicans*, thereby promoting the development of novel diverse antifungal drugs.

Keywords: *Candida albicans*; antagonistic microorganisms; biocontrol strategy; bioactive compounds

check for updates

Citation: Li, H.; Yang, J.; Zhang, X.; Xu, X.; Song, F.; Li, H. Biocontrol of *Candida albicans* by Antagonistic Microorganisms and Bioactive Compounds. *Antibiotics* **2022**, *11*, 1238. https://doi.org/10.3390/antibiotics11091238

Academic Editor: William N. Setzer

Received: 30 July 2022
Accepted: 9 September 2022
Published: 12 September 2022

1. Introduction

Fungal infection is a common global problem affecting humans and its incidence is on the rise [1]. Among them, *Candida* has been a life-threatening pathogen for a long time, accounting for almost 80% of fungal infections. Recently *C. albicans* infection causes more than 400,000 cases of blood infection each year, with a mortality rate of about 42% [2–4]. *C. albicans*, a small number in the normal body, is a part of healthy flora. It can exist in the oral cavity, intestinal tract, upper respiratory tract, and other parts. When growing in the state of unicellular yeast cells, it does not cause disease. However, when the normal flora interacts with each other out of balance or the body's immune function and defenses decline, *C. albicans* proliferates and grows into hyphae, invading cells and causing disease. It has been a major cause of morbidity and mortality in immunocompromised populations [5].

In host, the pathogenicity of *C. albicans* is caused by the decline of immune function, the change of conventional flora and the destruction of the epithelial protective barrier. During infection, the formation of *C. albicans* biofilm and the morphological switch from yeast-like to hyphal-like are considered to be two significant pathogenic characteristics of *C. albicans*. First of all, its morphological plasticity is crucial to the pathogenicity of fungi, as the hyphal form has a key role in the infection process [6–8]. In addition, the pathogenicity of *Candida* is greatly enhanced by the formation of biofilms [9]. Biofilms are microbial communities that irreversibly attach to surfaces. Biofilms behave very differently from planktonic cells, and once formed, they can increase resistance to existing antibiotics and immune responses [10]. Therefore, inhibition of hyphal development and inhibition of biofilm formation are considered to be an effective strategy against *C. albicans* infection.

Currently, there are very few drugs for the treatment and prevention of *Candidiasis* in clinic. The polyene antibiotic is the earliest specific drug isolated from *Streptomyces nodosus* in the 1950s to treat yeast infection. Since then, many antifungal agents have been developed [11,12]. There are four types of antifungal agents for *C. albicans* infection [13]. The most commonly used antifungal drugs and the mechanism of action include:(1) The widest range and most effective is polyene (Amphotericin B), which can kill most fungi. Polyenes bind to ergosterol in fungal cell membranes, creating stomata and causing cell death [14,15]. (2) Triazole antifungal drugs (fluconazole, voriconazole and itraconazole). Azoles can inhibit lanosterol 14α demethylase, which is an important enzyme in ergosterol biosynthesis [16–19]. (3) 5-fluorocytosine, it inhibits fungal DNA synthesis by inhibiting thymidylate synthetase [12,20]. (4) There are also some echinocandin antifungal drugs (anidulafungin, micafungin, and caspofungin) [21–27]. The mechanisms of these bioactive compounds against *C. albicans* are mainly related to inhibition of biofilm formation, inhibition of virulence factors and destruction of cell wall integrity. With the increasing drug resistance of *C. albicans*, it is compelling to find new antifungal methods and reagents to solve this complex medical problem. Biological control is considered to be a more effective and safe strategy [1,28,29].

Novel natural compounds produced by microorganisms, due to their complex structures, may exhibit novel antibacterial mechanisms and different modes of action. Moreover, they were considered as candidates to reduce drug resistance. People have been trying to find unique antifungal drugs from nature, which has led to important advances in the development of new antifungal drugs.

In recent years, there have been some reviews on natural products that could inhibit *C. albicans* [2,30–37]. In this paper, we have reviewed the antagonistic microorganisms against *C. albicans* considered in recent years and have also reviewed the active natural products produced by microorganisms that inhibit *C. albicans*. Researchers focus on the study of antagonistic microorganisms in order to use probiotics to inhibit *C. albicans*. Through the review of secondary metabolites, it can provide a reference for clinical drug development.

2. Antagonistic Microbes against *C. albicans*

Traditional azoles and their derivatives have poor effect on preventing recurrence of pathogenic fungus. In some patients, fluconazole can cause some side effects such as headache, discomfort, dizziness, gastrointestinal and, rash [38]. Bacteria, yeast, and fungus all can develop resistance to antibiotics and bactericidal chemicals [39]. Biological control of microbial infections is an alternative approach that utilizes antagonistic microorganisms to prevent the growth and infection of harmful microorganisms. Diverse microorganisms, including fungi (such as non-toxic *Aspergillus*, *Trichoderma*, *Penicillium*), yeast strains, and bacteria, have been studied as potential antagonistic organisms for the control of *C. albicans*. In this review, the microorganisms that inhibit *C. albicans* and their secondary metabolites are introduced from the perspective of antagonistic microorganisms. The microorganisms that have potential antagonism against *C. albicans* are listed in Table 1. The main species and inhibition activities of these antagonistic strains are also discussed. We have reviewed the antagonistic microorganisms against *C. albicans* in recent years with the aim to develop a new natural material, using beneficial bacteria or fungus, that would be useful for inhibiting the growth of pathogenic *C. albicans* in the human body.

As shown in Figure 1, the article reporting *Bacillus* spp. antagonists were dominant (40%) compared with the article reporting antagonistic *Bifidobacterium* (20%), antagonistic *Lactobacillus genus* (13.33%), antagonistic yeast (6.67%) and other antagonistic strains (20%).

Table 1. Antagonistic Microbes against *C. albicans*.

Antagonists	Species	Activity	References
Bacillus spp.	*B. sphaericus* A16, *B. circulans* M142, *B. brevis* M166, *B. brevis* T122	Strains showed extensive inhibition against *C. albicans*.	[40]
	B. subtilis spizizenii DK1-SA11	Cell-free supernatant had significant inhibitory activity against *C. albicans*.	[41]
	B. velezensis DTU001	Significantly inhibited the proliferation of *C. albicans*, and the inhibition ability of the strain was better than that of a single lipopeptide.	[42]
Bifidobacterium	*B. amyloliquefaciens* SYBC H47	Cell-free supernatant and Cell suspension had obvious inhibition against *C. albicans*.	[43]
	B. velezensis 1B-23	Inhibited *C. albicans* growth in vitro.	[44]
	B. longum BB536	The supernatant of fermented broccoli could inhibit the growth of *C. albicans* in vitro.	[45,46]
Lactobacillus genus	*L. johnsonii* MT4	Inhibited planktonic growth and biofilm formation of *C. albicans*	[47]
	Lactobacillus	Regulated growth and virulence of *C. albicans* through niche competition.	[48]
Yeast	*Metschnikowia pulcherrima*	Strong antagonistic activity against *C. albicans*.	[49]
Other strains	*Enterococcus*	Regulated growth and virulence of *C. albicans* through niche competition.	[48]
	Pseudomonas fluorescens	The strain showed extensive inhibition against *C. albicans*.	[40]
	Salivarius MG242	The strain had significant inhibitory effect on *C. albicans*.	[50]

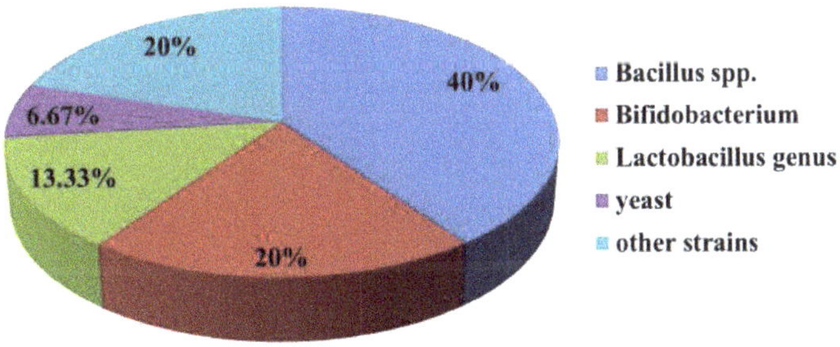

Figure 1. Percentages of different antagonistic microbes of *C. albicans*.

2.1. *Antagonistic Effect of Bacillus* spp. *against C. albicans*

Some beneficial bacteria or fungus are widely used in biocontrol. In particular, it is well known that *Bacillus* spp. is an excellent source of antifungal drugs, thus *Bacillus* spp. is widely used as a biological control agent [51–53]. Bacillus species are Gram-positive bacteria that can survive in different environments. They could form endospores and produce a large number of metabolites [53].

Researchers isolated four strains of *Bacillus* A16 (*B. sphaericus*), M142 (*B. circulans*), M166 (*B. brevis*) and T122 (*B. brevis*) from soil samples. These *Bacillus* showed extensive inhibitory activity against *C. albicans* [40]. Among them, *B. brevis* M166 showed antifungal activity against all tested microorganisms (*Sclerotium rolfsii*, *Rhizoctonia solani*, *Fusarium oxysporum*, *Staphylococcus aureus* and *C. albicans*), with a relatively wide antimicrobial spectrum. *B. circulans* M142 had strong antibacterial activity against *C. albicans* and *S. aureus*, while *B. brevis* T122 only had antibacterial activity against *C. albicans*. To our knowledge, no specific compounds inhibiting *C. albicans* had been identified.

In addition to the antifungal activity of *Bacillus* spp. from soil samples, *Bacillus* spp. from marine samples was also found to have inhibitory activity against *C. albicans*. *B. subtilis spizizenii* DK1-SA11 was isolated from Bay of Yellow Sea in China [41]. The cell-free supernatant had significant inhibitory activity against *C. albicans*. The inhibitory active ingredient had not been identified but was stable in nature, while the enzymatic hydrolysis of lipase, trypsin and papain made it lose activity. Antimicrobial activity tests against pathogens indicated that this strain could be used as a source of antibiotics, synbiotics, and probiotics.

B. velezensis was widespread in the environments and produced abundant lipopeptides with good bacteriostatic effect. Some researchers have studied on the inhibitory spectrum of *B. velezensis* DTU001 against 20 different species of human and/or plant pathogenic fungi [42]. The results showed that *B. velezensis* DTU001 was superior to a single lipopeptide (fengycin and iturin) in inhibiting the selected fungi. Co-culture of *B. velezensis* DTU001 and *C. albicans* significantly inhibited *C. albicans* proliferation, which further supported the biological control properties of *B. velezensis* DTU001.

B. amyloliquefaciens SYBC H47 was isolated from honey [43]. The cultured cell-free supernatant had significant inhibitory activity against *C. albicans*. The main antibacterial substances were surfactin, fengycin and bacillomycin. Three compounds had an inhibitory effect on spore germination of *Botryosphaeria dothidea*. However, compounds that inhibit *C. albicans* had not been identified.

Bacillus velezensis 1B-23 had inhibitory effect on the growth of *C. albicans* in vitro. It had a certain application prospect as a biological agent for biological control of fungal pathogens [44].

2.2. Antagonistic Effect of Bifidobacterium *spp. against C. albicans*

Bacillus spp. has been used clinically because of its bacteriostatic activity. Another probiotic, *Bifidobacterium*, can also be used to prevent and treat intestinal flora disorders in clinic. *Bifidobacterium* is a vital member of the normal human gut microbiota. Some strains of *Bifidobacterium* can be used as probiotics in food, medicine and feed [54,55]. *Bifidobacteria* could produce acetic acid and/or lactic acid during metabolism. Moreover, the action of lactic acid would reduce intestinal pH. Thereby, *Bifidobacterium* could inhibit the proliferation of pathogenic microorganisms [56,57].

Bifidobacterium longum BB536 which was isolated from the feces of healthy infants had been commercially used in various food applications and was considered safe [45,46,58]. The researchers fermented broccoli using *B. longum*. The supernatant could inhibit the growth of *C. albicans* and some other pathogenic bacteria in vitro. Researchers used beneficial bacteria such as *bifidobacteria* and used broccoli as a substrate for the growth of beneficial bacteria to develop substances. Maybe, we can use beneficial microorganisms and their secondary metabolites to develop products that inhibit the growth of pathogenic microorganisms. For example, as a daily oral care preparation, it can prevent the growth of *C. albicans* in human oral cavity [59].

2.3. Antagonistic Effect of Lactobacillus *spp. against C. albicans*

Lactobacillus johnsonii is a probiotic with wide antimicrobial characteristics and can be used as an antiallergic drug. Recent studies have shown that *L. johnsonii* also has inhibitory effects on *C. albicans*. *L. johnsonii* MT4 was isolated from the oral cavity of healthy mice. The strain affected the *C. albicans* growth in both biofilm and planktonic conditions. *L. johnsonii* MT4 showed an antagonistic effect on *C. albicans*, thus inhibiting the biofilm formation of *C. albicans* and planktonic growth. The study on the strain genome had shown that it produced metabolites with anti-*C. albicans* activity, but no active substances against fungi have been reported so far. The antibacterial mechanism needed to be further explored [47].

In addition to producing secondary metabolites that antagonize *C. albicans*, the competition for ecological niches of different strains during the growth process would also cause antagonism among strains, such as *C. albicans* and lactic acid bacteria in the gastrointestinal (GI) tract [48,60]. Non-pathogenic colonization of the human GI tract by *C. albicans* was common. *C. albicans* could regulate bacterial community in mice treated with broad-spectrum antibiotics. One of the most striking features was the significant change in the lactic acid bacteria (LAB) levels. *C. albicans* and *Lactobacillus* species shared a metabolic niche throughout the GI tract. LAB could antagonize *Enterococcus* and *C. albicans* in the GI tract. *C. albicans* and *Lactobacillus* could mutually regulate each other's growth and virulence in the GI tract [48].

2.4. Antagonistic Effect of Yeast against *C. albicans*

In addition to the bacteria mentioned above, yeast can also be used for biological control. *Metschnikowia* could accumulate pigments in cells and growth media. It was a highly effective biocontrol yeast. Antagonism of *M. pulcherrima* against phytopathogens had been demonstrated [49]. The researcher investigated three new strains of *Metschnikowia* which were isolated from grapes. The strain had strong antagonistic activity against *C. albicans*. The three strains produced the same amount of nevus pigments, but there were significant differences in antifungal activities against different microorganisms [61,62].

2.5. Antagonistic Effect of Other Strains against *C. albicans*

Salivarius MG242 isolated from human vagina presented a potential application in the biological control of *C. albicans*. MG242 had an obvious inhibitory impact on *C. albicans*, and the strain had the possibility to be developed into a probiotic product for the treatment of *C. albicans*. In order to develop stable living cell products, it was necessary to maintain anti-*Candida* activity and preserve cell viability during lyophilization. Lower storage temperature extended shelf life to 8.31 months [50]. Strains of K124 (*P. fluorescens*) was also isolated from soil samples, e.g., *B. sphaericus* A16, *B. circulans* M142, *B. brevis* M166 and *B. brevis* T122. *P. fluorescens* K124 showed extensive inhibitory activity against *C. albicans* [40]. *P. fluorescens* K124 only had antifungal activity against *C. albicans*. At present, no inhibitory compounds produced by the strain have been identified.

2.6. A Conclusion of Antagonistic Microbes

In conclusion, *Bacillus*, *Bifidobacterium*, *Lactobacillus*, and yeast strains can antagonize the growth of *C. albicans*. In particular, many strains of *Bacillus* have obvious advantages to exert antagonistic strains. Most of the strains exert antagonistic effects by producing active compounds. Moreover, some inhibit the growth of *C. albicans* through niche competition. We should intensify research on strains with inhibitory activity, especially probiotics. Research on different strains, especially probiotics, with antifungal activity is helpful to develop the agent for inhibiting *C. albicans*. Since the effective components of some strains against *C. albicans* are not clear, the compounds with obvious inhibitory activity should be further analyzed.

3. Inhibitory Nature Metabolites Produced by Diverse Antagonists

Secondary metabolites derived from many plants and microorganisms are valuable natural compounds. Many natural products have significant biological activities, such as anti-tumor activity, antibacterial activity [63–65]. The antagonistic effect of the strain is mainly due to the production of natural secondary metabolites, such as antibiotics and antimicrobial peptides [66–68]. The antifungal compounds reviewed in this paper are secondary metabolites derived from microorganisms for biological control of *C. albicans* and have strong inhibition against *C. albicans*. Table 2 lists the various antagonistic microbial strains, the characteristics of the active compounds produced, and their inhibition mechanism against *C. albicans*. Table 3 lists the structure and the activity of these inhibitory compounds.

Table 2. Inhibitory nature metabolites produced by antagonists against *C. albicans*.

Sources	Inhibitory Compounds	Main Characteristics of the Compounds	Other Inhibitory Actions	References
Bacteria				
Bacillus subtilis	5HM2F	Inhibit morphological transition	Reduced levels of secreted virulence factors and ergosterol to reduce the main sources of biofilms.	[69]
Pantoea agglomerans C9-1	2-amino-3-(oxane-2,3-dicarboxamido) propanoyl-valine	Inhibit growth	None	[70]
Tenacibaculum discolor sv11	Dipyrrolepyridines A and B	Inhibit growth		[71]
Yeast				
Saccharomyces boulardii	Capric acid	Inhibit hyphal formation, adhesion and biofilm development	Transcriptional levels of *HWP1*, *INO1* and *CSH1* genes were decreased.	[72]
Eendophytic fungi				
Biatriospora sp.	Biatriosporin D	Inhibit adhesion, biofilm formation and hyphal morphogenesis	Regulated Ras1-CAMP-Efg1 pathway, disrupted morphological transition and attenuated virulence	[73]
Drechmeria sp.	Drechmerin B	Inhibit growth	None	[74]
Phoma sp. SYSU-SK-7	Colletotric A	Inhibit growth	None	[75]
Stachybotrys chartarum	Atranone Q	Inhibit growth	None	[76]
Xylaria sp. YM 311647	Sesquiterpenes and Isomatanic diterpenes	Inhibit growth	None	[77]
Marine fungi				
Aspergillus isolates from Waikiki Beach	Waikialoid A and Waikialide A	Inhibit biofilm formation	None	[78]
Penicillium meleagrinum var. *viridiflavum*	PF1163A and B	Inhibit growth	None	[79]
Penicillium minioluteum ZZ1657	Purpurides E and F	Inhibit growth	None	[80]
Marine actinomycetes				
Actinoalloteichus cyanogriseus WH1-2216-6	Caerulomycin A and C	Inhibit growth	None	[81]
Streptomyces sp.	Bahamaolides A	Inhibit isocitrate lyase	None	[82]
Streptomyces sp. ZZ741	Streptoglutarimides A-J and Streptovitacin A	Inhibit growth	None	[83]
Lichen				
lichens	Usnic acid	Reduce the thickness of mature biofilms and Inhibit biofilm adhesion.		[84]
lichens	Retigeric acid B	Inhibit hyphal formation	RAB regulated the Ras1-cAMP-Efg1 pathway and inhibited hyphal formation	[85]
Lichens with *Talaromyces funiculosu*	Funiculosone	Inhibit growth	None	[86]
Other strains				
Acremonium sp. PSU-MA70	8-Deoxytrichocin and trichodermol	Inhibit growth	None	[87]
Aspergillus micronesiensis	Cyschalasins A and B	Inhibit growth	None	[88]
Curvularia hawaiiensis TA26-15	Moriniafungins B-G	Inhibit growth	None	[89]
Fusarium and *Gibberella* species	Zearalenone	Inhibit biofilm formation of and hyphal morphogenesis	None	[90–92]
Fusarium spp.	Deoxynivalenol	Inhibit biofilm formation and reduce metabolic activity	DON and its derivatives interplayed with lanosterol 14a-demethylase	[93]
Penicillium fuscum and *Penicillium camembertii/clavigerum*	Berkleyolactone A	Inhibit growth	A new mode of action that had not been resolved	[94]
Ustilago maydis	Ustilagic acid B and C	Inhibit growth	None	[95]

5HM2F: 5-hydroxymethyl-2-furaldehyde.

Table 3. The structures and activity of compounds against *C. albicans*.

Inhibitory Compounds	Compound Structure	Activity	References
Terpenoids			
Isomatanic diterpenes	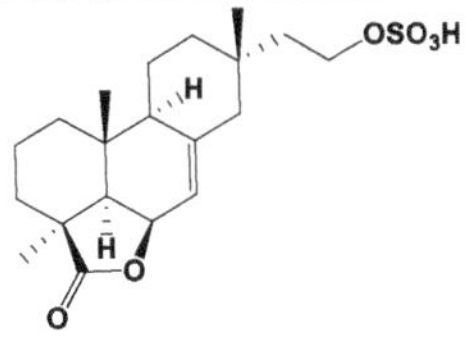	The MIC value was 16 μg/mL	[77]
Purpurides E and F	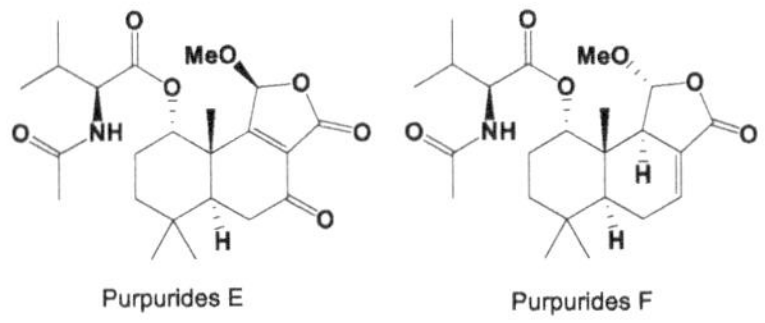	The MIC values were 12 and 6 μg/mL, respectively.	[80]
Usnic acid	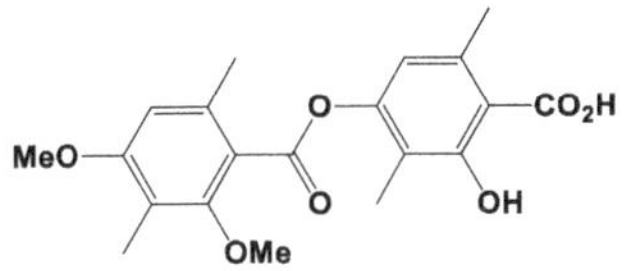	The MBIC value was 100 μg/mL.	[84]
Moriniafungins E	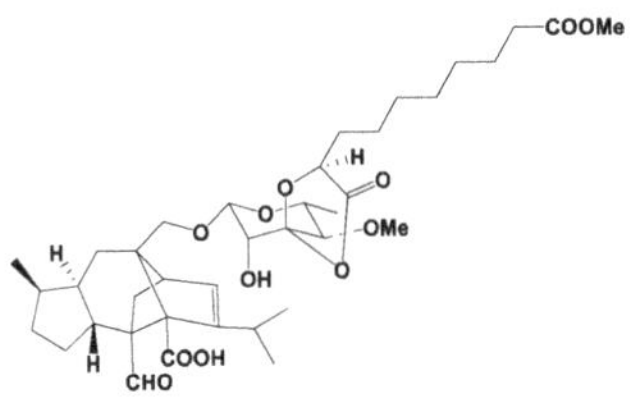	The MIC value was 2.9 μM.	[89]
Macrolides			
PF1163 A and B	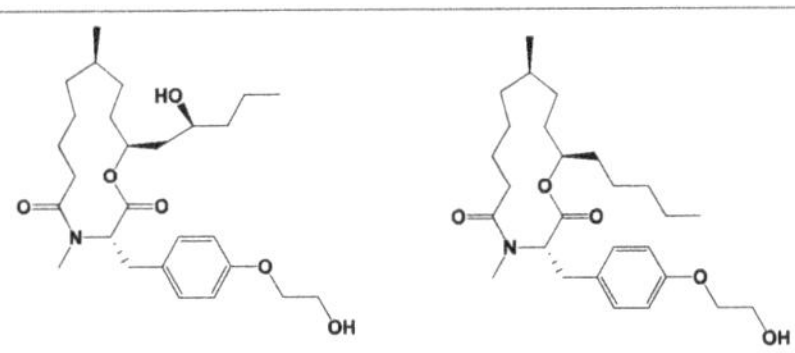	The inhibitory concentrations were 1 and 2 μg/mL, respectively.	[79]
Bahamaolides A	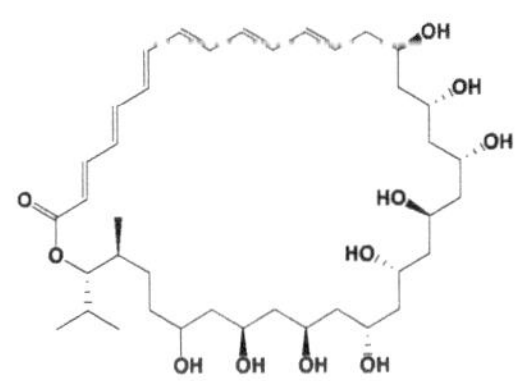	The MIC value was 12.5 μg/mL.	[82]
Berkleyolactone A	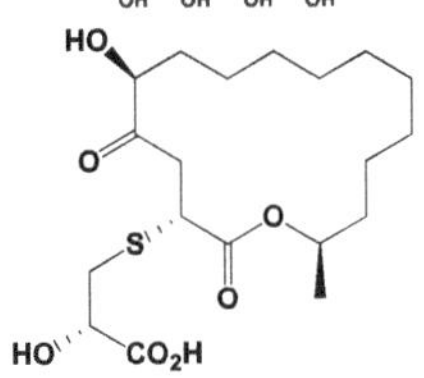	The MIC value was 1–2 μg/mL.	[94]

Table 3. *Cont.*

Inhibitory Compounds	Compound Structure	Activity	References
Organic acids			
Capric acid	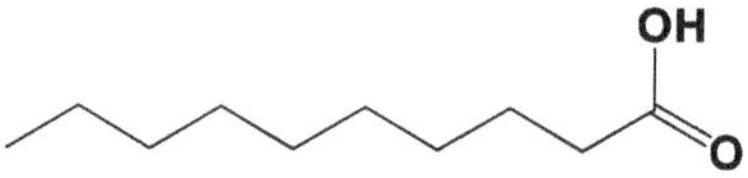	The inhibitory concentration was 45.3 μg/mL.	[72]
Retigeric acid B	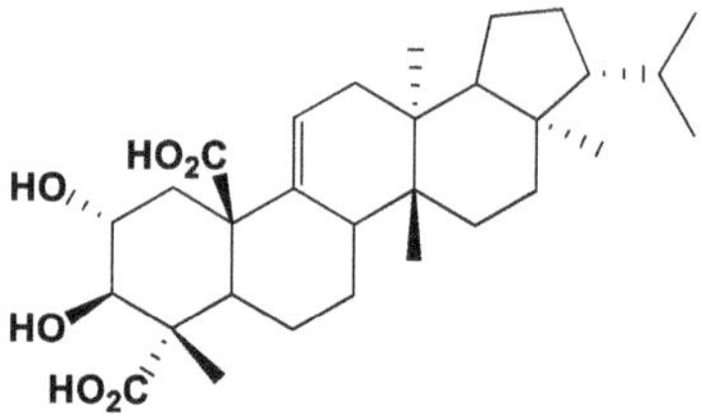	The MIC_{80} value was 8 μg/mL.	[85]
Ustilagic acid B and C	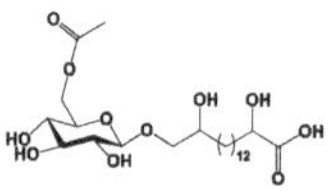 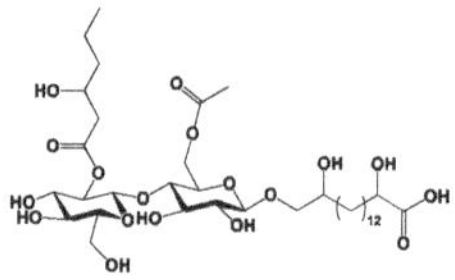	The MIC values were 50 and 100 μg/mL, respectively.	[95]
Alkaloids			
Ketones			
Colletotric A	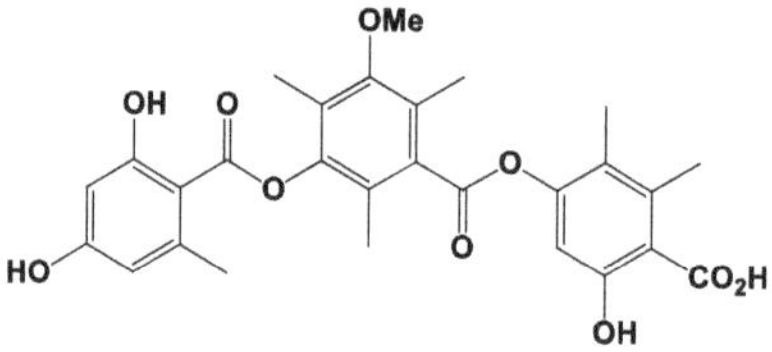	The MIC value was 3.27 μg/mL.	[75]
Atranone Q	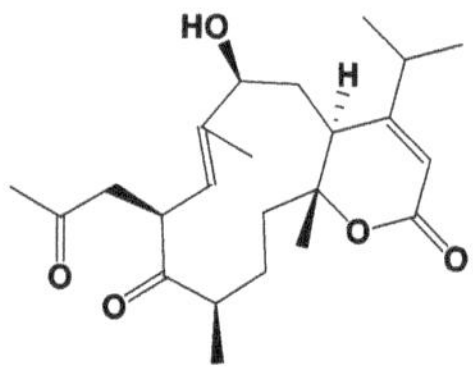	The MIC value was 8 μg/mL	[76]
Waikialoid A and Waikialide A	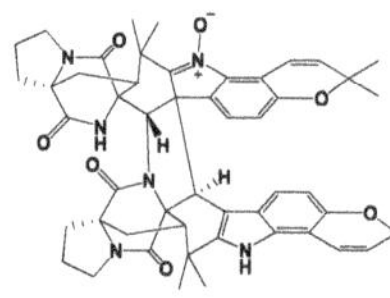 Waikialoid A 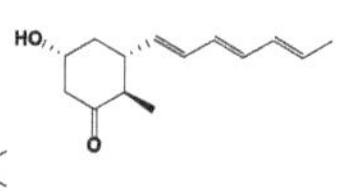 Waikialide A	The IC_{50} values were 1.4 and 32.4 μM, respectively.	[78]
Caerulomycin A and C	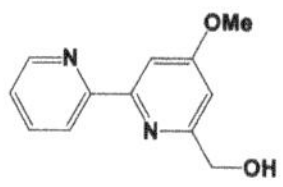 Caerulomycin A 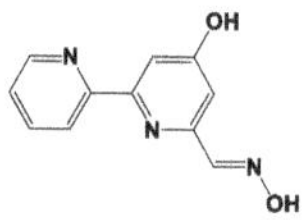 Caerulomycin C	The MIC values were 21.8 and 19.3 μM, respectively.	[81]

Table 3. *Cont.*

Inhibitory Compounds	Compound Structure	Activity	References
Cyschalasins A and B	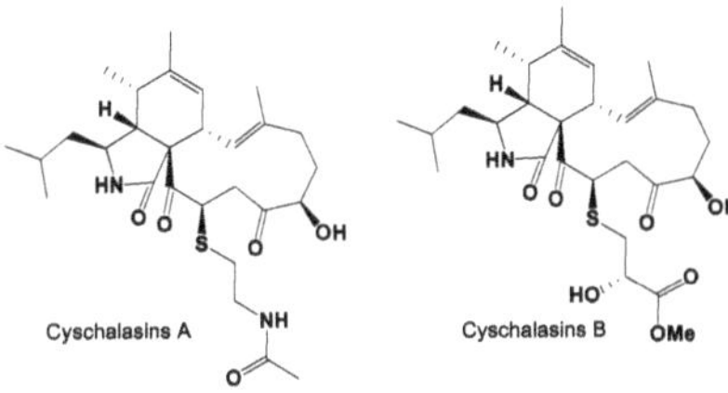	The MIC_{50} values were 43.3 ± 1.5 and 94.7 ± 1.3 μg/mL, respectively.	[88]
Zearalenone	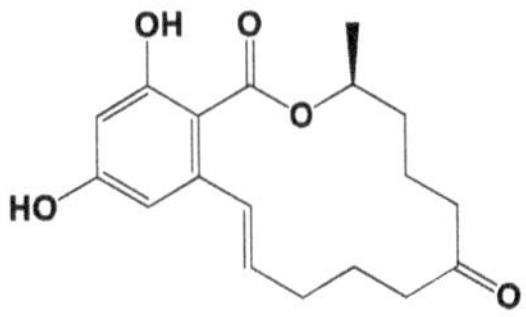	The inhibitory concentration was 100 μg/mL	[90–92]
Alcohols			
8-Deoxytrichothecin and trichodermol	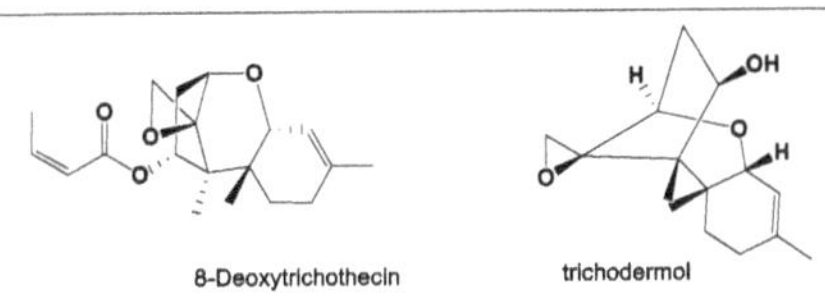	The MIC values were 16 and 64 μg/mL, respectively.	[87]
Deoxynivalenol and 3-acetyl-DON	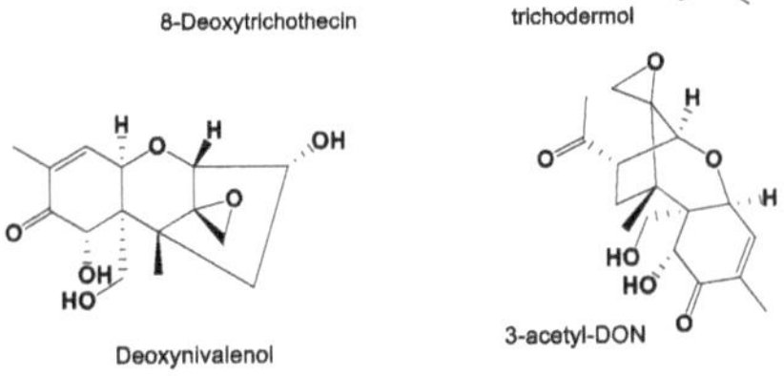	All inhibitory concentrations were 50 μg/mL.	[93]
Other structural compounds			
5HM2F	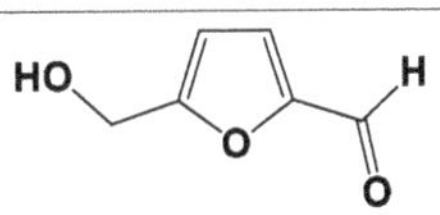	The MBIC value was 400 μg/mL.	[69]
2-amino-3-(oxane-2,3-dicarboxamido) propanoyl-valine	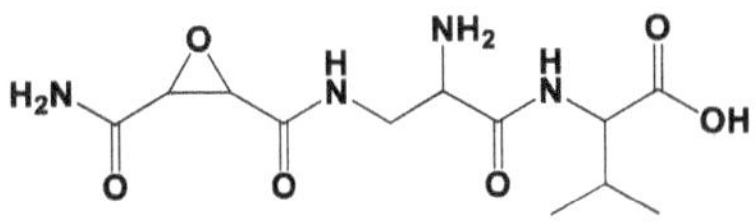	The inhibitory concentration was 1.5 μg/mL.	[70]
Dipyrrolepyridines A and B	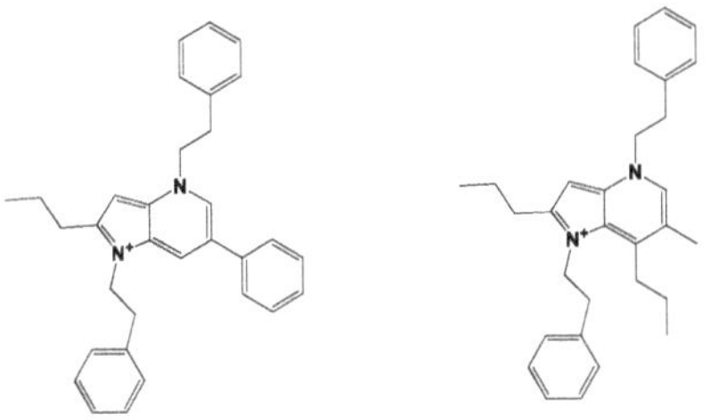	Certain antibacterial activity.	[71]

Table 3. *Cont.*

Inhibitory Compounds	Compound Structure	Activity	References
Biatriosporin D	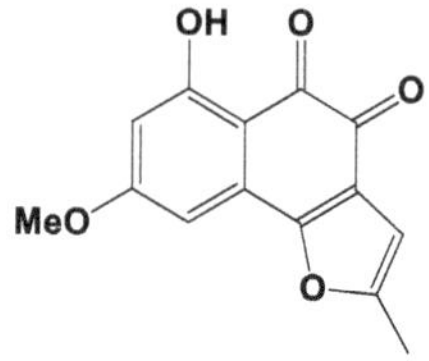	The inhibitory concentration was 2 μg/mL	[73]
Drechmerin B	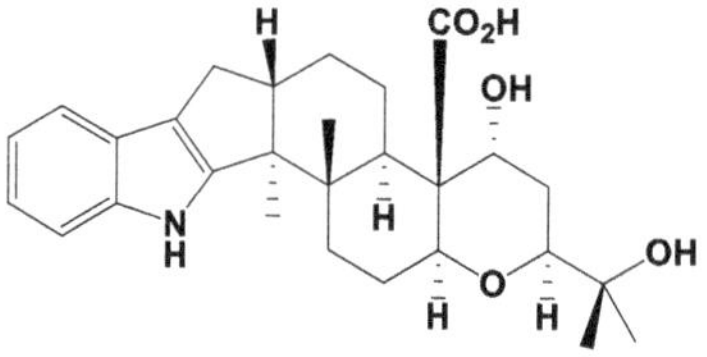	The MIC value was 12.5 μg/mL.	[74]
Streptoglutarimides D	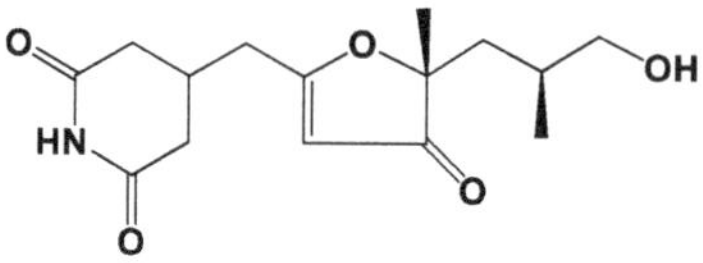	The MIC value was 4 μg/mL.	[83]
Funiculosone	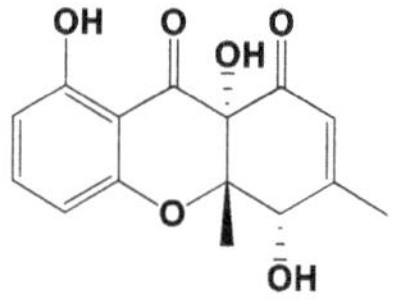	The IC_{50} value was 35 μg/mL.	[86]

BEC_{80}: 80% of biofilm-eradicating concentration; MBIC: maximum biofilm inhibitory concentration; 5HM2F: 5-hydroxymethyl-2-furaldehyde.

3.1. Nature Products Produced by Bacteria

Bacillus produces diverse active compounds, such as proteases, amylases, surfactants, and antibiotics [66,96–99]. Due to the high yield of antifungal active substances and the advantage of releasing peptides directly into the extracellular, *Bacillus subtilis* is a potential strain for the production of antifungal compounds [100–102]. The *B.subtilis* isolated from marine had antifungal membrane effect on *C. albicans*. It was found that 5-hydroxymethyl-2-furaldehyde (5HM2F) was one of the main components that inhibited *C. albicans* in the fermentation broth [69]. 5HM2F effectively disrupted the hyphal-like morphological transition of *C. albicans* and prevented the initial adhesion process. Further studies showed that 5HM2F reduced the main source of biofilms by reducing the levels of secreted virulence factors and ergosterol. In addition, the combination of 5HM2F with azole antifungal drugs effectively enhanced the anti-*C. albicans* activity of the tested drugs. Transcriptional level studies showed that 5HM2F increased the sensitivity of *C. albicans* to antifungal drugs by negatively regulating the expression levels of genes related to drug resistance mechanisms. As an antagonist, 5HM2F effectively inhibited the biofilm formation and reduced the resistance of *C. albicans* to traditional antifungal drugs.

Pantoea agglomerans are widespread in the environment [103,104]. *P. agglamerans* strain C9-1 was used as a biological control agent (BlightBan C9-1). A peptide antibiotic was isolated. The compound was 2-amino-3-(oxane-2,3-dicarboxamido)propanoyl-valine. This compound showed effectively inhibition on the growth of *C. albicans* [70].

Six novel alkaloids containing phenethylamine (PEA) were isolated from the culture medium of *Tenacibaculum discolor* sv11. Among them, Dipyrrolepyridines A and B had certain inhibitory activity against *C. albicans* FH2173 [71].

3.2. Nature Products Produced by Yeast

The researchers found that *S. boulardii* had inhibitory activity on *C. albicans*. The fermentation broth extracts inhibited hyphae formation, adhesion and biofilm development of *C. albicans* [72]. Further analysis showed that the fermentation broth contained 2-phenylethanol, capric, caprylic and caproic acid. The fermentation broth and the isolated pure compounds were tested for biological activity against *C. albicans*. Capric acid inhibited hyphae formation in *C. albicans* and also reduced adhesion and biofilm formation. However, compared with *S. boulardii* extract, the inhibitory effect on *C. albicans* was reduced by three times in the case of capric acid alone, so other compounds were contained to inhibit the adhesion of *C. albicans*. The transcriptional levels of *CSH1*, *INO1*, and *HWP1* genes were decreased in *C. albicans* treated with *B. boulardii* extract and capric acid.

3.3. Nature Products Produced by Endophytic Fungi

Biatriosporin D (BD), A phenolic compound, was isolated from *Biatriospora* spp. [73]. The compound inhibited biofilm formation, hyphal morphogenesis and adhesion of *C. albicans*. Notably, BD efficiently inhibited hyphal formation at doses lower than MIC value. Further studies showed that BD regulated the Ras1-cAMP-Efg1 pathway through reducing the cAMP level. As a prodrug, BD showed potential action against *C. albicans*. This provided possible application prospects for BD against clinically opportunistic fungi by targeting fungal virulence.

A fungus *Drechmeria* sp. was isolated from the roots of Panax notoginseng. Four known analogs and seven new indole diterpenoids, drechmerins A-G, were isolated from the fermentation broth. The MIC value of Drechmerin B against *C. albicans* was 12.5 μg/mL [74].

Five new polyketides and four known analogs were isolated from the *Phoma* sp. SYSU-SK-7 [75]. Among them, the polyketide colletotric B had strong antifungal activity against *C. albicans*, and the MIC value of colletotric A was 3.27 μg/mL. The MIC value of 3-hydroxy-5-methoxy-2, 4, 6-trimethylbenzoic acid was 2.62 μg/mL, and the MIC value of orsellinic acid was 2.10 μg/mL.

Three new monomers were isolated from the marine strain *Stachybotrys Chartarum*. The MIC value of compound Atranone Q was 8 μg/mL [76].

Nine sesquiterpenes and three diterpenes were isolated from the fermentation broth of the *Xylaria* sp. YM 311647 [77]. The MIC values of nine sesquiterpenes against *C. albicans* were different, while the activity of diterpenes was higher. One of the sesquiterpenes had the highest antibacterial activity against *C. albicans*, with an MIC value 16 μg/mL.

3.4. Nature Products Produced by Marine Fungi

One of the prenylated indole alkaloids, waikialoid A, was isolated from a metabolite-rich *Aspergillus* strain near Waikiki Beach. IC_{50} value of the natural product was 1.4 μM in inhibiting biofilm formation. Another compound, waikialide A, could inhibit the formation of *C. albicans* biofilm with a weaker IC_{50} value of 32.4 μM [78].

Two new 13-membered macrolide compounds and known PF1163A, B, D, H and F were isolated from *penicillium* strain. All of them had inhibitory activity against *C. albicans* when used in conjunction with fluconazole [79].

Three drimane sesquiterpene purpurides E-G were isolated from *P. minioluteum* ZZ1657. Purpurides E exhibited inhibitory activity against *C. albicans* with MIC values of 6–12 μg/mL, and Purpurides F was 3–6 μg/mL [80].

3.5. Nature Products Produced by Marine Source Actinomycetes

One new phenylpyridinealkaloid, five known analogues and five new bipyridine alkaloids were isolated from *Actinoalloteichus cyanogriseus* WH1-2216-6. The MICs of caerulomycin A and C against *C. albicans* were 21.8 and 19.3 μM, respectively [81].

Two new 36-membered macrolides, Bahamaolides A and B, were isolated from sediments of marine actinomycetes (*Streptomyces* sp.) on the North Cat Reef, Bahamas. Bahamaolides A obviously inhibited isocitrate lyase of *C. albicans* [82].

Streptovitacin A and new Streptoglutarimides A-J were isolated from marine actinomycetes *Streptomyces* sp. ZZ741. The MIC values of the obtained compounds against *C. albicans* were 8–20 μg/mL, and Streptoglutarimides D had a better inhibitory effect with 8 μg/mL [83].

3.6. Nature Products Produced by Lichen

Usnic acid, a secondary metabolite of lichens, effectively inhibited the hyphal switching of *C. albicans*. Usnic acid significantly reduced the thickness of mature biofilms and prevented the adhesion of biofilms. At the biofilm inhibitory concentration (BIC), the inhibitory effect of usnic acid on *C. albicans* biofilm could reach 65% [84].

As an inhibitor, Retigeric acid B (RAB) derived from *lichen* significantly inhibited the hyphae formation of *C. albicans* [105–107]. RAB prolonged the survival time of nematodes infected by *C. albicans*. RAB regulated the Ras1-CAMP-Efg1 pathway by reducing cAMP levels and inhibitd hyphal formation. By inhibiting the interruption of yeast-hyphal morphological transition and weakening the virulence of *C. albicans*, it provided a potential application for the treatment of *C. albicans* infection [85].

Funiculosone, a substituted dihydroxanthene-1, 9-dione, was isolated from the lichens of the *Trichosporaceae* fungus *T. funiculosus*. The IC_{50} value of *T. funiculosus* was 35 μg/mL [86].

3.7. Nature Products Produced by Other Fungal Sources

8-deoxytrichothecin and trichodermol, isolated from the *Acremonium* sp. PSU-MA70, exhibited moderate antifungal activity against *C. albicans* [87]. Two compounds cyschalasins A and B were isolated from *Aspergillus Micronesiensis* and showed antifungal activity against *C. albicans* [88]. Moriniafungins B-G, a new tetracyclic diterpene glycoside of Sordarincin, was isolated from *Curvularia hawaiiensis* TA26-15. Moriniafungins B-G had antifungal activity against *C. albicans* with an MIC value of 2.9 μM [89].

The F2 mycotoxin zearalenone (ZEN) produced by *Fusarium* and *Gibberella* species exhibited in vitro inhibitory effects on different microbial strains [90,91]. 100 μg/mL ZEN treatment significantly inhibited *C. albicans* hyphal morphogenesis and biofilm formation. Similarly, ZEN effectively destroyed established *C. albicans* biofilms without disturbing the planktonic cells. In vivo, ZEN prominently inhibited *C. albicans* infection in *Caenorhabditis elegans* [92].

Deoxynivalenol (DON), produced by *Fusarium* spp., was an epoxide sesquiterpene compound [93,108–110]. DON and 3-acetyl-DON exhibited a dose-dependent inhibitory effect on *C. albicans* in vitro. DON obviously reduced *C. albicans* metabolic activity, disrupted pre-formed biofilms, inhibited biofilm formation and inhibited hyphal that embedded in free-living planktonic cells and colonies. DON and 3-acetyl-DON mimicked the mechanism of through interplaying with lanosterol 14α-demethylase that was like the action of azole drugs. DON exhibited antifungal filament and antifungal membrane potential against *C. albicans* [111].

A carefully scheduled fermentation of *P. camembertii/clavigerum* and *P. fuscum* yielded eight novel 16-membered ring macrolides, Berkelilactone A exhibited the most potent antifungal activity in the macrolide series. It had low micromolar inhibitory activity against *C. albicans* (MIC = 1–2 μg/mL). Berkelilactone A did not inhibit protein synthesis and did not target ribosomes, suggesting a new mode of mechanism for its antibiotic activity, but the specific mechanism had not yet been elucidated [94].

U. maydis secreted a large amount of the glycolipid biosurfactant ustilagic acid. The new glycolipid ustilagic acid C and B were induced under special culture conditions. And the two compounds showed weak antifungal activity against *C. albicans* [95].

3.8. A Conclusion of Inhibitory Compounds Produced by Antagonistic Microbes

Many natural products that obtained from diverse microbial sources have been successfully applied in many fields. To overcome the increasing drug resistance of *C. albicans*, the discovery of new natural antifungal compounds is necessary. This review summarizes about 30 different compounds produced by microorganisms that have been found to have inhibitory effects on *C. albicans*. These compounds are derived from different bacteria and fungi, including bacteria such as *Bacillus*, *T. discolor* sv11 and *P. agglomerans*; yeast such as *S. bombicola* and *S. boulardii*; *Phoma* spp. SYSU-SK-7, *Biatriospora* sp.; marine-derived fungi such as *Aspergillus*, *P. minioluteum* ZZ1657; *Streptomyces* sp.; *A. cyanogriseus* WH1-2216-6; *Streptomyces* sp. ZZ741 and *Actinomycetes* of marine origin; other fungal sources: *Fusarium*, *Gibberella* species, *P. brown*, *P. camembertii/clavigerum*, *C. Hawaiian ensis* TA26-15, *U. maydis*; *A. micronesiensis*, *Acremonium* sp. PSU-MA70 and other fungi. It can be seen from Table 2 that the antifungal mechanisms of most isolated known or unknown compounds have not been clearly analyzed. Only a few compounds have been studied at the transcriptional level. These microorganisms produce compounds with different structures to inhibit *C. albicans* in different ways, such as inhibiting biofilm formation and hyphal morphological transformation.

4. Conclusions

With the emergence of *C. albicans* resistance against conventional antifungal therapies, new strategies to treat *C. albicans* infection are important [112]. Considering that *C. albicans* could threaten human life and health when the body's immune function declines or the normal flora is out of balance. both *Bacillus licheniformis* and *Bifidobacterium* can be used in clinic to prevent and treat intestinal microbiota disorders. This article reviews the different antagonistic microorganisms of *C. albicans* and various bioactive secondary metabolites produced by microorganisms, which are expected to achieve biological control of human pathogenic fungus *C. albicans*.

Biological control of microbial infections is an alternative approach that utilizes antagonistic microorganisms to prevent the growth and infection of harmful microorganisms. Antagonistic microbes, such as bacteria, yeast, and fungus, have been studied as potential antagonistic organisms for the control of *C. albicans*. Through the study on diverse strains with antifungal activity, it is helpful to develop the agent for inhibiting *C. albicans*. This is a potential strategy for biological control of *C. albicans*. On the other hand, secondary metabolites derived from microorganisms are valuable natural compounds. Many natural products have diverse structures and can exhibit significant biological activities. The structures of these compounds include: macrolides, terpenoids, alkaloids, organic acids, and other heterocyclic compounds. The secondary metabolites introduced in Tables 2 and 3 can significantly inhibit *C. albicans*. They are produced by diverse microorganisms. However no identified compounds are currently used as a drug against *C. albicans*. There are still four types of antifungal agents for *C. albicans* infection: polyene, triazole, 5-fluorocytosine, and echinocandin antifungal drugs [13]. Through the study of these active compounds, it is expected to obtain new drugs for the treatment and prevention of *C. albicans* infection, thereby maintaining human health.

Author Contributions: Writing—original draft preparation, H.L. (Honghua Li); writing—review and editing, F.S., X.Z., J.Y., X.X. and H.L. (Hehe Li); supervision, F.S. and H.L. (Hehe Li); funding acquisition, H.L. (Honghua Li) and F.S. All authors have read and agreed to the published version of the manuscript.

Funding: This research was funded by grants from the General Projects of Science and Technology Program of Beijing Municipal Education Commission (KM202210011008), the Key Lab of Marine Bioactive Substance and Modern Analytical Technique, SOA (MBSMAT-2019-06), Research Foundation for Youth Scholars of Beijing Technology and Business University (QNJJ2022-21), and Research Foundation for Advanced Talents of Beijing Technology and Business University (19008021176).

Institutional Review Board Statement: Not applicable for studies not involving humans or animals.

Informed Consent Statement: Not applicable.

Data Availability Statement: Not applicable.

Conflicts of Interest: The authors declare no conflict of interest.

References

1. Ramchandran, R.; Ramesh, S.; Thakur, R.; Chakrabarti, A.; Roy, U. Improved Production of Two Anti-*Candida* Lipopeptide Homologues Co-Produced by the Wild-Type *Bacillus subtilis* RLID 12.1 under Optimized Conditions. *Curr. Pharm. Biotechnol.* **2020**, *21*, 438–450. [CrossRef] [PubMed]
2. Karpiński, T.M.; Ożarowski, M.; Seremak-Mrozikiewicz, A.; Wolski, H.; Adamczak, A. Plant Preparations and Compounds with Activities against Biofilms Formed by *Candida* spp. *J. Fungi* **2021**, *7*, 360. [CrossRef] [PubMed]
3. Brown, G.D.; Denning, D.W.; Gow, N.A.; Levitz, S.M.; Netea, M.G.; White, T.C. Hidden killers: Human fungal infections. *Sci. Transl. Med.* **2012**, *4*, 165rv13. [CrossRef] [PubMed]
4. Moran, C.; Grussemeyer, C.A.; Spalding, J.R.; Benjamin, D.K., Jr.; Reed, S.D. Comparison of costs, length of stay, and mortality associated with *Candida glabrata* and *Candida albicans* bloodstream infections. *Am. J. Infect Control.* **2010**, *38*, 78–80. [CrossRef]
5. Vila, T.; Sultan, A.S.; Montelongo-Jauregui, D.; Jabra-Rizk, M.A. Oral Candidiasis: A Disease of Opportunity. *J. Fungi* **2020**, *6*, 15. [CrossRef]
6. Ruben, S.; Garbe, E.; Mogavero, S.; Albrecht-Eckardt, D.; Hellwig, D.; Häder, A.; Krüger, T.; Gerth, K.; Jacobsen, I.D.; Elshafee, O.; et al. Ahr1 and Tup1 Contribute to the Transcriptional Control of Virulence-Associated Genes in *Candida albicans*. *mBio* **2020**, *11*, e00206-20. [CrossRef]
7. Zakikhany, K.; Naglik, J.R.; Schmidt-Westhausen, A.; Holland, G.; Schaller, M.; Hube, B. In vivo transcript profiling of Candida albicans identifies a gene essential for interepithelial dissemination. *Cell. Microbiol.* **2007**, *9*, 2938–2954. [CrossRef]
8. Mavor, A.L.; Thewes, S.; Hube, B. Systemic fungal infections caused by *Candida* species: Epidemiology, infection process and virulence attributes. *Curr. Drug Targets* **2005**, *6*, 863–874. [CrossRef]
9. Nobile, C.J.; Johnson, A.D. *Candida albicans* Biofilms and Human Disease. *Annu. Rev. Microbiol.* **2015**, *69*, 71–92. [CrossRef]
10. Chandra, J.; Kuhn, D.M.; Mukherjee, P.K.; Hoyer, L.L.; McCormick, T.; Ghannoum, M.A. Biofilm formation by the fungal pathogen *Candida albicans*: Development, architecture, and drug resistance. *J. Bacteriol.* **2001**, *183*, 5385–5394. [CrossRef]
11. Nett, J.E.; Andes, D.R. Antifungal Agents: Spectrum of Activity, Pharmacology, and Clinical Indications. *Infect. Dis. Clin. N. Am.* **2016**, *30*, 51–83. [CrossRef] [PubMed]
12. Peyclit, L.; Yousfi, H.; Rolain, J.M.; Bittar, F. Drug Repurposing in Medical Mycology: Identification of Compounds as Potential Antifungals to Overcome the Emergence of Multidrug-Resistant Fungi. *Pharmaceuticals* **2021**, *14*, 488. [CrossRef] [PubMed]
13. Barantsevich, N.; Barantsevich, E. Diagnosis and Treatment of Invasive Candidiasis. *Antibiotics* **2022**, *11*, 718. [CrossRef]
14. Anderson, T.M.; Clay, M.C.; Cioffi, A.G.; Diaz, K.A.; Hisao, G.S.; Tuttle, M.D.; Nieuwkoop, A.J.; Comellas, G.; Maryum, N.; Wang, S.; et al. Amphotericin forms an extramembranous and fungicidal sterol sponge. *Nat. Chem. Biol.* **2014**, *10*, 400–406. [CrossRef] [PubMed]
15. Robbins, N.; Caplan, T.; Cowen, L.E. Molecular Evolution of Antifungal Drug Resistance. *Annu. Rev. Microbiol.* **2017**, *71*, 753–775. [CrossRef]
16. Hassanmoghadam, F.; Shokohi, T.; Hedayati, M.T.; Aslani, N.; Haghani, I.; Nabili, M.; Lotfali, E.; Davari, A.; Moazeni, M. High prevalence of itraconazole resistance among *Candida parapsilosis* isolated from Iran. *Curr. Med. Mycol.* **2019**, *5*, 43–46. [CrossRef]
17. Galia, L.; Pezzani, M.D.; Compri, M.; Callegari, A.; Rajendran, N.B.; Carrara, E.; Tacconelli, E. The Combacte Magnet Epi-Net Network. Surveillance of Antifungal Resistance in Candidemia Fails to Inform Antifungal Stewardship in European Countries. *J. Fungi* **2022**, *8*, 249. [CrossRef]
18. Berkow, E.L.; Lockhart, S.R. Fluconazole resistance in Candida species: A current perspective. *Infect. Drug Resist.* **2017**, *10*, 237–245. [CrossRef]
19. Whaley, S.G.; Berkow, E.L.; Rybak, J.M.; Nishimoto, A.T.; Barker, K.S.; Rogers, P.D. Azole Antifungal Resistance in *Candida albicans* and Emerging Non-*albicans Candida* Species. *Front. Microbiol.* **2017**, *7*, 2173. [CrossRef]
20. Delma, F.Z.; Al-Hatmi, A.M.S.; Brüggemann, R.J.M.; Melchers, W.J.G.; de Hoog, S.; Verweij, P.E.; Buil, J.B. Molecular Mechanisms of 5-Fluorocytosine Resistance in Yeasts and Filamentous Fungi. *J. Fungi* **2021**, *7*, 909. [CrossRef]
21. Ham, Y.Y.; Lewis, J.S.; Thompson, G.R. Rezafungin: A novel antifungal for the treatment of invasive candidiasis. *Future Microbiol.* **2021**, *16*, 27–36. [CrossRef] [PubMed]
22. Miesel, L.; Lin, K.Y.; Ong, V. Rezafungin treatment in mouse models of invasive candidiasis and aspergillosis: Insights on the PK/PD pharmacometrics of rezafungin efficacy. *Pharmacol. Res. Perspect.* **2019**, *7*, e00546. [CrossRef] [PubMed]
23. Miesel, L.; Cushion, M.T.; Ashbaugh, A.; Lopez, S.R.; Ong, V. Efficacy of Rezafungin in Prophylactic Mouse Models of Invasive Candidiasis, Aspergillosis, and Pneumocystis Pneumonia. *Antimicrob. Agents Chemother.* **2021**, *65*, e01992-20. [CrossRef]
24. Lepak, A.J.; Zhao, M.; Andes, D.R. Determination of Pharmacodynamic Target Exposures for Rezafungin against *Candida tropicalis* and *Candida dubliniensis* in the Neutropenic Mouse Disseminated Candidiasis Model. *Antimicrob. Agents Chemother.* **2019**, *63*, e01556-19. [CrossRef] [PubMed]

25. Pfaller, M.A.; Carvalhaes, C.; Messer, S.A.; Rhomberg, P.R.; Castanheira, M. Activity of a Long-Acting Echinocandin, Rezafungin, and Comparator Antifungal Agents Tested against Contemporary Invasive Fungal Isolates (SENTRY Program, 2016 to 2018). *Antimicrob. Agents Chemother.* **2020**, *64*, e00099-20. [CrossRef] [PubMed]
26. Farhadi, Z.; Farhadi, T.; Hashemian, S.M. Virtual screening for potential inhibitors of β(1,3)-D-glucan synthase as drug candidates against fungal cell wall. *J. Drug Assess.* **2020**, *9*, 52–59. [CrossRef] [PubMed]
27. Szymański, M.; Chmielewska, S.; Czyżewska, U.; Malinowska, M.; Tylicki, A. Echinocandins-structure, mechanism of action and use in antifungal therapy. *J. Enzyme Inhib. Med. Chem.* **2022**, *37*, 876–894. [CrossRef]
28. Sandrin, C.; Peypoux, F.; Michel, G. Coproduction of surfactin and iturin A, lipopeptides with surfactant and antifungal properties, by *Bacillus subtilis*. *Biotechnol. Appl. Biochem.* **1990**, *12*, 370–375.
29. Ahimou, F.; Jacques, P.; Deleu, M. Surfactin and iturin A effects on Bacillus subtilis surface hydrophobicity. *Enzyme Microb. Technol.* **2000**, *27*, 749–754. [CrossRef]
30. Newman, D.J.; Cragg, G.M. Natural Products as Sources of New Drugs over the Nearly Four Decades from 01/1981 to 09/2019. *J. Nat. Prod.* **2020**, *83*, 770–803. [CrossRef]
31. Sun, F.J.; Li, M.; Gu, L.; Wang, M.L.; Yang, M.H. Recent progress on anti-*Candida* natural products. *Chin. J. Nat. Med.* **2021**, *19*, 561–579. [CrossRef]
32. Guimarães, R.; Milho, C.; Liberal, Â.; Silva, J.; Fonseca, C.; Barbosa, A.; Ferreira, I.C.F.R.; Alves, M.J.; Barros, L. Antibiofilm Potential of Medicinal Plants against *Candida* spp. Oral Biofilms: A Review. *Antibiotics* **2021**, *10*, 1142. [CrossRef] [PubMed]
33. Espinel-Ingroff, A. Commercial Methods for Antifungal Susceptibility Testing of yeasts: Strengths and Limitations as Predictors of Resistance. *J. Fungi* **2022**, *8*, 309. [CrossRef] [PubMed]
34. Murphy, S.E.; Bicanic, T. Drug Resistance and Novel Therapeutic Approaches in Invasive Candidiasis. *Front. Cell. Infect. Microbiol.* **2021**, *11*, 759408. [CrossRef]
35. Li, B.; Pan, L.; Zhang, H.; Xie, L.; Wang, X.; Shou, J.; Qi, Y.; Yan, X. Recent Developments on Using Nanomaterials to Combat *Candida albicans*. *Front. Chem.* **2021**, *9*, 813973. [CrossRef] [PubMed]
36. Khan, F.; Bamunuarachchi, N.I.; Tabassum, N.; Jo, D.M.; Khan, M.M.; Kim, Y.M. Suppression of hyphal formation and virulence of *Candida albicans* by natural and synthetic compounds. *Biofouling* **2021**, *37*, 626–655. [CrossRef]
37. Owen, M.K.; Clenney, T.L. Management of vaginitis. *Am. Fam. Physician* **2004**, *70*, 2125–2132. [PubMed]
38. Siikala, E.; Rautemaa, R.; Richardson, M.; Saxen, H.; Bowyer, P.; Sanglard, D. Persistent *Candida albicans* colonization and molecular mechanisms of azole resistance in autoimmune polyendocrinopathy-candidiasis-ectodermal dystrophy (APECED) patients. *J. Antimicrob. Chemother.* **2010**, *65*, 2505–2513. [CrossRef]
39. Zida, A.; Bamba, S.; Yacouba, A.; Ouedraogo-Traore, R.; Guiguemdé, R.T. Anti-Candida albicans natural products, sources of new antifungal drugs: A review. *J. Mycol. Med.* **2017**, *27*, 1–19. [CrossRef]
40. Ghai, S.; Sood, S.S.; Jain, R.K. Antagonistic and antimicrobial activities of some bacterial isolates collected from soil samples. *Indian J. Microbiol.* **2007**, *47*, 77–80. [CrossRef]
41. Khan, M.N.; Lin, H.; Li, M.; Wang, J.; Mirani, Z.A.; Khan, S.I.; Buzdar, M.A.; Ali, I.; Jamil, K. Identification and growth optimization of a Marine *Bacillus* DK1-SA11 having potential of producing broad spectrum antimicrobial compounds. *Pak. J. Pharm. Sci.* **2017**, *30*, 839–853. [PubMed]
42. Devi, S.; Kiesewalter, H.T.; Kovács, R.; Frisvad, J.C.; Weber, T.; Larsen, T.O.; Kovács, Á.T.; Ding, L. Depiction of secondary metabolites and antifungal activity of *Bacillus velezensis* DTU001. *Synth. Syst. Biotechnol.* **2019**, *4*, 142–149. [CrossRef] [PubMed]
43. Li, X.; Zhang, Y.; Wei, Z.; Guan, Z.; Cai, Y.; Liao, X. Antifungal Activity of Isolated Bacillus amyloliquefaciens SYBC H47 for the Biocontrol of Peach Gummosis. *PLoS ONE* **2016**, *11*, e0162125. [CrossRef] [PubMed]
44. Li, M.S.M.; Piccoli, D.A.; McDowell, T.; MacDonald, J.; Renaud, J.; Yuan, Z.C. Evaluating the biocontrol potential of *Canadian* strain *Bacillus velezensis* 1B-23 via its surfactin production at various pHs and temperatures. *BMC Biotechnol.* **2021**, *21*, 31. [CrossRef]
45. Bonfrate, L.; Di Palo, D.M.; Celano, G.; Albert, A.; Vitellio, P.; De Angelis, M.; Gobbetti, M.; Portincasa, P. Effects of *Bifidobacterium longum* BB536 and *Lactobacillus rhamnosus* HN001 in IBS patients. *Eur. J. Clin. Invest.* **2020**, *50*, e13201. [CrossRef]
46. Lau, A.S.; Yanagisawa, N.; Hor, Y.Y.; Lew, L.C.; Ong, J.S.; Chuah, L.O.; Lee, Y.Y.; Choi, S.B.; Rashid, F.; Wahid, N.; et al. *Bifidobacterium longum* BB536 alleviated upper respiratory illnesses and modulated gut microbiota profiles in Malaysian pre-school children. *Benef. Microbes* **2018**, *9*, 61–70. [CrossRef]
47. Vazquez-Munoz, R.; Thompson, A.; Russell, J.T.; Sobue, T.; Zhou, Y.; Dongari-Bagtzoglou, A. Insights From the *Lactobacillus johnsonii* Genome Suggest the Production of Metabolites With Antibiofilm Activity Against the Pathobiont *Candida albicans*. *Front. Microbiol.* **2022**, *13*, 853762. [CrossRef]
48. Zeise, K.D.; Woods, R.J.; Huffnagle, G.B. Interplay between *Candida albicans* and Lactic Acid Bacteria in the Gastrointestinal Tract: Impact on Colonization Resistance, Microbial Carriage, Opportunistic Infection, and Host Immunity. *Clin. Microbiol. Rev.* **2021**, *34*, e0032320. [CrossRef]
49. Sipiczki, M. Metschnikowia strains isolated from botrytized grapes antagonize fungal and bacterial growth by iron depletion. *Appl. Environ. Microbiol.* **2006**, *72*, 6716–6724. [CrossRef]
50. Kang, C.H.; Han, S.H.; Kim, Y.; Paek, N.S.; So, J.S. In Vitro Probiotic Properties of Lactobacillus salivarius MG242 Isolated from Human Vagina. *Probiotics Antimicrob. Proteins* **2018**, *10*, 343–349. [CrossRef]
51. Chae, G.P.; Shoda, M.; Kubota, H. Suppressive effect of Bacillus subtilis and it's products on phytopathogenic microorganisms. *J. Ferment. Bio. Eng.* **1990**, *69*, 1–7. [CrossRef]

52. Ongena, M.; Jacques, P. Bacillus lipopeptides: Versatile weapons for plant disease biocontrol. *Trends Microbiol.* **2008**, *16*, 115–125. [CrossRef] [PubMed]
53. Abriouel, H.; Franz, C.M.; Ben Omar, N.; Gálvez, A. Diversity and applications of Bacillus bacteriocins. *FEMS Microbiol. Rev.* **2011**, *35*, 201–232. [CrossRef] [PubMed]
54. Abou-Kassem, D.E.; Elsadek, M.F.; Abdel-Moneim, A.E.; Mahgoub, S.A.; Elaraby, G.M.; Taha, A.E.; Elshafie, M.M.; Alkhawtani, D.M.; Abd El-Hack, M.E.; Ashour, E.A. Growth, carcass characteristics, meat quality, and microbial aspects of growing quail fed diets enriched with two different types of probiotics (*Bacillus toyonensis* and *Bifidobacterium bifidum*). *Poult. Sci.* **2021**, *100*, 84–93. [CrossRef] [PubMed]
55. Kadja, L.; Dib, A.L.; Lakhdara, N.; Bouaziz, A.; Espigares, E.; Gagaoua, M. Influence of Three Probiotics Strains, *Lactobacillus rhamnosus* GG, *Bifidobacterium animalis* subsp. Lactis BB-12 and *Saccharomyces boulardii* CNCM I-745 on the Biochemical and Haematological Profiles and Body Weight of Healthy Rabbits. *Biology* **2021**, *10*, 1194. [CrossRef] [PubMed]
56. Lau, A.S.; Liong, M.T. Lactic Acid Bacteria and Bifidobacteria-Inhibited *Staphylococcus epidermidis*. *Wounds.* **2014**, *26*, 121–131.
57. Fukuda, S.; Toh, H.; Hase, K.; Oshima, K.; Nakanishi, Y.; Yoshimura, K.; Tobe, T.; Clarke, J.M.; Topping, D.L.; Suzuki, T.; et al. *Bifidobacteria* can protect from enteropathogenic infection through production of acetate. *Nature* **2011**, *469*, 543–547. [CrossRef]
58. El-Zahar, K.M.; Hassan, M.F.Y.; Al-Qaba, S.F. Protective Effect of Fermented Camel Milk Containing *Bifidobacterium longum* BB536 on Blood Lipid Profile in Hypercholesterolemic Rats. *J. Nutr. Metab.* **2021**, *2021*, 1557945. [CrossRef]
59. Suido, H.; Miyao, M. *Bifidobacterium longum*-fermented broccoli supernatant inhibited the growth of *Candida albicans* and some pathogenic bacteria in vitro. *Biocontrol. Sci.* **2008**, *13*, 41–48. [CrossRef]
60. Chevalier, M.; Ranque, S.; Prêcheur, I. Oral fungal-bacterial biofilm models in vitro: A review. *Med Mycol.* **2018**, *56*, 653–667. [CrossRef]
61. Kluyver, A.J.; van der Walt, J.P.; van Triet, A.J. Pulcherrimin, The Pigment of *Candida pulcherrima*. *Proc. Natl. Acad. Sci. USA* **1953**, *39*, 583–593. [CrossRef] [PubMed]
62. Türkel, S.; Ener, B. Isolation and characterization of new *Metschnikowia pulcherrima* strains as producers of the antimicrobial pigment pulcherrimin. *Zeitschrift für Naturforschung* **2009**, *64*, 405–410. [CrossRef] [PubMed]
63. Renda, G.; Kadıoğlu, M.; Kılıç, M.; Korkmaz, B.; Kırmızıbekmez, H. Anti-inflammatory secondary metabolites from *Scrophularia kotschyana*. *Hum. Exp. Toxicol.* **2021**, *40*, S676–S683. [CrossRef] [PubMed]
64. Muhaj, F.F.; George, S.J.; Nguyen, C.D.; Tyring, S.K. Antimicrobials and resistance part II: Antifungals, antivirals, and antiparasitics. *J. Am. Acad. Dermatol.* **2022**, *86*, 1207–1226. [CrossRef]
65. Shala, A.; Singh, S.; Hameed, S.; Khurana, S.M.P. Essential Oils as Alternative Promising Anti-Candidal Agents: Progress and Prospects. *Curr. Pharm. Des.* **2022**, *28*, 58–70. [CrossRef]
66. Dehghanifar, S.; Keyhanfar, M.; Emtiazi, G. Production and partial purification of thermostable bacteriocins from *Bacillus pumilus* ZED17 and DFAR8 strains with antifungal activity. *Mol. Biol. Res. Commun.* **2019**, *8*, 41–49. [CrossRef]
67. Boparai, J.K.; Sharma, P.K. Mini Review on Antimicrobial Peptides, Sources, Mechanism and Recent Applications. *Protein Pept. Lett.* **2020**, *27*, 4–16. [CrossRef]
68. Bin Hafeez, A.; Jiang, X.; Bergen, P.J.; Zhu, Y. Antimicrobial Peptides: An Update on Classifications and Databases. *Int. J. Mol. Sci.* **2021**, *22*, 11691. [CrossRef]
69. Subramenium, G.A.; Swetha, T.K.; Iyer, P.M.; Balamurugan, K.; Pandian, S.K. 5-hydroxymethyl-2-furaldehyde from marine bacterium *Bacillus subtilis* inhibits biofilm and virulence of *Candida albicans*. *Microbiol. Res.* **2018**, *207*, 19–32. [CrossRef]
70. Sammer, U.F.; Völksch, B.; Möllmann, U.; Schmidtke, M.; Spiteller, P.; Spiteller, M.; Spiteller, D. 2-amino-3-(oxirane-2,3-dicarboxamido)-propanoyl-valine, an effective peptide antibiotic from the epiphyte *Pantoea agglomerans* 48b/90. *Appl. Environ. Microbiol.* **2009**, *75*, 7710–7717. [CrossRef]
71. Wang, L.; Linares-Otoya, V.; Liu, Y.; Mettal, U.; Marner, M.; Armas-Mantilla, L.; Willbold, S.; Kurtán, T.; Linares-Otoya, L.; Schäberle, T.F. Discovery and Biosynthesis of Antimicrobial Phenethylamine Alkaloids from the Marine Flavobacterium *Tenacibaculum discolor* sv11. *J. Nat. Prod.* **2022**, *85*, 1039–1051. [CrossRef] [PubMed]
72. Murzyn, A.; Krasowska, A.; Stefanowicz, P.; Dziadkowiec, D.; Łukaszewicz, M. Capric acid secreted by *S. boulardii* inhibits *C. albicans* filamentous growth, adhesion and biofilm formation. *PLoS ONE* **2010**, *5*, e12050. [CrossRef] [PubMed]
73. Zhang, M.; Chang, W.; Shi, H.; Zhou, Y.; Zheng, S.; Li, Y.; Li, L.; Lou, H. Biatriosporin D displays anti-virulence activity through decreasing the intracellular cAMP levels. *Toxicol. Appl. Pharmacol.* **2017**, *322*, 104–112. [CrossRef] [PubMed]
74. Zhao, J.C.; Wang, Y.L.; Zhang, T.Y.; Chen, Z.J.; Yang, T.M.; Wu, Y.Y.; Sun, C.P.; Ma, X.C.; Zhang, Y.X. Indole diterpenoids from the endophytic fungus *Drechmeria* sp. as natural antimicrobial agents. *Phytochemistry* **2018**, *148*, 21–28. [CrossRef] [PubMed]
75. Chen, Y.; Yang, W.; Zou, G.; Chen, S.; Pang, J.; She, Z. Bioactive polyketides from the mangrove endophytic fungi *Phoma* sp. SYSU-SK-7. *Fitoterapia* **2019**, *139*, 104369. [CrossRef] [PubMed]
76. Yang, B.; He, Y.; Lin, S.; Zhang, J.; Li, H.; Wang, J.; Hu, Z.; Zhang, Y. Antimicrobial Dolabellanes and Atranones from a Marine-Derived Strain of the Toxigenic Fungus *Stachybotrys chartarum*. *J. Nat. Prod.* **2019**, *82*, 1923–1929. [CrossRef]
77. Wu, S.H.; He, J.; Li, X.N.; Huang, R.; Song, F.; Chen, Y.W.; Miao, C.P. Guaiane sesquiterpenes and isopimarane diterpenes from an endophytic fungus *Xylaria* sp. *Phytochemistry* **2014**, *105*, 197–204. [CrossRef]
78. Wang, X.; You, J.; King, J.B.; Powell, D.R.; Cichewicz, R.H. Waikialoid A suppresses hyphal morphogenesis and inhibits biofilm development in pathogenic *Candida albicans*. *J. Nat. Prod.* **2012**, *75*, 707–715. [CrossRef]

79. Okabe, M.; Sugita, T.; Kinoshita, K.; Koyama, K. Macrolides from a Marine-Derived Fungus, *Penicillium meleagrinum* var. *viridiflavum*, Showing Synergistic Effects with Fluconazole against Azole-Resistant *Candida albicans*. *J. Nat. Prod.* **2016**, *79*, 1208–1212. [CrossRef]
80. Ma, M.Z.; Ge, H.J.; Yi, W.W.; Wu, B.; Zhang, Z.Z. Bioactive drimane sesquiterpenoids and isocoumarins from the marine-derived fungus *Penicillium minioluteum* ZZ1657. *Tetrahedron Lett.* **2019**, *7*, 13. [CrossRef]
81. Fu, P.; Wang, S.; Hong, K.; Li, X.; Liu, P.; Wang, Y.; Zhu, W. Cytotoxic bipyridines from the marine-derived actinomycete *Actinoalloteichus cyanogriseus* WH1-2216-6. *J. Nat. Prod.* **2011**, *74*, 1751–1756. [CrossRef]
82. Kim, D.G.; Moon, K.; Kim, S.H.; Park, S.H.; Park, S.; Lee, S.K.; Oh, K.B.; Shin, J.; Oh, D.C. Bahamaolides A and B, antifungal polyene polyol macrolides from the marine actinomycete *Streptomyces* sp. *J. Nat. Prod.* **2012**, *75*, 959–967. [CrossRef] [PubMed]
83. Zhang, D.; Yi, W.; Ge, H.; Zhang, Z.; Wu, B. Bioactive Streptoglutarimides A-J from the Marine-Derived *Streptomyces* sp. ZZ741. *J. Nat. Prod.* **2019**, *82*, 2800–2808. [CrossRef] [PubMed]
84. Nithyanand, P.; Beema Shafreen, R.M.; Muthamil, S.; Karutha Pandian, S. Usnic acid inhibits biofilm formation and virulent morphological traits of *Candida albicans*. *Microbiol. Res.* **2015**, *179*, 20–28. [CrossRef]
85. Chang, W.; Li, Y.; Zhang, L.; Cheng, A.; Lou, H. Retigeric acid B attenuates the virulence of *Candida albicans* via inhibiting adenylyl cyclase activity targeted by enhanced farnesol production. *PLoS ONE* **2012**, *7*, e41624. [CrossRef]
86. Padhi, S.; Masi, M.; Cimmino, A.; Tuzi, A.; Jena, S.; Tayung, K.; Evidente, A. Funiculosone, a substituted dihydroxanthene-1,9-dione with two of its analogues produced by an endolichenic fungus *Talaromyces funiculosus* and their antimicrobial activity. *Phytochemistry* **2019**, *157*, 175–183. [CrossRef] [PubMed]
87. Rukachaisirikul, V.; Rodglin, A.; Sukpondma, Y.; Phongpaichit, S.; Buatong, J.; Sakayaroj, J. Phthalide and isocoumarin derivatives produced by an *Acremonium* sp. isolated from a mangrove Rhizophora apiculata. *J. Nat. Prod.* **2012**, *75*, 853–858. [CrossRef] [PubMed]
88. Wu, Z.; Zhang, X.; Anbari, W.H.A.; Zhou, Q.; Zhou, P.; Zhang, M.; Zeng, F.; Chen, C.; Tong, Q.; Wang, J.; et al. Cysteine Residue Containing Merocytochalasans and 17,18-seco-Aspochalasins from *Aspergillus micronesiensis*. *J. Nat. Prod.* **2019**, *82*, 2653–2658. [CrossRef]
89. Zhang, M.Q.; Xu, K.X.; Xue, Y.; Cao, F.; Yang, L.J.; Hou, X.M.; Wang, C.Y.; Shao, C.L. Sordarin Diterpene Glycosides with an Unusual 1,3-Dioxolan-4-one Ring from the Zoanthid-Derived Fungus *Curvularia hawaiiensis* TA26-15. *J. Nat. Prod.* **2019**, *82*, 2477–2482. [CrossRef]
90. Geng, Z.; Zhu, W.; Su, H.; Zhao, Y.; Zhang, K.Q.; Yang, J. Recent advances in genes involved in secondary metabolite synthesis, hyphal development, energy metabolism and pathogenicity in *Fusarium graminearum* (teleomorph *Gibberella zeae*). *Biotechnol. Adv.* **2014**, *32*, 390–402. [CrossRef]
91. Hidy, P.H.; Baldwin, R.S.; Greasham, R.L.; Keith, C.L.; McMullen, J.R. Zearalenone and some derivatives: Production and biological activities. *Adv. Appl. Microbiol.* **1977**, *22*, 59–82. [CrossRef] [PubMed]
92. Rajasekharan, S.K.; Lee, J.H.; Zhao, Y.; Lee, J. The Mycotoxin Zearalenone Hinders *Candida albicans* Biofilm Formation and Hyphal Morphogenesis. *Indian J. Microbiol.* **2018**, *58*, 19–27. [CrossRef] [PubMed]
93. Rajasekharan, S.K.; Byun, J.; Lee, J. Inhibitory effects of deoxynivalenol on pathogenesis of *Candida albicans*. *J. Appl. Microbiol.* **2018**, *125*, 1266–1275. [CrossRef] [PubMed]
94. Stierle, A.A.; Stierle, D.B.; Decato, D.; Priestley, N.D.; Alverson, J.B.; Hoody, J.; McGrath, K.; Klepacki, D. The Berkeleylactones, Antibiotic Macrolides from Fungal Coculture. *J. Nat Prod.* **2017**, *80*, 1150–1160. [CrossRef]
95. Yang, X.L.; Takayoshi, A.; Toshiyuki, W.; Ikuro, A. Induced production of the novel glycolipid ustilagic acid C in the plant pathogen *Ustilago maydis*. *Tetrahedron Lett.* **2013**, *54*, 3655–3657. [CrossRef]
96. Caetano, T.; Krawczyk, J.M.; Mösker, E.; Süssmuth, R.D.; Mendo, S. Heterologous expression, biosynthesis, and mutagenesis of type II lantibiotics from *Bacillus licheniformis* in *Escherichia coli*. *Chem. Biol.* **2011**, *18*, 90–100. [CrossRef]
97. Fira, D.; Dimkić, I.; Berić, T.; Lozo, J.; Stanković, S. Biological control of plant pathogens by *Bacillus* species. *J. Biotechnol.* **2018**, *285*, 44–55. [CrossRef] [PubMed]
98. Chen, W.; Wang, J.; Huang, D.; Cheng, W.; Shao, Z.; Cai, M.; Zheng, L.; Yu, Z.; Zhang, J. Volatile Organic Compounds from *Bacillus aryabhattai* MCCC 1K02966 with Multiple Modes against Meloidogyne incognita. *Molecules* **2021**, *27*, 103. [CrossRef]
99. Chu, J.; Wang, Y.; Zhao, B.; Zhang, X.M.; Liu, K.; Mao, L.; Kalamiyets, E. Isolation and identification of new antibacterial compounds from *Bacillus pumilus*. *Appl. Microbiol. Biotechnol.* **2019**, *103*, 8375–8381. [CrossRef]
100. Fu, L.; Wang, Y.; Ju, J.; Cheng, L.; Xu, Y.; Yu, B.; Wang, L. Extracellular production of active-form *Streptomyces mobaraensis* transglutaminase in *Bacillus subtilis*. *Appl. Microbiol. Biotechnol.* **2020**, *104*, 623–631. [CrossRef]
101. Zhang, K.; Su, L.; Wu, J. Enhancing Extracellular Pullulanase Production in *Bacillus subtilis* through dltB Disruption and Signal Peptide Optimization. *Appl. Biochem. Biotechnol.* **2022**, *194*, 1206–1220. [CrossRef] [PubMed]
102. Xia, Y.; Zhao, J.; Chen, H.; Liu, X.; Wang, Y.; Tian, F.; Zhang, H.P.; Zhang, H.; Chen, W. Extracellular secretion in *Bacillus subtilis* of a cytoplasmic thermostable beta-galactosidase from *Geobacillus stearothermophilus*. *J. Dairy Sci.* **2010**, *93*, 2838–2845. [CrossRef] [PubMed]
103. Nissan, G.; Gershovits, M.; Morozov, M.; Chalupowicz, L.; Sessa, G.; Manulis-Sasson, S.; Barash, I.; Pupko, T. Revealing the inventory of type III effectors in *Pantoea agglomerans* gall-forming pathovars using draft genome sequences and a machine-learning approach. *Mol. Plant Pathol.* **2018**, *19*, 381–392. [CrossRef] [PubMed]

104. Nissan, G.; Chalupowicz, L.; Sessa, G.; Manulis-Sasson, S.; Barash, I. Two *Pantoea agglomerans* type III effectors can transform nonpathogenic and phytopathogenic bacteria into host-specific gall-forming pathogens. *Mol. Plant Pathol.* **2019**, *20*, 1582–1587. [CrossRef]
105. Sun, L.M.; Cheng, A.X.; Wu, X.Z.; Zhang, H.J.; Lou, H.X. Synergistic mechanisms of retigeric acid B and azoles against *Candida albicans*. *J. Appl. Microbiol.* **2010**, *108*, 341–348. [CrossRef]
106. Chang, W.Q.; Wu, X.Z.; Cheng, A.X.; Zhang, L.; Ji, M.; Lou, H.X. Retigeric acid B exerts antifungal effect through enhanced reactive oxygen species and decreased cAMP. *Biochim. Biophys. Acta* **2011**, *1810*, 569–576. [CrossRef]
107. Chang, W.; Li, Y.; Zhang, L.; Cheng, A.; Liu, Y.; Lou, H. Retigeric acid B enhances the efficacy of azoles combating the virulence and biofilm formation of *Candida albicans*. *Biol Pharm Bull.* **2012**, *35*, 1794–1801. [CrossRef]
108. Sobrova, P.; Adam, V.; Vasatkova, A.; Beklova, M.; Zeman, L.; Kizek, R. Deoxynivalenol and its toxicity. *Interdiscip. Toxicol.* **2010**, *3*, 94–99. [CrossRef]
109. Audenaert, K.; Vanheule, A.; Höfte, M.; Haesaert, G. Deoxynivalenol: A major player in the multifaceted response of *Fusarium* to its environment. *Toxins* **2013**, *6*, 1–19. [CrossRef]
110. May, H.D.; Wu, Q.; Blake, C.K. Effects of the *Fusarium* spp. mycotoxins fusaric acid and deoxynivalenol on the growth of *Ruminococcus albus* and *Methanobrevibacter ruminantium*. *Can. J. Microbiol.* **2000**, *46*, 692–699. [CrossRef]
111. Berthiller, F.; Dall'Asta, C.; Schuhmacher, R.; Lemmens, M.; Adam, G.; Krska, R. Masked mycotoxins: Determination of a deoxynivalenol glucoside in artificially and naturally contaminated wheat by liquid chromatography-tandem mass spectrometry. *J. Agric Food Chem.* **2005**, *53*, 3421–3425. [CrossRef] [PubMed]
112. Ellepola, A.N.; Samaranayake, L.P.; Khan, Z.U. Extracellular phospholipase production of oral *Candida albicans* isolates from smokers, diabetics, asthmatics, denture wearers and healthy individuals following brief exposure to polyene, echinocandin and azole antimycotics. *Braz. J. Microbiol.* **2016**, *47*, 911–916. [CrossRef] [PubMed]

antibiotics

MDPI

Article

Mixing Propolis from Different Apiaries and Harvesting Years: Towards Propolis Standardization?

Marta Peixoto [1], Ana Sofia Freitas [1,2,3], Ana Cunha [1,2,3], Rui Oliveira [1,2,3,*] and Cristina Almeida-Aguiar [1,2,3,*]

1 Department of Biology, School of Sciences, University of Minho, Campus de Gualtar, 4710-057 Braga, Portugal
2 CITAB, Centre for the Research and Technology of Agro-Environmental and Biological Sciences, University of Minho, 4710-057 Braga, Portugal
3 CBMA, Centre of Molecular and Environmental Biology, University of Minho, 4710-057 Braga, Portugal
* Correspondence: ruipso@bio.uminho.pt (R.O.); cristina.aguiar@bio.uminho.pt (C.A.-A.)

Abstract: Global demand for safe, effective and natural products has been increasing in parallel with consumers' concerns about personal and environmental health. Propolis, a traditional and potentially medicinal product with several health benefits, is a beehive product with a worldwide reputation. However, despite the bioactivities reported, the low productivity and high chemical heterogeneity have been extensively hampering broader industrial uses. To assist in overcoming some of these problems, we prepared and characterized mixtures of ethanol extracts of a heterogeneous propolis sample (Pereiro) collected over a five-year period (2011–2015) and, additionally, we mixed two different propolis samples from distinct regions of Portugal (Pereiro and Gerês), also harvested at different times. An investigation of the antimicrobial and antioxidant properties, as well as characterization of the chemical composition of the eleven propolis blends were performed in this work. The antioxidant and antimicrobial activities of such blends of propolis samples, either from different localities and/or different years, were maintained, or even enhanced, when a comparison of the individual extracts was conducted. The differences in the chemical composition of the original propolis samples were also diluted in the mixtures. The results reemphasize the great potential of propolis and suggest that mixing different samples, regardless of provenance or harvesting date, can contribute to propolis standardization while simultaneously increasing its availability and adding value to this beehive byproduct.

Keywords: propolis; ethanol extracts; blends of propolis; phenolic compounds; antimicrobial activity; antioxidant potential

Citation: Peixoto, M.; Freitas, A.S.; Cunha, A.; Oliveira, R.; Almeida-Aguiar, C. Mixing Propolis from Different Apiaries and Harvesting Years: Towards Propolis Standardization? *Antibiotics* **2022**, *11*, 1181. https://doi.org/10.3390/antibiotics11091181

Academic Editors: Fuhang Song and Yunjiang Feng

Received: 15 July 2022
Accepted: 25 August 2022
Published: 31 August 2022

1. Introduction

The global market for natural products has been growing due to increasing consumer concern about personal health and the widespread use of synthetic chemicals with potentially harmful side effects and environmental impacts. The rising need for safe, effective and natural alternatives has given attention to propolis, widely recognized as a traditional and potentially medicinal product with several health benefits [1,2]. Propolis or "bee glue" is a sticky product composed of resinous and balsamic materials collected by bees from several plant sources and also of other substances resulting from bees' metabolism [3]. Interest in propolis has mostly arisen due to its broad range of valuable bioactivities, such as its antioxidant and antimicrobial properties, which have been mainly ascribed to phenolic compounds [4–6].

Worldwide demand for propolis has been rising, with China and India being the fastest-growing markets back in 2015 [7]. Leading the propolis market since 2018, North America is expected to be at the top by 2024 [8]. According to the most recent projections, the global propolis market size is expected to reach around USD 700 million, with a CAGR (Compound Annual Growth Rate) of 6% between 2019 and 2024 [8].

Despite the difficulty in accurately estimating the total sales of propolis and its byproducts, as beekeeping is essentially a home-made industry, US propolis sales were estimated at 46,000 € in 1996 [9]. Japan's interest in propolis has triggered a price increase from 4.4 to 176.4 €/ Kg in recent years, revealing the economic potential of bee glue [10]. However, in other countries, such as Portugal, propolis has still been an undervalued and cheaper beehive product [11], not because of a lower quality but rather due to its low level of exploitation [12,13]. This low level of commercial exploitation by most Portuguese beekeepers is mainly due to their lack of awareness and lack of technical knowledge, along with the rooted tradition of honey production, making propolis production practically neglected in the context of national apiculture [14]. Additionally, propolis productivity is very low: a European hive can produce between 50–150 g propolis per year [15] and, almost similarly, the estimated annual production of Portuguese propolis is around 100 g/ hive [15]. Lower productivities, such as 15.7 g propolis/ hive [16] or 24.2 ± 22.5 g/ hive [17] have been reported worldwide too, contributing to a reduced interest in propolis production as a considerable volume is often required by its main target, the pharmaceutical industry. In addition, until adequate quality parameters are developed, propolis will remain an alternative treatment without acceptance in medicine. The complex and variable propolis chemistry makes propolis quality standardization very challenging. More than 800 different chemical compounds were identified but the link between marker compound(s) for propolis and their respective therapeutic potential is still missing [18].

Some factors have been identified as determinants in propolis production and quality —the resin botanical origin, the genetics of the honey bee, the hive structure and material, food availability, environmental factors, and disease (see [19] for a review)—but the lack of quality standards and proper legislation still hinders the introduction of propolis to the world market. Therefore, a compilation of the parameters that need to be ensured to commercialize propolis for specific applications is critical and regulatory agencies should establish the quality parameters for propolis in a certain country.

The low productivity of propolis and its great chemical variability allied to the lack of standardization are the main obstacles to propolis applications, for example, in the food industry as a food preservative but especially in the pharmaceutical industry for therapeutic purposes [10,20–22]. In this framework, and given the great potential of this natural product [6,23,24], the aim of this work is the evaluation of the antimicrobial and antioxidant activities of mixtures of propolis, as well as the characterization of its chemical composition through in vitro methodologies. We previously showed that mixing ethanol extracts of propolis collected over five years from an apiary (Gerês; G) results in the maintenance or improvement of bioactivities [13]. However, propolis from Gerês seems rather unique, as it shows chemical and biological constancy along harvests in different years [25], which is different to what is largely known for bee glue. As a result, we intended to go further in evaluating (i) the outcome of mixing a very different but more typical propolis sample collected in different years from another apiary (Pereiro; P) and showing different bioactive and chemical profiles over the years [26]; and (ii) the effect of blending the two completely different propolis samples collected from the two apiaries (G and P) in different years.

Thus, in this study we prepared blends of (i) the ethanol extracts of propolis collected in Pereiro over a 5-year period (2011–2015); and of (ii) the ethanol extracts of propolis collected from two distinct apiaries/ regions—Pereiro and Gerês—in selected years. By mixing propolis from different apiaries and harvesting years, we expect to be able to overcome the limitations of the low yield per hive diluting the differences found between individual samples and increasing the available propolis for the market, in this way contributing towards a standardization of propolis biochemical profiles and propolis value.

2. Results

2.1. Characterization of Mixtures Obtained from Propolis Samples Harvested from the Same Apiary over Different Years

2.1.1. Total Polyphenols and Flavonoids Contents

In order to make a broad chemical characterization of individual extracts and blends of propolis from Pereiro, total polyphenols contents (TPC) and total flavonoid contents (TFC) were determined (Table 1).

Table 1. Total polyphenols (TPC) and flavonoids contents (TFC) of the ethanol extracts of propolis harvested at Pereiro in the years 2011 to 2015 and of the mixtures of P.EEs prepared in this work. The results are presented as mean ± standard deviation of mg of gallic acid equivalents (GAE) and mg of quercetin equivalents (QE) per g extract (mg GAE/ g extract and mg QE/ g extract per g of extract), respectively. Significant differences ($p < 0.05$) between TPC or between TFC of single extracts (P.EEs) and mixtures of extracts (mP.EEs) are noted with different lowercase letters. Significant differences between TPC or between TFC of mP.EEs ($p < 0.05$) are represented with different uppercase letters.

Samples	TPC (mg GAE/ g Extract)	TFC (mg QE/ g Extract)
P11.EE	224.6 ± 11.5^{a}	43.4 ± 0.7^{a}
P12.EE	173.2 ± 10.1^{c}	32.4 ± 1.5^{b}
mP(P11.EE+P12.EE)	$\mathbf{206.6 \pm 4.5^{b;\,A}}$	$\mathbf{43.9 \pm 2.0^{a;\,C}}$
P11.EE	224.6 ± 11.5^{a}	43.4 ± 0.7^{a}
P13.EE	217.6 ± 6.6^{a}	38.1 ± 2.0^{b}
mP(P11.EE+P13.EE)	$\mathbf{200.3 \pm 6.3^{b;\,A}}$	$\mathbf{46.2 \pm 0.8^{a;\,C}}$
P11.EE	224.6 ± 11.5^{a}	43.4 ± 0.7^{b}
P14.EE	174.4 ± 4.2^{c}	33.4 ± 2.0^{c}
mP(P11.EE+P14.EE)	$\mathbf{198.5 \pm 9.3^{b;\,A}}$	$\mathbf{47.7 \pm 0.5^{a;\,C}}$
P12.EE	173.2 ± 10.1^{b}	32.4 ± 1.5^{c}
P13.EE	217.6 ± 6.6^{a}	38.1 ± 2.0^{b}
mP(P12.EE+P13.EE)	$\mathbf{194.9 \pm 12.2^{b;\,A}}$	$\mathbf{44.6 \pm 2.7^{a;\,C}}$
P13.EE	217.6 ± 6.6^{a}	38.1 ± 2.0^{b}
P14.EE	174.4 ± 4.2^{b}	33.4 ± 2.0^{b}
mP(P13.EE+P14.EE)	$\mathbf{202.5 \pm 10.5^{a;\,A}}$	$\mathbf{47.8 \pm 2.1^{a;\,C}}$
P13.EE	217.6 ± 6.6^{b}	38.1 ± 2.0^{c}
P15.EE	262.2 ± 4.3^{a}	78.4 ± 1.7^{a}
mP(P13.EE+P15.EE)	$\mathbf{204.6 \pm 15.5^{b;\,A}}$	$\mathbf{67.9 \pm 2.0^{b;\,A}}$
P14.EE	174.4 ± 4.2^{c}	33.4 ± 2.0^{c}
P15.EE	262.2 ± 4.3^{a}	78.4 ± 1.7^{a}
mP(P14.EE+P15.EE)	$\mathbf{209.2 \pm 16.5^{b;\,A}}$	$\mathbf{67.0 \pm 2.2^{b;\,A}}$
P11.EE	224.6 ± 11.5^{a}	43.4 ± 0.7^{a}
P12.EE	173.2 ± 10.1^{b}	32.4 ± 1.5^{c}
P13.EE	217.6 ± 6.6^{a}	38.1 ± 2.0^{b}
mP(P11.EE−P13.EE)	$\mathbf{202.6 \pm 12.4^{a;\,A}}$	$\mathbf{46.5 \pm 2.6^{a;\,C}}$
P11.EE	224.6 ± 11.5^{b}	43.4 ± 0.7^{c}
P12.EE	173.2 ± 10.1^{c}	32.4 ± 1.5^{e}
P13.EE	217.6 ± 6.6^{b}	38.1 ± 2.0^{d}
P14.EE	174.4 ± 4.2^{c}	33.4 ± 2.0^{e}
P15.EE	262.2 ± 4.3^{a}	78.4 ± 1.7^{a}
mP(P11.EE−P15.EE)	$\mathbf{215.1 \pm 11.3^{b;\,A}}$	$\mathbf{55.3 \pm 1.7^{b;\,B}}$

Note: mP—mixtures of P.EEs. Different lowercase letters (a, b, c, d, e) were used for significant differences ($p < 0.05$) between TPC or between TFC of single extracts (P.EEs) and mixtures of extracts (mP.EEs). Different uppercase letters (**A**, **B**, **C**) were used for significant differences between TPC or between TFC of mP.EEs ($p < 0.05$).

The total polyphenols content of mixtures of ethanol extracts of Pereiro propolis (Table 1) range between 194.9 and 215.1 mg GAE/ g propolis extract, displayed by the mixture of P12.EE and P13.EE ($mP_{(P12.EE+P13.EE)}$) and the mixture of all of the five EEs of propolis collected in different years, respectively. The TPC of mP.EEs is frequently amid the contents of the individual EEs of Pereiro propolis. Interestingly, the exception was for the mixtures $mP_{(P11.EE+P13.EE)}$ and $mP_{(P13.EE+P15.EE)}$, with a slightly lower content when compared with the respective individual extracts, which are the ones with the highest TPC. The nature of the method for estimation of the phenolic content might account for such results, as the TPC methodology is based on the reduction of the Folin–Ciocalteu reagent, which could be affected by synergisms and/ or antagonisms of the phenolics that came together when the mixtures were prepared. The method was reported as giving higher values for polyphenols than the sum of the individual compounds as measured by HPLC [27]. Oxidation of multiple phenolic groups may generate products that are themselves reducing agents, thus giving a higher Folin value but the same can potentially occur in vivo and thus the Folin–Ciocalteu measurement may be relevant [28]. However, despite the differences observed between the individual extracts, the TPC of the mixtures were very similar ($p > 0.05$; ^{A}). In addition, the TPC of the mP.EEs were similar to the ones found in other single EEs of European propolis samples [29,30], meaning that even when mixing propolis samples with different TPC values, the phenolics contents of the blends remain within the values generally found for propolis from other sources. Moreover, the mixtures obtained from individual extracts with high and low TPC produced blends with intermediate TPC (see the pairings) or high TPC (compare, for instance, $mP(_{P11.EE\,-\,P15.EE})$ and $mP_{(P11.EE-P13.EE)}$ with their respective individual PEEs), highlighting not only that blending diluted the content differences between samples, but also that blending samples of several years did not significantly reduce the high TPC found in the samples of particular years.

The total flavonoids content (Table 1) varied between 43.9 and 67.9 mg QE/ g propolis extract in the mP.EEs, with the higher contents being measured in P15.EE-containing mixtures (from 55.3 to 67.9 mg QE/ g extract), with this individual extract being the one displaying the highest TFC (78.4 mg QE/ g extract). The TFC of the mixtures generally maintained the values of the individual extracts or showed a slight increase. Still, the values were within the range of contents found in other European individual samples [31,32]. In the case of the TFC, the pairings obtained with an extract of high TFC (again those from odd years) always resulted in mixtures retaining this high value or with a significantly higher TFC, but nevertheless within the range of contents found in other European individual samples [31,32]. This increased TFC of the mixtures may be due to the occurrence of some kind of synergism between compounds from different extracts and suggests a higher potential of blends in a broad spectrum of applications. In addition, as shown by Chang et al. [33], this method does not detect flavanones since these compounds do not form stable complexes with Al^{3+}. Nevertheless, the contents in flavones and flavonols detected by the assay are reliable indicators of the antioxidant activity of the samples (see further discussion in Figure 1b).

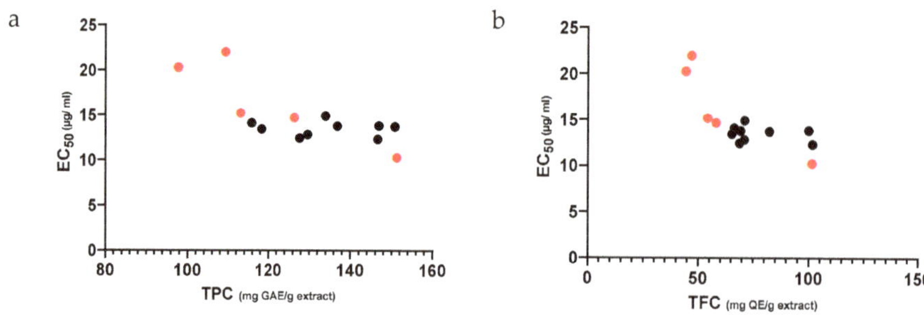

Figure 1. Correlation between antioxidant potential (EC_{50} values)) and TPC (**a**) or TFC (**b**) of individual ethanol extracts of propolis from Pereiro (•) and of mP.EEs (•).

2.1.2. Antioxidant Potential of Propolis

Antioxidant activity, expressed by the EC_{50} parameter, varied between 12.3 and 14.9 µg/mL (Table 2) for Pereiro propolis-containing mixtures. Interestingly, even when the EC_{50} values were very different among the individual extracts of the blend, the antioxidant potential of the majority of mP.EEs was similar to the one of the most-active individual extracts present in the blend (see, for instance, $mP_{(P11.EE+P12.EE)}$ and $mP_{(P11.EE+P14.EE)}$), except for P15.EE-containing blends, which showed an intermediate EC_{50} value. Together, these results highlight the advantage of mixing multiple propolis samples: higher homogeneity between blends/lots, thereby contributing to its standardization, and maintenance of the higher phenolic and flavonoid contents and antioxidant capacity, meaning an improved, more constant and predictable quality of the blends.

Table 2. P.EEs' and mP.EEs' ability to capture DPPH• free radicals. Antioxidant potential is expressed as a mean ± standard deviation of EC_{50} values (µg/mL). Significant differences between single P.EEs and mP.EEs ($p < 0.05$) are noted with different lowercase letters. Significant differences between mP.EEs ($p < 0.05$) are represented with different uppercase letters.

Samples	EC_{50} (µg/mL)
P11.EE	14.7 ± 2.7^{a}
P12.EE	22.0 ± 0.4^{b}
$mP_{(P11.EE+P12.EE)}$	**$13.5 \pm 0.3^{a;\ A,B,C}$**
P11.EE	14.7 ± 2.7^{a}
P13.EE	15.2 ± 2.3^{a}
$mP_{(P11.EE+P13.EE)}$	**$12.5 \pm 0.2^{a;\ A,B,C}$**
P11.EE	14.7 ± 2.7^{b}
P14.EE	20.3 ± 0.3^{a}
$mP_{(P11.EE+P14.EE)}$	**$12.8 \pm 0.3^{b;\ A,B,C}$**
P12.EE	22.0 ± 0.4^{a}
P13.EE	15.2 ± 2.3^{b}
$mP_{(P12.EE+P13.EE)}$	**$14.1 \pm 0.7^{b;\ A,B}$**
P13.EE	15.2 ± 2.3^{b}
P14.EE	20.3 ± 0.3^{a}
$mP_{(P13.EE+P14.EE)}$	**$14.9 \pm 0.4^{b;\ A}$**
P13.EE	15.2 ± 2.3^{a}
P15.EE	10.3 ± 1.7^{b}
$mP_{(P13.EE+P15.EE)}$	**$12.3 \pm 0.2^{a,b;\ C}$**
P14.EE	20.3 ± 0.3^{a}
P15.EE	10.3 ± 1.7^{c}
$mP_{(P14.EE+P15.EE)}$	**$13.8 \pm 0.6^{b;\ A,B}$**
P11.EE	14.7 ± 2.7^{b}
P12.EE	22.0 ± 0.4^{a}
P13.EE	15.2 ± 2.3^{b}
$mP_{(P11.EE-P13.EE)}$	**$13.8 \pm 0.7^{b;\ A,B}$**
P11.EE	14.7 ± 2.7^{c}
P12.EE	22.0 ± 0.4^{a}
P13.EE	$15.2 \pm 2.3^{c,b}$
P14.EE	$20.3 \pm 0.3^{a,b}$
P15.EE	10.3 ± 1.7^{d}
$mP_{(P11.EE-P15.EE)}$	**$13.7 \pm 0.4^{c;\ B}$**

Note: mP—mixtures of P.EEs. Different lowercase letters (a, b, c, d) were used for significant differences ($p < 0.05$) between TPC or between TFC of single extracts (P.EEs) and mixtures of extracts (mP.EEs). Different uppercase letters (**A**, **B**, **C**) were used for significant differences between TPC or between TFC of mP.EEs ($p < 0.05$).

In general, all of the mixtures displayed a similar antioxidant potential (Table 2), once again in the range reported both for single European (26.45 µg/mL Ireland and 27.72 µg /mL Czech Republic) [29,34] and for other individual Portuguese propolis samples [35], even being propolis diverse in this activity as well [36,37]. A lower DPPH• scavenging activity ($IC_{50} = 32.35 \pm 2.84$ µg mL^{-1}) was reported for an EE of Chinese propolis [38], whereas Brazilian green propolis appears to have the best antioxidant potential [39].

Figure 1 shows a negative correlation between both the TPC (Figure 1a) and TFC (Figure 1b) values (Table 1) of the P.EEs and their antioxidant capacity, measured by the radical scavenging activity (Table 2); the higher the former, the higher the latter. In addition, it shows that, when mixing propolis individual extracts, the TPC, TFC and EC_{50} values of the mixtures narrowed down its range, clustering near the most active and phenolic-rich individual extracts. Yet, when mixing propolis, the TPC, TFC and radical scavenging activity of the mixtures stood between the values of the individual samples or similar to the most active individual propolis extract. In fact, mP.EEs displayed roughly the same EC_{50} value (Figure 1a,b; black dots) regardless of the TPC (Figure 1a) or the TFC (Figure 1b) of the individual extracts (red dots).

These observations suggest that mixing different known propolis samples will ensure that the bioactivities of the mixtures will be in the range of the individual samples, making the choice of blends more predictable in terms of their bioactivity, thereby contributing to normalizing the antioxidant potential.

2.1.3. Antimicrobial Potential of Propolis

Antibacterial properties of EEs and of mixtures of EEs of propolis from Pereiro samples are presented in Table 3. In most cases mP.EEs showed antibacterial activity that was identical to the most-active single extract, as seen in $mP_{(P11.EE-P13.EE)}$ against *Bacillus subtilis* and *Staphylococcus aureus*, or an even higher activity such as in $mP_{(P11.EE+P12.EE)}$ against *B. subtilis* or $mP_{(P13.EE+P15.EE)}$ against *S. aureus*.

Table 3. MIC values (µg/mL) of the ethanol extracts of propolis from Pereiro collected from 2011 to 2015 and of the mixtures of P.EEs against *Bacillus subtilis*, *Propionibacterium acnes*, *Staphylococcus aureus*, methicillin-resistant *Staphylococcus aureus* (MRSA) and *Escherichia coli*. The results show the mean ± standard deviation of three assays with three replicates each.

	Gram-Positive				Gram-Negative
Samples	***B. subtilis***	***P. acnes***	***S. aureus***	**MRSA**	***E. coli***
P11.EE	500	—	>750	—	—
P12.EE	500	—	500	—	>1000
$mP_{(P11.EE+P12.EE)}$	**200**	**500**	**500**	**1500**	**1500**
P11.EE	500	—	>750	—	—
P13.EE	200	—	750	—	—
$mP_{(P11.EE+P13.EE)}$	**200**	**500**	**750**	**1500**	**1500**
P11.EE	500	—	>750	—	—
P14.EE	100	—	500	—	—
$mP_{(P11.EE+P14.EE)}$	**200**	**500**	**500**	**1500**	**1500**
P12.EE	500	—	500	—	>1000
P13.EE	200	—	750	—	—
$mP_{(P12.EE+P13.EE)}$	**200**	**500**	**500**	**1000**	**1000**
P13.EE	200	—	750	—	—
P14.EE	100	—	500	—	—
$mP_{(P13.EE+P14.EE)}$	**200**	**500**	**500**	**1500**	**1500**
P13.EE	200	—	750	—	—
P15.EE	500	200	750	>1250	>1250
$mP_{(P13.EE+P15.EE)}$	**200**	**500**	**500**	**1000**	**1000**
P14.EE	100	—	500	—	—
P15.EE	500	200	750	>1250	>1250
$mP_{(P14.EE+P15.EE)}$	**200**	**500**	**500**	**>1500**	**>1500**
P11.EE	500	—	>750	—	—
P12.EE	500	—	500	—	>1000
P13.EE	200	—	750	—	—
$mP_{(P11.EE-P13.EE)}$	**200**	**500**	**500**	**1500**	**1500**
P11.EE	500	—	>750	—	—
P12.EE	500	—	500	—	>1000
P13.EE	200	—	750	—	—
P14.EE	100	—	500	—	—
P15.EE	500	200	750	>1250	>1250
$mP_{(P11.EE-P15.EE)}$	**200**	**500**	**500**	**1500**	**1500**

Note: mP—mixtures of P.EEs.

The same type of result was observed against more-resistant bacteria, such as methicillin-resistant *Staphylococcus aureus* (MRSA) or *Escherichia coli*, with $mP_{(P12.EE+P13.EE)}$ and $mP_{(P13.EE+P15.EE)}$ as the most active mixtures (MIC = 1000 μg/mL). mP.EEs displayed a higher antibacterial activity against *E. coli* than individual Portuguese propolis samples studied by [40], though were similar in their activity against *S. aureus*. The remaining mixtures showed a similar antibacterial activity, displaying identical MIC values and being more active against Gram-positive than towards Gram-negative bacteria, as generally reported [21]. Additionally, MIC values of propolis mixtures against *S. aureus*, *B. subtilis* and *E. coli* were very close to the average MIC calculated by [41] from several studies (mean MIC values of EEs of propolis were 457, 180 and 784 μg/mL respectively).

According to Saraiva et al. [42], plant extracts with MIC ranging from 100 to 500 μg/mL are active extracts and moderately active if MIC varies from 500 to 1000 μg/mL while extracts with MIC ranging from 1000 to 2000 μg/mL are considered to have low activity. Transposing this classification to propolis extracts, mP.EEs were revealed to have particularly interesting activities against *B. subtilis*, *Propionibacterium acnes* and *S. aureus*. Moreover, although no data were obtained for the individual extracts against *P. acnes*, for instance, the MIC of the mixtures and the findings regarding its behavior against the other indicator strains support the choice of blends over individual extracts and open perspectives for clinical applications in the case of this acne-causing strain.

The results obtained for mP.EEs anti-yeast activity followed the same pattern as it did for antibacterial activity. The mixtures mostly displayed an MIC value similar to the most-active extract or, in some blends, even lower MIC values (Table 4). For *S. cerevisiae*, for example, almost all of the mixtures displayed the MIC of the most-active extract, except $mP_{(P11.EE+P13.EE)}$, which was even more active than the single extracts.

A curious observation from all of these results is that P14.EE, one of the extracts with the lowest TPC and TFC (Table 1) and antioxidant capacity (Table 2) was the one with the highest antibacterial activity (Table 3) but did not have a high anti-yeast activity (Table 4), where typically P15.EE and P15.EE-mixtures excel. This seems to suggest that propolis mode of action against bacteria is not as tightly related to its level of polyphenols and flavonoids contents per se, but possibly is related to certain combinations of compounds or specific synergisms, as has been reported [43].

2.2. *Characterization of Mixtures of Propolis Obtained from Two Apiaries and Different Harvesting Years*

It is well known that propolis composition depends on a myriad of factors and that chemical and biological diversity are propolis signatures [21,22,44]. Over the years, propolis from Gerês, in the north of Portugal, has shown great consistency in terms of its biological activity and phenolic profiles [25]. On the contrary, propolis from Pereiro, in the center of the country, has shown great heterogeneity over the years [26]. According to our findings in this study, blending may potentiate propolis bioactivities and somehow allow us to standardize its characteristics by diluting year-dependent variability. Similar findings were previously reported for propolis from Gerês [13], which has been showing remarkably constant chemical and biological profiles [25]. This constant behaviour may be explained by the surrounding vegetation of the Gerês apiary, which belongs to a protected area of a National Park, by the apiculture practices (organic beekeeping) as well as the standardized way of propolis production and harvesting (from grids, rather than being harvested by scraping during the annual cleaning of the beehive, as in Pereiro). Therefore, we hypothesized that mixing samples from different apiaries and harvesting years could result in a similar outcome in terms of potency. Hence, we investigated the bioactivities of blends based on propolis samples from Gerês and Pereiro.

Table 4. MIC values (µg/mL) of the ethanol extracts of propolis from Pereiro collected from 2011 to 2015 and of the mixtures of P.EEs against *Saccharomyces cerevisiae* and *Candida albicans*. The results show the mean ± standard deviation of three assays with three replicates each.

Samples	*S. cerevisiae*	*C. albicans*
P11.EE	750	500
P12.EE	>1000	750
mP$_{(P11.EE+P12.EE)}$	**750**	**750**
P11.EE	750	500
P13.EE	750	750
mP$_{(P11.EE+P13.EE)}$	**500**	**750**
P11.EE	750	500
P14.EE	750	>750
mP$_{(P11.EE+P14.EE)}$	**750**	**750**
P12.EE	>1000	750
P13.EE	750	750
mP$_{(P12.EE+P13.EE)}$	**750**	**750**
P13.EE	750	750
P14.EE	750	>750
mP$_{(P13.EE+P14.EE)}$	**750**	**750**
P13.EE	750	750
P15.EE	500	500
mP$_{(P13.EE+P15.EE)}$	**500**	**500**
P14.EE	750	>750
P15.EE	500	500
mP$_{(P14.EE+P15.EE)}$	**500**	**750**
P11.EE	750	500
P12.EE	>1000	750
P13.EE	750	750
mP$_{(P11.EE-P13.EE)}$	**750**	**750**
P11.EE	750	500
P12.EE	>1000	750
P13.EE	750	750
P14.EE	750	>750
P15.EE	500	500
mP$_{(P11.EE-P15.EE)}$	**500**	**750**

Note: mP—mixtures of P.EEs.

2.2.1. Total Polyphenols and Flavonoids Contents

When EEs of propolis from the two apiaries—Gerês and Pereiro—were mixed in two different blends, the TPC of these mixtures mP + G were similar (Table 5) and comparable to the TPC values previously observed for mP.EEs (Table 1) ($p > 0.05$), though it seemed that G.EE was superimposed in the mixture. Thus, mixing propolis with very different characteristics, collected from different apiaries and regions of the country as well as from different harvesting years does not appear to alter the range of the TPC found in mixtures of propolis from the same apiary (Table 1). In addition, as previously observed for blends of propolis from Gerês and from different years [13], the TPC values became closer, with no significant differences between the mixtures, suggesting that mixing propolis samples can contribute to standardization.

Table 5. Total polyphenols and flavonoids contents of the mixtures prepared with ethanol extracts of propolis from Pereiro and from Gerês, considering the most and the least active extract with regard to its antioxidant activity. The TPC and TFC of each EE are included for ease of analysis. The results are presented as mean ± standard deviation of mg of gallic acid equivalents per g of extract (mg GAE/ g extract) and mg of quercetin equivalents per g extract (mg QE/ g extract) for the TPC and TFC, respectively. Significant differences between single EEs (P and G) and mixtures (mP + G) ($p < 0.05$) are noted with different lowercase letters. Significant differences ($p < 0.05$) between mP + G are represented with different uppercase letters.

Samples	TPC (mg GAE/ g Extract)	TFC (mg QE/ g Extract)
P14.EE	174.4 ± 4.2^{b}	33.4 ± 2.0^{b}
G15.EE	207.9 ± 7.5^{a}	51.7 ± 0.9^{a}
mP + G$_{(P14.EE+G15.EE)}$	**$207.2 \pm 6.8^{a; A}$**	**$38.5 \pm 3.7^{b; B}$**
P15.EE	262.2 ± 4.3^{a}	78.4 ± 1.7^{a}
G13.EE	205.8 ± 3.5^{b}	32.6 ± 0.8^{c}
mP + G$_{(P15.EE+G13.EE)}$	**$217.1 \pm 9.1^{b; A}$**	**$53.7 \pm 4.1^{b; A}$**

Note: mP + G—mixtures of a P.EE and a G.EE. Different lowercase letters (a, b, c) were used for significant differences ($p < 0.05$) between TPC or between TFC of single extracts (P.EEs) and mixtures of extracts (mP.EEs). Different uppercase letters (**A**, **B**) were used for significant differences between TPC or between TFC of mP.EEs ($p < 0.05$).

On the contrary, the two tested mP + G mixtures showed different TFC values, the highest was found in the P15.EE-containing mixture (Table 5), yet an intermediate value between the ones of the respective individual extracts. Osés et al. [45] also found a high but diverse TFC in 13 samples of propolis from different American and European regions, with values ranging between 18.48 and 83.76 mg QE/ g extract, notwithstanding that all had a strong antioxidant activity, despite their TFC variation.

2.2.2. Antioxidant Potential

In the blends made with EEs of propolis from two different localities, there was also an improvement of antioxidant capacity relative to the individual extracts, particularly in m$_{(P14.EE+G15.EE)}$ (Table 6). Despite the EC_{50} differences between the individual extracts, both mixtures showed a similar antioxidant potential (Table 6), once again suggesting that mixing propolis may contribute to the standardization of this natural product.

Table 6. The ability to capture DPPH• free radicals of the P.EE and G.EE mixtures is expressed as a mean ± standard deviation of EC_{50} values (µg/ mL). Significant differences between single EEs (P and G) and mixtures (mP + G) ($p < 0.05$) are noted with different lowercase letters. Significant differences between mP + G ($p < 0.05$) are represented with different uppercase letters.

Samples	EC_{50} (µg/ mL)
P14.EE	20.3 ± 0.3^{a}
G15.EE	19.7 ± 8.8^{a}
mP + G$_{(P14.EE+G15.EE)}$	**$15.4 \pm 1.5^{b; A}$**
P15.EE	10.3 ± 1.7^{c}
G13.EE	25.2 ± 2.5^{a}
mP + G$_{(P15.EE+G13.EE)}$	**$15.3 \pm 2.0^{b; A}$**

Note: mP + G—mixtures of a P.EE and a G.EE. Different lowercase letters (a, b, c) were used for significant differences ($p < 0.05$) between TPC or between TFC of single extracts (P.EEs) and mixtures of extracts (mP.EEs). Different uppercase letters (**A**) were used for significant differences between TPC or between TFC of mP.EEs ($p < 0.05$).

Thus, the use of mixtures appears to be beneficial, allowing us to make use of even the least-active propolis samples without a loss of bioactivity or even their potentiation. Mixing propolis, whether from a single apiary and different harvesting years, or from different apiaries and harvesting years, can contribute to the standardization of bioactivity, as different mixtures show very close EC_{50}, regardless of either TPC or TFC values or the EC_{50} values of the individual samples (Figure 2), although in these mixtures the best correlation with antioxidant capacity was found in the TFC.

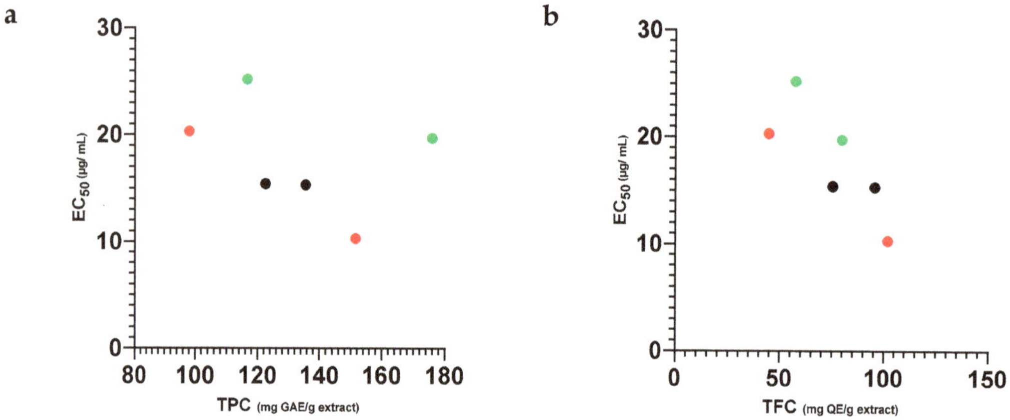

Figure 2. Correlation between antioxidant potential (EC_{50} values) and TPC (**a**) and TFC (**b**) of individual ethanol extracts of propolis from Pereiro (•) and from Gerês (•) and of propolis blends (mP + G) (•).

2.2.3. Antimicrobial Potential of Propolis Blends

Propolis antibacterial activity can occur directly against microorganisms or indirectly by stimulating the immune system [41,46]. Mixtures made with propolis from Pereiro and Gerês showed MIC values that were generally different to the ones displayed by the individual extracts (Table 7), albeit no loss of activity was detected. For *B. subtilis*, for example, the mixtures were found to have the lowest MIC or an intermediate value between the two individual extracts. Al-Waili [47] recently reported that mixtures of propolis from two different regions of Iraq displayed greater antimicrobial activity and faster wound-healing than individual propolis samples, also showing a greater potential of the mixtures compared with the individual propolis extracts.

Table 7. MIC values (µg/ ml) of P.EEs and G.EEs blends against *Bacillus subtilis*, *Propionibacterium acnes*, *Staphylococcus aureus*, methicillin-resistant *Staphylococcus aureus* (MRSA) and *Escherichia coli*. The results show the mean ± standard deviation of three assays with three replicates each.

	Gram-Positive				Gram-Negative
Samples	*B. subtilis*	*P. acnes*	*S. aureus*	MRSA	*E. coli*
P14.EE	100	—	500	—	—
G15.EE	50	50	>750	>1250	>1250
mP + G$_{(P14.EE+G15.EE)}$	**50**	**500**	**500**	**1000**	**1000**
P15.EE	500	200	750	>1250	>1250
G13.EE	50	—	200	>2000	>2000
mP + G$_{(P15.EE+G13.EE)}$	**200**	**500**	**750**	**1000**	**1500**

Note: mP + G—mixtures of a P.EE and a G.EE.

Regarding their anti-yeast activity (Table 8), the mixtures mP + G are less active than P14.EE and P15.EE, but more active (or as active as) the EEs of propolis from Gerês. Although antibacterial activity is the most prominent antimicrobial propolis property, propolis mixtures can display antifungal activity to the same extent as individual propolis samples against *C. albicans*, for instance. Antifungal activity has been correlated with several propolis constituents, such as chrysin and cinnamic acid derivatives, and this knowledge can contribute to the selection of molecules or propolis samples with higher activity and effectiveness in antifungal treatments [48].

Table 8. MIC values (μg/mL) of P.EEs and G.EEs mixtures against *Saccharomyces cerevisiae* and *Candida albicans*. The results show the mean ± standard deviation of three assays with three replicates each.

Samples	*S. cerevisiae*	*C. albicans*
P14.EE	750	>750
G15.EE	>1500	1000
mP + G(P14.EE+G15.EE)	**1500**	**1000**
P15.EE	500	500
G13.EE	>2000	>2000
mP + G(P15.EE+G13.EE)	**1500**	**1500**

Note: mP + G—mixtures of a P.EE and a G.EE.

The antimicrobial activity either of extracts or of mixtures of propolis falls within the variability described for this bioactivity in European propolis [25,26,38,41,44]. More importantly, our results indicate that mixing propolis extracts—either from different regions of the country and/ or from different harvesting years—does not promote the loss of antimicrobial activity. Instead, this activity can even be improved in some cases [13]. Furthermore, all of the ethanol extracts used in this study were prepared in the years that the propolis samples were collected and did not lose either antioxidant or antimicrobial activities when used some years later (not shown). Evidence has also been given that propolis in an ethanol solution over a period of 10–15 years can increase its antibacterial activity [49] and that samples of fresh and aged propolis have similar qualitative composition, radical scavenging and antimicrobial properties [50]. In a recent work, we suggested that aged propolis should not be discarded but explored for alternative applications, as a propolis leftover (with more than a year of storage) showed antimicrobial activity [51]. Furthermore, although collected and prepared over four different years (2011–2014), G.EEs chemical profiles showed a huge similarity regarding the type of phenolic compounds, with variations being mostly quantitative [25]. This maintenance of propolis properties for long periods is another asset in propolis conservation and the production of propolis mixtures from samples that are stored for a long time.

3. Materials and Methods

3.1. Propolis Origin and Ethanol Extraction

Propolis samples were collected over a period of five years (between 2011 and 2015), in apiaries from two different regions of Portugal. One of the apiaries is located in Guarda district, in Pereiro (40°44′57.135″ N; 7°0′59.403″ O) and the other is in Gerês (41°45′41.62″ N; 7°58′03.34″ W). Samples were collected annually between August and September and were identified by the letters P or G, according to their origin (Pereiro and Gerês, respectively) followed by two digits corresponding to the harvest year (for example, P12 corresponds to propolis of Pereiro collected in 2012). Sample P was obtained by scraping whereas G was collected from grids.

All propolis samples were extracted with ethanol in the same year of collection, following the methodology reported [25]. Briefly, 15 g of propolis was incubated for 24 h with 80 mL of absolute ethanol, in the dark at room temperature (RT) at 110 revolutions per minute (rpm). The suspension was filtered and the resulting residue was further extracted as described above but with 50 mL of absolute ethanol. Filtrates resulting from the two-step extraction process were pooled and the solvent was evaporated on a rotavapor (Buchi, Flawil, Switzerland) connected to a bath at 38–40 °C and at 47 rpm. The ethanol extracts (EE) prepared with propolis from Pereiro (P.EEs)—P11.EE, P12.EE, P13.EE, P14.EE and P15.EE—and from Gerês (G.EEs)—G11.EE, G12.EE, G13.EE, G14.EE and G15.EE—were stored in the dark at 4 °C, until needed.

3.2. *Preparation of Blends of Propolis Ethanol Extracts*

3.2.1. Mixtures of Ethanol Extracts of Propolis from Pereiro

Mixtures of EEs of propolis from the apiary Pereiro (mP.EE) were prepared considering the antioxidant, antibacterial and antifungal activities of each individual extract. Firstly, each EE was classified into three categories labeled as: "most active", "least active" and "intermediate" (the remaining extracts) for each of the bioactivities, similar to that which was previously conducted for propolis from Gerês [13]. Based on this categorization, mP.EEs were prepared according to the following criteria: (i) the most and the least active extracts, (ii) an intermediate and the least active, (iii) an intermediate and the most active, (iv) all the intermediate P.EEs and (v) a mixture of all of the P.EEs. Thirteen mixtures were planned according to these five criteria but only nine different mixtures were prepared, as some of the blends shared the same composition (Table 9).

Table 9. Composition of blends of P.EEs based on five criteria and taking into account the classification of each P.EE regarding antioxidant, antifungal and antibacterial activities.

Criteria used in P.EEs Mixtures	Antioxidant Activity	Antifungal Activity	Antibacterial Activity
Most active + least active	$mP_{(P15.EE + P14.EE)}$	$mP_{(P11.EE+P12.EE)}$	$mP_{(P14.EE+P11.EE)}$
Intermediate + least active	$mP_{(P13.EE+P14.EE)}{}^{\beta}$	$mP_{(P13.EE+P12.EE)}{}^{\gamma}$	$mP_{(P13.EE+P11.EE)}{}^{\alpha}$
Most active + intermediate	$mP_{(P13.EE+P15.EE)}$	$mP_{(P13.EE+P11.EE)}{}^{\alpha}$	$mP_{(P13.EE+P14.EE)}{}^{\beta}$
Mixture of intermediates	$mP_{(P11.EE-P13.EE)}$	$mP_{(P13.EE+P14.EE)}{}^{\beta}$	$mP_{(P13.EE+P12.EE)}{}^{\gamma}$
Mixture of all the extracts		$mP_{(P11.EE-P15.EE)}$	

Note: mP—mixtures of P.EEs; β, α, γ—mixtures with the same formulation.

3.2.2. Mixtures of Ethanol Extracts of Propolis from Pereiro and from Gerês

Two blends were prepared with propolis from the two apiaries. For this purpose, P.EEs and G.EEs were used, taking into account the most and least active extracts in terms of antioxidant capacity (Table 10).

Table 10. Mixtures of ethanol extracts of propolis from Pereiro (P) and from Gerês (G) prepared with the most and least active extracts of each apiary regarding antioxidant potential.

Criteria Underlying the Mixtures	Mixtures
Most active (G) + least active (P)	$mP + G_{(P14.EE+G15.EE)}$
Most active (P) + least active (G)	$mP + G_{(P15.EE+G13.EE)}$

Note: mP + G—mixtures of a P.EE and a G.EE.

All blends were prepared from P.EEs (Table 9) or from P.EEs and G.EEs (Table 10) at the same concentration, 10 mg/ mL, and adding equal volumes of each individual extract of the blend. The mixtures were then used for chemical characterization and biological activity assays.

3.3. *Determination of Total Poliphenols Contents*

The total polyphenols content (TPC) was determined following an adaptation of the Folin–Ciocalteu colorimetric method [5,52]. A volume of 50 µL of propolis ethanol extracts or mixtures—prepared in a concentration range of 10 to 200 µg/mL in absolute ethanol—were added to 50 µL of 10% (*v*/*v*) Folin–Ciocalteu reagent and 40 µL of 7.5% (*w*/*v*) Na_2CO_3. After 1 h incubation in the dark at RT, the absorbance was measured at 760 nm. The results are expressed in milligrams of gallic acid equivalents (GAE) per gram of propolis extract (mg GAE/ g extract), upon a calibration curve performed with gallic acid at concentrations between 5 and 50 µg/mL.

3.4. Determination of Total Flavonoids Contents

The total flavonoids content (TFC) was determined following the Woisky and Salatino method [53]. Propolis EEs or mixtures were prepared in the concentrations of 100 and 1400 μg/mL in absolute ethanol and mixed (50 μL) with 50 μL of 2% (*w*/*v*) $AlCl_3$. The absorbance was measured at 420 nm after 1 h incubation in the dark at RT. Quercetin varying from 5 to 200 μg/ ml was used as a standard. The results are presented as milligrams of quercetin equivalents (QE) per gram of propolis extract (mg QE/ g extract).

3.5. Determination of of DPPH• Scavenging Potential

The radical *2,2-diphenyl-1-picrylhydrazyl* (DPPH•) is a stable free radical which is reduced in the presence of hydrogen-donating antioxidants, promoting a color change that can be measured by spectrophotometry [54]. The DPPH• scavenging activity of propolis was determined by mixing 50 μL of samples at concentrations of 1 to 50 μg/mL (in absolute ethanol) and 100 μL of 0.004% (*w*/*v*) DPPH•, followed by 20 min incubation at RT, in the dark, and an absorbance measurement at 517 nm. The percentage reduction for each tested concentration was calculated using the following equation:

$$Inhibition\ (\%) = \times 100 \quad (1)$$

where A_{Sample} is the absorbance of the extract with DPPH• and $A_{Control}$ the absorbance of the control (DPPH• and ethanol). EC_{50} (μg/mL), which defines the propolis concentration needed to scavenge 50% of the free radical, was calculated by interpolation from those values. Gallic acid was used as a standard.

3.6. Evaluation of Antimicrobial Activity

To evaluate propolis antimicrobial activity, a panel of microorganisms was selected, taking into consideration their susceptibility and their clinical and pharmaceutical importance. Gram-negative (*Escherichia coli* CECT 423) and Gram-positive bacteria (*Bacillus subtilis* 48886, *Bacillus cereus* ATCC 7064, *Bacillus megaterium*, *Propionibacterium acnes* H60803, *Staphylococcus aureus* ATCC 6538 and *Methicillin-resistant Staphylococcus aureus* M746665 (MRSA)), as well as the yeast *Candida albicans* 53B and *Saccharomyces cerevisiae* BY4741, were all provided by the culture collection of the Department of Biology of the University of Minho. Bacteria were cultured in LB medium (Difco) for 24 h at 37 °C and 200 rpm and yeast in YPD medium (Difco) for 48 h at 30 °C and 200 rpm. Agar 2% (*w*/*v*) was added to each recipe to prepare solid media (LBA and YPDA).

An adaptation of the agar dilution method was used to determine the antibacterial and antifungal activities of all of the propolis mixtures. Each mixture was incorporated in LBA and YPDA media at various concentrations (25, 50, 100, 200, 500, 750, 1000, 1500 or 2000 μg/ ml) depending on the strain under study. Subsequently, 5 μL drops of exponential phase microbial cultures (OD_{600} = 0.4–0.6) were transferred to the Petri dishes, in triplicate, with incubation for 24 h at 37 °C in the case of bacteria, or 48 h at 30 °C in the case of yeast. Minimum inhibitory concentration (MIC) values were obtained upon observation of the presence/absence of microbial growth. Experiments were repeated three times.

3.7. Statistical Analysis

All of the assays for chemical characterization and antioxidant potential were analyzed and the results were presented as mean ± standard deviation from a variable number of assays, always equal to or greater than three ($n \geq 3$). GraphPad Prism for Windows (version 8.0.1, GraphPad Software, San Diego, California USA, www.graphpad.com, accessed on 2 December 2021 was used in the statistical analysis of the results. The results were analyzed using analysis of variance (ANOVA) followed by the Tukey test. Differences considered statistically significant ($p \leq 0.05$) were expressed with the alphabetical notation system, using different letters (lowercase when comparing mixtures and individual extracts, uppercase when comparing mixtures).

4. Conclusions

This study is a first approach to mixing propolis samples from different years and regions, and consequently with different bio and chemical profiles. We found that when mixing propolis from the same apiary but collected over different years, or when combining propolis samples collected from different regions and years, the antimicrobial and antioxidant activities of the most active of the individual extracts were either preserved or enhanced. Considering the results regarding the chemical characterization (TPC and TFC) and bioactivities (antioxidant and antimicrobial activities) of the mixtures, we can also conclude that the differences between individual propolis samples can be attenuated and a reduction in heterogeneity was obtained, thereby contributing to propolis standardization. These findings support the great potential of propolis and add even more value to this hive resource. Such valorization is also related to a greater use of the product, since samples from different years can be used without any loss of bioactivity. This efficiency in combining different propolis extracts/ samples can contribute to increasing beekeepers' interest in this product and enable them to face larger demands for this natural product [55]. Together, these main outcomes are two important starting points for the valorization and standardization of propolis.

Characterization and quantification of propolis bioactive molecules, such as polyphenols, work as a fingerprint of propolis samples, being of interest in medicine and nutraceuticals [56] and several other applications. With the possible standardization of propolis, its application in combating several health problems, such as obesity and associated metabolic disorders becomes easier [57], or as an antioxidant and anti-inflammatory agent in the prevention and care of various diseases [58,59].

Author Contributions: Conceptualization, C.A.-A. and R.O.; methodology, M.P. and A.S.F.; formal analysis, M.P., A.S.F., A.C., R.O. and C.A.-A.; investigation, M.P. and A.S.F.; resources, A.C., R.O. and C.A.-A.; writing—original draft preparation, M.P. writing—review and editing, M.P., A.S.F., A.C., R.O. and C.A.-A.; supervision, R.O. and C.A.-A.; project administration, C.A.-A.; funding acquisition, A.C., R.O. and C.A.-A. All authors have read and agreed to the published version of the manuscript.

Funding: This research was funded by European Investment Funds by FEDER/COMPETE/POCI—Operational Competitiveness and Internationalization Programme, under Project POCI-01-0145-FEDER-006958 and National Funds by FCT—Portuguese Foundation for Science and Technology, under the project UID/AGR/04033/2019. Ana Freitas acknowledges the financial support provided by national funds through FCT—Portuguese Foundation for Science and Technology (PD/BD/128276/2017), under the Doctoral Programme "Agricultural Production Chains—from fork to farm" (PD/00122/2012) and from the European Social Funds and the Regional Operational Programme Norte 2020. This study was also supported by CITAB research unit (UIDB/04033/2020) and by CBMA research unit (UIDB/04050/2020) funded by national funds through the FCT I.Pg.

Institutional Review Board Statement: Not applicable.

Informed Consent Statement: Not applicable.

Data Availability Statement: Not applicable.

Acknowledgments: The authors thank beekeepers Amadeu Fortunas (Casa do Couto; Montalegre, Portugal and Pedro Fernandes (Encostas do Côa; Pinhel, Quinta Nova, Portugal for the propolis samples used in this work.

Conflicts of Interest: The authors declare no conflict of interest.

References

1. Stan, L.; Mărghitaş, L.A.; Dezmirean, D. Quality criteria for propolis standardization. *Sci. Pap. Anim. Sci. Biotechnol.* **2011**, *44*, 137–140.
2. Osés, S.M.; Melgosa, L.; Pascual-Maté, A.; Fernández-Muiño, M.A.; Sancho, M.T. Design of a food product composed of honey and propolis. *J. Apic. Res.* **2015**, *54*, 461–467. [CrossRef]
3. Sforcin, J.M. Biological properties and therapeutic applications of propolis. *Phytother. Res.* **2016**, *30*, 894–905. [CrossRef]

4. Santos, F.A.; Bastos, E.M.A.; Uzeda, M.; Carvalho, M.A.R.; Farias, L.M.; Moreira, E.S.A.; Braga, F.C. Antibacterial activity of Brazilian propolis and fractions against oral anaerobic bacteria. *J. Ethnopharmacol.* **2002**, *80*, 1–7. [CrossRef]
5. Kumazawa, S.; Hamasaka, T.; Nakayama, T. Antioxidant activity of propolis of various geographic origins. *Food Chem.* **2004**, *84*, 329–339. [CrossRef]
6. Anjum, S.I.; Ullah, A.; Khan, K.A.; Attaullah, M.; Khan, H.; Ali, H.; Bashir, M.A.; Tahir, M.; Ansari, M.J.; Ghramh, H.A.; et al. Composition and functional properties of propolis (bee glue): A review. *Saudi J. Biol. Sci.* **2018**, *26*, 1695–1703. [CrossRef] [PubMed]
7. Srivastava, S.; Sharma, P.K.; Kumara, S. Nutraceuticals: A Review. *J. Chronother. Drug Deliv.* **2015**, *6*, 1–10.
8. Market Research Future. *Propolis Market Research Report—Global Forecast Till 2024*; ID: MRFR/F-B & N/0298-HCR 140; Market Research Future: Pune, India, 2019.
9. Burdock, G.A. Review of the biological properties and toxicity of bee propolis (propolis). *Food Chem. Toxicol.* **1998**, *36*, 347–363. [CrossRef]
10. Pereira, A.S.; Seixas, F.R.M.S.; Aquino Neto, F.R. Própolis: 100 years of research and future perspectives. *Química Nova* **2002**, *25*, 321–326. [CrossRef]
11. Pardal, P.; Casalta, F.; Godinho, J. Própolis: Avaliação quantitativa da produção de própolis. *Rev. Unidade Investig. Inst. Politécnico St.* **2014**, *2*, 182–191.
12. Dias, L.G.; Pereira, A.P.; Estevinho, L.M. Comparative study of different Portuguese samples of propolis: Pollinic, sensorial, physicochemical, microbiological characterization and antibacterial activity. *Food Chem. Toxicol.* **2012**, *50*, 4246–4253. [CrossRef]
13. Peixoto, M.; Freitas, A.S.; Cunha, A.; Oliveira, R.; Almeida-Aguiar, C. Antioxidant and antimicrobial activity of blends of propolis samples collected in different years. *LWT Food Sci. Technol.* **2021**, *145*, 111311. [CrossRef]
14. FNAP. Manual de apicultura em modo de produção biológico. In *Lisboa: FNAP—Federação Nacional dos Apicultores de Portugal*; FNAP: Lisbon, Portugal, 2008.
15. FNAP. *Manual de Produção de Polén e Própolis*; FNAP—Federação Nacional dos Apicultores de Portugal: Lisbon, Portugal, 2010.
16. Abu Fares, R.; Nazer, I.K.; Darwish, R.M.; Abu Zarqa, M. Honey bee hive modification for propolis collection. *Jordan J. Agric. Sci.* **2008**, *4*, 138–147.
17. Nuru, A.; Hepburn, H.R.; Radloff, S.E. Induction of propolis production by *Apis mellifera bandasii* in traditional basket and Langstroth movable-frame hives in Ethiopia. *J. Apic. Res.* **2002**, *41*, 101–106. [CrossRef]
18. Kasote, D.; Bankova, V.; Viljoen, A.M. Propolis: Chemical diversity and challenges in quality control. *Phytochem. Rev.* **2022**, 1–25. [CrossRef] [PubMed]
19. Mountford-McAuley, R.; Prior, J.; Clavijo McCormick, A. Factors affecting propolis production. *J. Apic. Res.* **2021**, 1–9. [CrossRef]
20. Feás, X.; Pacheco, L.; Iglesias, A.; Estevinho, L.M. Use of Propolis in the Sanitization of Lettuce. *Int. J. Mol. Sci.* **2014**, *15*, 12243–12257. [CrossRef]
21. Silva-Carvalho, R.; Baltazar, F.; Almeida-Aguiar, C. Propolis: A Complex Natural Product with a Plethora of Biological Activities That Can Be Explored for Drug Development. *Evid. Based Complementary Altern. Med.* **2015**, *2015*, 1–29. [CrossRef]
22. Graikou, K.; Popova, M.; Gortzi, O.; Bankova, V.; Chinou, I. Characterization and biological evaluation of selected Mediterranean propolis samples. Is it a new type? *LWT Food Sci. Technol.* **2016**, *65*, 261–267. [CrossRef]
23. Pasupuleti, V.R.; Sammugam, L.; Ramesh, N.; Gan, S.H. Honey, propolis, and royal jelly: A comprehensive review of their biological actions and health benefits. *Oxidative Med. Cell. Longev.* **2017**, *2017*, 1259510. [CrossRef]
24. Zulhendri, F.; Chandrasekaran, K.; Kowacz, M.; Ravalia, M.; Kripal, K.; Fearnley, J.; Perera, C.O. Antiviral, Antibacterial, Antifungal, and Antiparasitic Properties of Propolis: A Review. *Foods* **2021**, *10*, 1360. [CrossRef] [PubMed]
25. Freitas, A.S.; Cunha, A.; Cardoso, S.M.; Oliveira, R.; Almeida-Aguiar, C. Constancy of the bioactivities of propolis samples collected on the same apiary over four years. *Food Res. Int.* **2019**, *119*, 622–633. [CrossRef] [PubMed]
26. Marques, R.A.C. Contributos para a Elucidação do Modo de Açao de Própolis Português: O Caso do Própolis do Pereiro. Master's Thesis, University of Minho, Braga, Portugal, 2015.
27. Mursu, J.; Voutilainen, S.; Nurmi, T.; Tuomainen, T.P.; Kurl, S.; Salonen, J.T. Flavonoid intake and the risk of ischaemic stroke and CVD mortality in middle-aged Finnish men: The Kuopio Ischaemic Heart Disease Risk Factor Study. *Br. J. Nutr.* **2008**, *100*, 890–895. [CrossRef] [PubMed]
28. Agbor, G.A.; Vinson, J.A.; Donnelly, P.E. Folin-Ciocalteau reagent for polyphenolic assay. *Int. J. Food Sci. Nutr. Diet.* **2014**, *3*, 147–156. [CrossRef]
29. Mašek, T.; Perin, N.; Racané, L.; Cindrić, M.; Čipčić Paljetak, H.; Perić, M.; Matijašić, M.; Verbanac, D.; Radić, B.; Šuran, J. Chemical Composition, Antioxidant and Antibacterial Activity of Different Extracts of Poplar Type Propolis. *Croat. Chem. Acta* **2018**, *91*, 81–88. [CrossRef]
30. Papotti, G.; Bertelli, D.; Bortolotti, L.; Plessi, M. Chemical and functional characterization of Italian propolis obtained by different harvesting methods. *J. Agric. Food Chem.* **2012**, *60*, 2852–2862. [CrossRef]
31. Ozdal, T.; Ceylan, F.D.; Eroglu, N.; Kaplan, M.; Olgun, E.O.; Capanoglu, E. Investigation of antioxidant capacity, bioaccessibility and LC-MS/MS phenolic profile of Turkish propolis. *Food Res. Int.* **2019**, *122*, 528–536. [CrossRef]
32. Silva, V.; Genta, G.; Möller, M.N.; Masner, M.; Thomson, L.; Romero, N.; Radi, R.; Fernandes, D.C.; Laurindo, F.R.; Heinzen, H. Antioxidant activity of Uruguayan propolis. In vitro and cellular assays. *J. Agric. Food Chem.* **2011**, *59*, 6430–6437. [CrossRef]

33. Chang, C.C.; Yang, M.H.; Wen, H.M.; Chern, J.C. Estimation of total flavonoid content in propolis by two complementary colometric methods. *J. Food Drug Anal.* **2002**, *10*, 3.
34. Al-Ani, I.; Zimmermann, S.; Reichling, J.; Wink, M. Antimicrobial activities of European propolis collected from various geographic origins alone and in combination with antibiotics. *Medicines* **2018**, *5*, 2. [CrossRef]
35. Falcão, S.I.; Freire, C.; Vilas-Boas, M. A proposal for physicochemical standards and antioxidant activity of Portuguese propolis. *J. Am. Oil Chem. Soc.* **2013**, *90*, 1729–1741. [CrossRef]
36. Miguel, M.G.; Nunes, S.; Dandlen, S.A.; Cavaco, A.M.; Antunes, M.D. Phenols and antioxidant activity of hydro-alcoholic extracts of propolis from Algarve, South of Portugal. *Food Chem. Toxicol.* **2010**, *48*, 3418–3423. [CrossRef]
37. Kurek-Górecka, A.; Keskin, Ş.; Bobis, O.; Felitti, R.; Górecki, M.; Otręba, M.; Stojko, J.; Olczyk, P.; Kolayli, S.; Rzepecka-Stojko, A. Comparison of the Antioxidant Activity of Propolis Samples from Different Geographical Regions. *Plants* **2022**, *11*, 1203. [CrossRef] [PubMed]
38. Zhang, J.; Cao, X.; Ping, S.; Wang, K.; Shi, J.; Zhang, C.; Zeng, H.; Hu, F. Comparisons of ethanol extracts of Chinese propolis (poplar type) and poplar gums based on the antioxidant activities and molecular mechanism. *Evid. Based Complementary Altern. Med.* **2015**, *2015*, 307594. [CrossRef]
39. de Figueiredo, S.M.; Binda, N.S.; Vieira-Filho, S.A.; de Moura Almeida, B.; Abreu, S.R.L.; Paulino, N.; Pastore, G.M.; Sato, H.H.; Theodoropoulos, V.C.T.; Tapia, E.V.; et al. Physicochemical Characteristics of Brazilian Green Propolis Evaluated During a Six-Year Period. *Curr. Drug Discov. Technol.* **2017**, *14*, 127–134. [CrossRef]
40. Silva, J.C.; Rodrigues, S.; Feás, X.; Estevinho, L.M. Antimicrobial activity, phenolic profile and role in the inflammation of propolis. *Food Chem. Toxicol.* **2012**, *50*, 1790–1795. [CrossRef]
41. Przybyłek, I.; Karpiński, T.M. Antibacterial properties of propolis. *Molecules* **2019**, *24*, 2047. [CrossRef]
42. Saraiva, A.M.; Castro, R.H.A.; Cordeiro, R.P.; Peixoto Sobrinho, T.J.S.; Castro, V.T.N.A.; Amorim, E.L.C.; Xavier, H.S.; Pisciottano, M.N.C. In vitro evaluation of antioxidant, antimicrobial and toxicity properties of extracts of *Schinopsis brasiliensis* Engl. (Anacardiaceae). *Afr. J. Pharm. Pharmacol.* **2011**, *5*, 1724–1731. [CrossRef]
43. Freitas, A.S.; Cunha, A.; Oliveira, R.; Almeida-Aguiar, C. Propolis antibacterial and antioxidant synergisms with gentamicin and honey. *J. Appl. Microbiol.* **2022**, *132*, 2733–2745. [CrossRef]
44. Falcão, S.I.; Vale, N.; Cos, P.; Gomes, P.; Freire, C.; Maes, L.; Vilas-Boas, M. In Vitro Evaluation of Portuguese Propolis and Floral Sources for Antiprotozoal, Antibacterial and Antifungal Activity. *Phytother. Res.* **2014**, *28*, 437–443. [CrossRef]
45. Osés, S.M.; Marcos, P.; Azofra, P.; de Pablo, A.; Fernández-Muíño, M.Á.; Sancho, M.T. Phenolic profile, antioxidant capacities and enzymatic inhibitory activities of propolis from different geographical areas: Needs for analytical harmonization. *Antioxidants* **2020**, *9*, 75. [CrossRef] [PubMed]
46. Sforcin, J.M.; Bankova, V. Propolis: Is there a potential for the development of new drugs? *J. Ethnopharmacol.* **2011**, *133*, 253–260. [CrossRef]
47. Al-Waili, N. Mixing two different propolis samples potentiates their antimicrobial activity and wound healing property: A novel approach in wound healing and infection. *Vet. World* **2018**, *11*, 1188. [CrossRef]
48. Papp, Z.; Bouchelaghem, S.; Szekeres, A.; Meszéna, R.; Gyöngyi, Z.; Papp, G. The Scent of Antifungal Propolis. *Sensors* **2021**, *21*, 2334. [CrossRef]
49. Bogdanov, S. Propolis: Biological properties and medical applications. In *The Propolis Book*; Bogdanov, S., Ed.; Bee Hexagon Knowledge Network: Muehlethurnen, Switzerland, 2017; Available online: http://www.bee-hexagon.net/propolis/ (accessed on 15 December 2015).
50. Schmidt, E.M.; Stock, D.; Chada, F.J.G.; Finger, D.; Sawaya, A.C.H.F.; Eberlin, M.N.; Felsner, M.L.; Quináia, S.P.; Monteiro, M.C.; Torres, Y.R. A comparison between characterization and biological properties of Brazilian fresh and aged propolis. *BioMed Res. Int.* **2014**, *2014*, 257617. [CrossRef]
51. Pereira, L.; Cunha, A.; Almeida-Aguiar, C. Portuguese propolis from Caramulo as a biocontrol agent of the apple blue mold. *Food Control* **2022**, *139*, 109071. [CrossRef]
52. Singleton, V.L.; Orthofer, R.; Lamuela-Raventós, R.M. Analysis of total phenols and other oxidation substrates and antioxidants by means of folin-ciocalteu reagent. *Methods Enzymol.* **1999**, *299*, 152–178.
53. Woisky, R.G.; Salatino, A. Analysis of propolis: Some parameters and procedures for chemical quality control. *J. Apic. Res.* **1998**, *37*, 99–105. [CrossRef]
54. Mitra, K.; Uddin, N. Total phenolics, flavonoids, proanthrocyanidins, ascorbic acid contents and *in-vitro* antioxidant activities of newly developed isolated soya protein. *Discourse J. Agric. Food Sci.* **2014**, *2*, 160–168. Available online: careproducts.htm (accessed on 12 December 2015).
55. Vilas-Boas, M. Própolis: Um negócio por explorar. *Rev. Raízes*; Escola Superior Agrária—Instituto Politécnico de Bragança, 2015. Available online: http://esa.ipb.pt/blogs/noticiasesa/2015/propolis-um-negocio-por-explorar/ (accessed on 12 December 2015).
56. Galeotti, F.; Capitani, F.; Fachini, A.; Volpi, N. Recent advances in analytical approaches for the standardization and quality of polyphenols of propolis. *J. Med. Plants Res.* **2019**, *13*, 487–500.
57. Cardinault, N.; Tourniaire, F.; Astier, J.; Couturier, C.; Perrin, E.; Dalifard, J.; Karkeni, E. Poplar Propolis Ethanolic Extract Reduces Body Weight Gain and Glucose Metabolism Disruption in High-fat Diet-fed Mice. *Mol. Nutr. Food Res.* **2020**, *64*, e2000275. [CrossRef] [PubMed]

58. Braakhuis, A. Evidence on the health benefits of supplemental propolis. *Nutrients* **2019**, *11*, 2705. [CrossRef] [PubMed]
59. Ali, A.M.; Kunugi, H. Propolis, bee honey, and their components protect against coronavirus disease 2019 (COVID-19): A review of in silico, in vitro, and clinical studies. *Molecules* **2021**, *26*, 1232. [CrossRef] [PubMed]

antibiotics

MDPI

Article

Formulation, In Vitro Characterization and Antibacterial Activity of Chitosan-Decorated Cream Containing Bacitracin for Topical Delivery

Rumana Zaib Khattak [1], Asif Nawaz [1,*], Maha Abdallah Alnuwaiser [2], Muhammad Shahid Latif [1], Sheikh Abdur Rashid [1], Asghar Ali Khan [3] and Soha A. Alamoudi [4]

1 Advanced Drug Delivery Lab, Gomal Centre of Pharmaceutical Sciences, Faculty of Pharmacy, Gomal University, Dera Ismail Khan 29050, Pakistan
2 Department of Chemistry, College of Science, Princess Nourah bint Abdulrahman University, P.O. Box 84428, Riyadh 11671, Saudi Arabia
3 Department of Agronomy, Faculty of Agriculture, Gomal University, Dera Ismail Khan 29050, Pakistan
4 Biological Sciences Department, College of Science and Arts, King Abdulaziz University, Rabigh 21911, Saudi Arabia
* Correspondence: asifnawaz676@gmail.com

Abstract: (1) Background: Bacitracin is a broad spectrum antibiotic that is used against various microorganisms. Chitosan is a natural polymer that has been widely investigated as an antimicrobial agent for preventing and treating infections owing to its intrinsic antimicrobial properties, as well as its ability to effectively deliver extrinsic antimicrobial compounds to infected areas. Topical drug delivery offers important benefits for improving the therapeutic effect and reducing systemic side effects of administered compounds/drugs. The topical use of chitosan-decorated bacitracin-loaded cream improves the permeation of the drug across the skin and enhances the drug bioavailability by prolonging the residence time of the drug when applied topically, as well as producing synergistic effects and reducing the side effects of the drug. Topical chitosan-decorated cream can be a promising approach to administer the drug more efficiently and enhance the efficacy of treatment in wound healing and antibacterial activity. (2) Methods: This study was conducted to prepare, assess and investigate the synergistic antibacterial activity of a chitosan-coated bacitracin cream. The results were compared to the antibacterial activity of simple bacitracin-loaded cream. The prepared cream was evaluated for various in vitro characteristics such as rheology, pH, viscosity, drug content and antibacterial activity studies. (3) Result: The formulations were found to be stable regarding color, liquefaction and phase separation at all accelerated conditions. It was observed that with time, substantial variations in the pH of the preparations were found. The introduction of chitosan results in controlled release of the drug from the formulations. The antibacterial activity of the formulated creams was assessed with the disc diffusion method against *Staphylococcus aureus (ATCC), Escherichia coli (STCC), Pseudomonas aeruginosa (ATCC)* and *Bacillus cereus (ATCC)*. The strains, *E. coli, S. aureus, P. aeruginosa* and *B. cereus* were susceptible to 50 μg chitosan-decorated bacitracin cream, showing inhibition zones of 10 ± 0.6, 34 ± 1.5, 31 ± 0.76 and 21 ± 2.02 mm, respectively. The zones of inhibition for simple bacitracin-loaded cream were significantly smaller than chitosan-decorated cream, at 2 ± 0.2, 28 ± 0.92, 15 ± 0.5 and 11 ± 1.25 mm (ANOVA; $p < 0.05$), respectively. (4) Conclusions: It was observed that the zones of inhibition of simple bacitracin-loaded cream were significantly smaller than those of chitosan-decorated bacitracin-loaded cream. Chitosan synergistically improves the antimicrobial activity of bacitracin. Hence, the developed formulation was effective and should be considered as a suitable candidate for topical management of skin infections and wound healing.

Keywords: bacitracin; cream; antibacterial activity; in vitro evaluation; chitosan

check for updates

Citation: Khattak, R.Z.; Nawaz, A.; Alnuwaiser, M.A.; Latif, M.S.; Rashid, S.A.; Khan, A.A.; Alamoudi, S.A. Formulation, In Vitro Characterization and Antibacterial Activity of Chitosan-Decorated Cream Containing Bacitracin for Topical Delivery. *Antibiotics* **2022**, *11*, 1151. https://doi.org/10.3390/antibiotics11091151

Academic Editors: Fuhang Song and Yunjiang Feng

Received: 30 July 2022
Accepted: 22 August 2022
Published: 25 August 2022

1. Introduction

Chitosan is a derivative of chitin. It is polysaccharide with molecular weight between 300 to 1000 kDa [1]. It is the second most commonly used natural polymer after cellulose [2]. The chemical structure of chitin is 1-4 linked 2-acetamido-2-deoxy- β-D-glucopyranose [3]. Since the 1970s, chitosan has been popular in science and industry due to its specific structure, chemical composition, compatibility, biodegradability and other intrinsic properties [4].

Chitosan has many therapeutic qualities such as antibacterial, antifungal, healing property, anti-cholesteric properties, anticancer ability, and an immune-system enhancing effect. The antibacterial activity of chitosan has been observed to be enhanced if its degree of deacetylation and molecular weight is increased, and vice versa [5]. Chitosan's antimicrobial properties, as well as its modes of action, demonstrate that it is a versatile substance with a wide range of therapeutic applications. Chitosan is also used as a wound-healing accelerator in medicine due to its property of enhancing the functions of inflammatory cells. Chitosan appears to have no adverse effects after implantation in tissues and, for this reason, it has been used for a wide range of biomedical applications [6]. Chitosan is used to inhibit fibroplasia in wound healing and to promote tissue growth and differentiation in culture. As a natural polymer, chitosan has been widely investigated as an antimicrobial agent for preventing and treating infections owing to its intrinsic antimicrobial properties, as well as its ability to effectively deliver extrinsic antimicrobial compounds into infected areas. Keeping in view all these benefits, chitosan has been added to topical formulations (cream) and is considered to be a good candidate for burn and wound management [7].Topical drug delivery offers important benefits for improving the therapeutic effect and reducing systemic side effects of the administered compounds/drugs [8]. The topical application of drugs using particulate systems consisting of chitosan is one of the most popular drug delivery routes. The aim of the topical use of chitosan particles is to improve the permeation of drug across the skin or enhance drug bioavailability by prolonging the residence time of drugs applied topically or to produce synergistic effects, as well as to reduce the side effects of the drugs [9].

Bacitracin is a topical antibiotic used to treat skin injuries such as cuts, scrapes and burns by medical personnel and the general population [10]. The United States FDA permitted the clinical use of bacitracin in 1948 for the cure and prevention of both acute and chronic skin infections. Intramuscular injection of bacitracin can also be given for the systemic treatment of staphylococcal pneumonia and emphysema. However, in 2020, the FDA requested the withdrawal of bacitracin for injections from the market [11]. To date, there is no evidence of any non-FDA permitted uses of topical bacitracin. Bacitracin is a combination of many closely related cyclic polypeptide antibiotics that have both bacteriostatic and bactericidal effects. Gram-positive bacteria such as Staphylococcus spp, Corynebacterium spp, Streptococcus spp, Actinomyces spp and Clostridium spp are susceptible to bacitracin. Gram-negative bacteria such as Neisseria spp. also possess susceptibility, while many gram-negative bacteria are resistant [12]. Bacitracin interferes with the dephosphorylation of C55-isoprenyl pyrophosphate and bactoprenol pyrophosphate. Both of these lipid's function as membrane carrier molecules that transport the building-blocks of the peptidoglycan bacterial cell wall outside of the inner membrane; thus, their dephosphorylation leads to membrane damage. Topical bacitracin prevents the transfer of mucopeptides into the cell walls of bacteria and is absorbed through burnt or granulated skin. Bacitracin inhibits the growth of microorganism by causing leaks in the cell wall of bacteria via the ionic surface interaction method, preventing protein synthesis and the formation of mRNA [13].

The need to develop new effective antimicrobial agents has become imperative due to the rapid development of antibiotic resistance. New synthetic compounds require long duration. Rapid way to overcome this problem is the use of natural compounds including natural polymers such as chitosan. Natural compounds play a key role as sources of new scaffolds for antibiotics. Wounds are a serious health issue all around the world. Wounds can cause major consequences as a result of subsequent microbial infections [14].

Simple cream formulations have low stability, drug entrapment efficacy and penetration rate. To overcome all these problems topical chitosan-decorated cream can be a promising approach to administer the drug more efficiently and enhance the efficacy of treatment in skin infections such as wound healing. The aim of the present study was to develop chitosan-decorated bacitracin-based cream for topical delivery.

2. Materials and Methods

2.1. Materials

The model drug used was bacitracin (Sigma-Aldrich, St. Louis, MO, USA). Chitosan (molecular weight; 15,000 Da and degree of deacetylation; 85%; Sigma-Aldrich, USA) of low molecular weight was used in this study. Liquid paraffin and white soft paraffin (The Dow Chemical Company., Washington, MD, USA) were used in the preparation of cream. Cetyl alcohol and beeswax (Sigma-Aldrich, St. Louis, MO, USA,) was used in preparation of the oil phase of the cream. All the chemicals used in this study were of analytical grade.

2.2. ATR-FTIR Analysis

The FTIR analysis was performed using an ATR-FTIR spectrometer (L1600300, PerkinElmer, Beacons field, MA, USA. The FTIR spectra of bacitracin, chitosan and formulations C1 and C2 were obtained. The recording range of the spectrum was 600–4000 cm^{-1} at 32 scans per minute with a resolution of 4 cm^{-1} in transmission mode. The samples were analyzed without further processing and directly placed on zinc selenide. Three spectra of each sample were taken and results were averaged.

2.3. Preparation of Base Cream

Chitosan-decorated bacitracin cream for topical application was prepared in two steps/phases, i.e., aqueous and oil phases. For the aqueous phase, 19 g of distilled water was taken in a 50 mL beaker and placed on a hot plate magnetic stirrer at 40 °C. 1 g of bacitracin was taken and dissolved in distilled water. Chitosan (1 g) was accurately weighed and added to the drug solution, then stirred at 500 rpm until completely dissolved. For the oil phase, 20 g liquid paraffin, 37 g white soft paraffin, 4 g cetyl alcohol and 18 g beeswax were accurately weighed with the help of a digital balance and placed in separate beakers. All the beakers were wrapped in aluminum foil and placed in a water bath at 70 °C for 30 min. Cetyl alcohol was placed on a hot plate magnetic stirrer and all the chemicals were added to it with continuous stirring at 500 rpm and temperature at 40 °C.

For the preparation of cream, the oil phase was placed on a hot plate magnetic stirrer at 40 °C and 500 rpm. The aqueous phase was gently added to the oil phase, with continuous stirring for 30 min at 500 rpm until the aqueous phase was completely dissolved. The prepared cream was taken from the hot plate magnetic stirrer and moved to a suitable container for cooling at room temperature. The prepared cream was then evaluated for various physical characteristics (see composition in Table 1).

Table 1. Composition of cream (*w*/*w*).

S. No.	Ingredients	C1	C2
1	Bacitracin	1 g	1 g
2	Chitosan	-	1 g
3	Beeswax	18 g	18 g
4	Liquid paraffin	20 g	20 g
5	Cetyl alcohol	4 g	4 g
6	Distilled water	20 g	19 g
7	White soft paraffin	37 g	37 g

2.4. *In Vitro Characterization of Cream*

2.4.1. Physical Appearance

The physical appearance of the prepared creams was observed in terms of color and roughness, and was graded.

2.4.2. Determination of pH

The formulated cream encoded as C1 and C2 were checked by digital pH meter (InoLab®, Xylem Analytics, Dr. Karl Slevogt Street 1. 82362 Weilheim, Germany) for the determination of pH. The pH was measured at different temperatures (8 °C, 25 °C, 40 °C, and 40 °C + 75% RH) during the study duration of 28 days. The pH values were calculated in triplicate and averaged as mean ± SD [15].

2.4.3. Homogeneity, Organoleptic and Smear Tests

To determine the composition of the prepared creams, a homogeneity test was performed. This test revealed the homogeneity of the prepared formulations by physical touch. A smear test was performed to check the run-off effect and nature of the cream. The test was performed by applying a small amount of cream on the skin to check if it was greasy or non- greasy in nature [16].

2.4.4. Viscosity

The viscosity of prepared creams were checked with the help of a digital viscometer (AMETEK Brookfield, 11 Commerce Blvd, Middleboro, MA 02346, United States) over the study period of 28 days at different temperatures (8 °C, 25 °C, 40 °C and 40 °C-RH). Spindle no. 2 was used for the determination of viscosity at 6, 12, 30 and 60 rpm. The result was computed in centipoise. All the reading were taken in triplicate and averaged as mean ±SD [17].

2.4.5. Spreadability

To test the spreadability of the prepared creams, the slip and drag method (parallel plate method) was utilized. In this technique, one glass slide is attached to a wooden block and a second one is put above the first one with similar dimensions. One gram of cream was poured on the fixed slide and distributed using a spreader. Cream was squeeze in between the slides. A 100 g solid support was placed on the glass slides to uniformly spread cream over the slide, and the top slide was freely moved under a 10 g weight linked to the upper slide's hook. The time it took for the top slide to glide over the bottom slide was recorded, and spreadability was computed in g.cm/s. The experiment was performed for all of the cream formulations [15].

2.4.6. Drug Content

Drug content was measured by taking 1 g of cream (C1 and C2) in a 100 mL flask containing distilled water. The conical flask was stirred continuously for 30 min until a clear solution was obtained. The obtained solution was filtered through Whatman filter paper (grade 42) and the filtrate was collected. To make dilutions, flasks were taken with 10 mL distilled water and 1 mL filtrate was added from the 100 mL solution. Absorbance of the resulting dilutions was checked at 215 nm by using a UV-spectrophotometer, and the content of bacitracin was calculated by using the following equation [18]:

$$\%\ \text{Drug Content} = (\text{Absorbance of sample}/\text{absorbance of standard}) \times 100\% \quad (1)$$

2.4.7. Stability Studies

These studies was done to determine the change in color, liquefaction and phase separation. The formulations coded as C1 and C2 were divided into four samples, each of which was stored at different temperatures, including 8 °C, 25 °C, 40 °C, and 40 °C, ±75% RH. The study was conducted for 60 days. Samples were thoroughly monitored with regards to color change, liquefaction and phase separation under various storage settings [19].

2.5. *In Vitro Drug Release*

The drug release pattern was observed by using a Franz diffusion cell (PermeGear, Hellertown, PA, USA). The hydrophilic cellulose acetate membrane (ADVANTEC C300A142C) was submerged in acetate buffer solution (pH 5.5; simulating skin pH) and inserted between the receptor and donor compartments of the Franz diffusion cell. One gram of cream was added to the donor compartment of the Franz diffusion cell while the cell was continuously stirred using a magnetic bar at 100 rpm. At predetermined time intervals (0, 0.5, 1, 1.5, 2, 4, 8, 12, 16, 20 and 24 h), 2 mL of samples were taken from the receptor compartment using a syringe and examined on a UV visible spectrophotometer at 215 nm for quantification of bacitracin [20].

2.6. *Antibacterial Assay*

Antibacterial activity assay of the formulations (C1 and C2) was performed against the following species: *Staphylococcus aureus, Pseudomonas aeruginosa, Escherichia coli* and *Bacillus cereus.*

2.6.1. Preparation of Nutrient Agar Media

The nutrient agar media was made by adding 0.5% peptone, 0.3% beef extract, 1.5% agar and 0.5% NaCl in 1 L distilled water. The complete dissolution of all ingredients was assured by continuous stirring and heat. The medium was autoclaved at 121 °C for 15 min and the solution was allowed to cool at room temperature. The agar media was taken carefully and poured in sterile petri dishes until it set on the surface; then, it was refrigerated by replacing the lid of the petri dish. Nutrient agar media was used for plating bacterial species for liquid overnight culture [21].

2.6.2. Disc Plate Method

The test was carried out using the disc plate diffusion method. The microorganisms were sub-cultured (the previous day) to ensure that the bacteria were in the log phase of growth. The bacterial inoculum was spread on the surface of a nutrient agar plate containing 25 mL of media with a sterile cotton swab. The dried inoculated plates were impregnated with sterile 5 mm paper discs (Difco). Petri plates were incubated for 24 h at a temperature of 37 °C. Within 15 min of inoculation, the samples were placed in the plates. The test was performed using 50 µg of chitosan-decorated bacitracin cream (C2) per disc as a microbiological susceptibility control, and the same procedure was repeated for non-chitosan-decorated bacitracin cream (C1). The inoculation plates were incubated at 37 °C for 24, 48 and 72 h before being checked for inhibition zones measured in mm using a ruler for each disc displaying inhibition zones [22]. The results of the chitosan-decorated bacitracin cream was compared with those of the non-chitosan-decorated bacitracin cream.

2.7. *Statistical Analysis*

All the experiments were conducted in triplicates and the results were averaged. The data were evaluated by using one-way analysis of variance (ANOVA) using SPSS. Post hoc multiple comparisons were applied when necessary. A p value of <0.05 was considered significant.

3. Results and Discussion

3.1. *ATR-FTIR Analysis*

The characteristic peaks of the drug (bacitracin) were 2961 cm^{-1} (stretching vibrations of C-H and C-C), 1644 cm^{-1} (stretching vibrations of C-O), 1520 cm^{-1} (stretching vibrations of C-H and C-C) and 1105 cm^{-1} (C-O alcohol bond). Chitosan's characteristic peaks were at 3403 cm^{-1} representing NH functional groups (primary amine). The absorption band at around 2978 cm^{-1} can be attributed to CH symmetric stretching and 1644 cm^{-1} can be attributed to stretching vibrations of C-O. The FTIR spectra of Formulation C1 contains the characteristic peaks of the drug, showing that the drug is free in the cream base. The characteristic peaks of the drug were weakened in Formulation C2, as shown in Figure 1d, and the chitosan characteristic peaks were dominant in Formulation C2, showing the successful coating of the drug with chitosan (Figure 1).

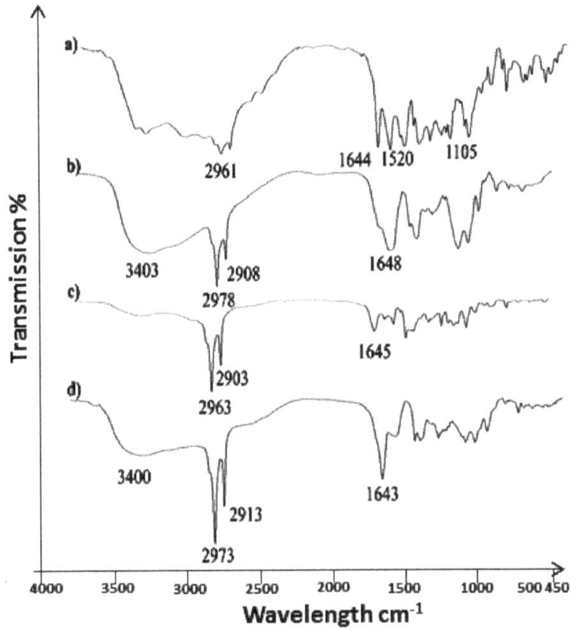

Figure 1. FTIR spectra of (**a**) drug, (**b**) chitosan, (**c**) Formulation C1 and (**d**) Formulation C2.

3.2. *Physicochemical Characterization of Cream*

3.2.1. Physical Appearance

The prepared cream was organoleptically evaluated for its color, which was observed as off white to light yellow (C1 and C2). The prepared cream had a homogeneous composition with semisolid consistency. The cream also possessed good moisturizing properties with a pleasant odor. The prepared cream showed no phase separation, passed all the physical tests and was considered best candidate to be used topically.

3.2.2. Determination of pH

The pH of cream is a significant factor to determine its efficiency. Ijaz et al. stated that the pH value of the prepared formulation must be in the range suitable for skin [15]. The pH of the formulation was adjusted to the normal range by adding NaOH dropwise under continuous stirring. The pH of the formulation in the current study was in the range of 4.5 to 6.0, as shown in Table 2, which was within the official limit and in accordance with the pH of the skin. The same phenomena were discussed by F. Rahmandari et al., stated that all formulations intended for application to the skin must be in the range suitable for skin, i.e., 4.0 to 6.0 [23]. It was observed that pH of the preparations kept at different environmental conditions (8 ± 2 °C, 25 ± 2 °C and 40 ± 2 °C) rose in the first week of study, but later the values were declined over the whole study, with significant variations. It was assumed that there were some acidic metabolites causing specific decline in the pH of the formulations [23]. Through statistical analysis, it was observed that the changes in the pH of the formulations were insignificant (ANOVA; $p > 0.05$) at different environmental temperatures and time intervals. The formulations showed no skin irritation and were appropriate for application to the skin.

Table 2. pH value of formulations at 8 ± 2 °C, 25 ± 2 °C, 40 ± 2 °C.

Time Period	8 ± 2 °C		25 ± 2 °C		40 ± 2 °C	
	C1	C2	C1	C2	C1	C2
Fresh	5.3	5.6	5.3	5.6	5.3	5.6
12 h	5.35	5.5	5.31	5.52	5.59	5.68
24 h	5.31	5.52	5.29	5.57	5.5	5.61
36 h	5.27	5.6	5.23	5.48	5.49	5.58
48 h	5.24	5.5	5.2	5.52	5.43	5.51
72 h	5.18	5.48	5.19	5.4	5.31	5.43
1 week	5.14	5.41	5.5	5.34	5.24	5.3
2 weeks	5.09	5.3	5.09	5.2	5.12	5.35
3 weeks	4.98	5.2	4.64	4.9	5.01	5.1
4 weeks	4.72	5.21	4.53	4.7	4.91	4.9

3.2.3. Homogeneity and Smear Test

The homogeneity and smear test were conducted to evaluate the uniformity of the ingredients used in the formulation. The test was conducted over a time period of four weeks at different storage conditions, as mentioned above. It was observed that the formulations (C1 and C2) were stable throughout the four weeks and no change occurred. The smear test revealed that C1 and C2 are greasy and have good moisturizing properties. Prepared creams were readily removed with tap water, indicating that they may be utilized without difficulty. The cream homogeneity test revealed that the formulations had a homogeneous composition. The colors of the prepared formulations were off white to pale yellow, as shown in Table 2, as per the organoleptic evaluation. All of the creams had a pleasant odor and a semisolid consistency. Neither of the formulations showed any phase separation (Table 3).

Table 3. Physical appearance of prepared creams.

Parameters	C1	C2
Color	Light yellow	Off white
Phase Separation	Nil	Nil
Homogeneity	V. Good	Excellent
Consistency	V. Good	V. Good
Smear Test	Greasy	Greasy
Spreadability (g × cm/s)	44.31 ± 1.24	41.34 ± 1.45
Drug Content (%)	95.16	96.12

3.2.4. Viscosity

The viscosity of any semisolid formulation usually indicates its consistency. One of the most important characteristics of topical formulations is the consistency of the semisolid mixture, as it is applied to the skin as a thin layer. It was observed that the viscosity of the cream was inversely related to the shear stress. When shear stress is increased, the viscosity reduces, showing non-Newtonian flow behavior [24]. This behavior is preferred, since it has a low resistance to flow when applied under high shear circumstances [25]. In the present study, the viscosity of the formulations was measured by using different shear stresses and different revolutions per min (6, 12, 30 and 60 rpm). It was observed that by increasing the shear stress, the viscosity of the formulations significantly declined, and vice versa. The viscosity of the developed cream formulation was measured with a Brookfield viscometer using spindle no. 2. The viscosity of creams is significant in their application, since relatively low viscosity leads to easy flow off the surface and high viscosity causes problems with spreading [26]. The viscosity of the formulations were in the order C2 > C1, but both were in the acceptable range and easily flowed off the skin (ANOVA; $p < 0.05$). It can be seen that introduction of chitosan causes a rise in the viscosity of the creams. This might be due to chitosan's swelling nature. Surfactant molecules, micelles and oil droplets form a network as the concentration in the external phase rises. The denser the network, the higher the viscosity and the closer the gap between the dispersed phase and the yield value. As shown in Figure 2, the viscosity is inversely proportional to the temperature; the greater the temperature, the lower the viscosity, and vice versa [27].

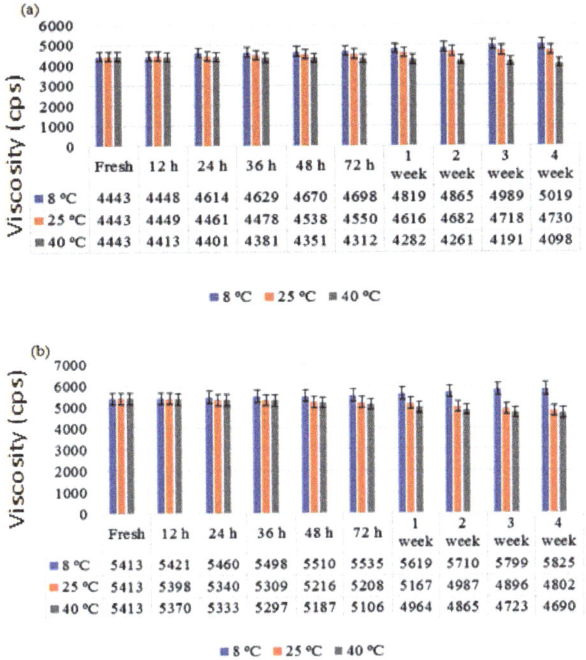

Figure 2. Viscosity of formulations (**a**) C1 and (**b**) C2 over 28 days.

3.2.5. Spreadability

Spreadability of creams is an important parameter that shows the efficacy and extent of spreading on the skin after their application [28]. The spreadability values of the prepared creams were in the suitable range [29]. The spreadability of the prepared creams was in the order of C1 > C2 (C1-44.31 ± 1.24 g·cm/s; C2-41.34 ± 1.45 g·cm/s) (ANOVA; $p < 0.05$). The prepared cream formulations (C1 and C2) exhibited a good spreadability rate over the surface of the skin. The reason might be due to the presence of an appropriate quantity of liquid paraffin and white soft paraffin. These excipients offer lubrication and give the formulated cream good spreadability. The addition of chitosan has an insignificant effect over the spreadability of the formulated cream preparations (C1 and C2). Various factors, including low and high temperatures, affect the spreadability coefficient of a cream formulation. Indeed, at low temperatures, the viscosity of cream formulations increased, resulting in decreased spreadability. Conversely, at high temperatures, the viscosity of cream formulations decreased, resulting in high spreadability [30]. The study showed that our prepared cream formulations (C1 and C2) exhibited good and uniform spreadability (Table 3) [31].

3.2.6. Drug Content

For semisolid preparations, drug content uniformity is required to ensure the homogeneity of the distributed drug throughout the formulation [32]. To ensure uniform drug distribution in the prepared cream formulations, the percentage drug content test was performed. Results showed that the percent drug contents of C1 and C2 were 95.16% and 96.12%, respectively (ANOVA; $p < 0.05$). It was also revealed that C2 has more drug entrapment in the formulation as compared to C1. This may be due to the use of chitosan as a polymer and coating agent in C2. As the concentration of polymer increased, the entrapment efficiency also increased, due to its stabilization effect [32]. The results verify that the drug content was within the approved range of 90–110% (Table 3). This demonstrates that the medication was evenly dispersed throughout the creams. As a result, the method adopted in this study appears to be suitable for the manufacturing of cream.

3.2.7. Stability Studies

Stability studies were conducted to evaluate the physical appearance of the prepared formulation. Different parameters such as color, odor, consistency, homogeneity and phase separation were observed over the time period of 60 days (Table 4 and Figure 3). Rodrigues et al. stated that these parameters must be taken in account, as they can compromise the efficacy and presentation of the formulation, which remains either changed or unchanged [33]. In the recent study, it was ensured that the prepared cream formulations (C1 and C2) passed the test of homogeneity and no significant changes occurred in the color and odor at different temperatures (8 ± 2 °C, 25 ± 2 °C and 40 °C) in an oven for a time period of 60 days. It was also observed that no significant phase separation occurred in the specified period of time (ANOVA; $p > 0.05$). However, there was a slight variation in the pH. Y. Pakzad et al., documented that the rate of degradation of cream depends upon two parameters: pH and temperature [34]. It was concluded that the prepared formulations (C1 and C2) passed the homogeneity, pH and phase separation test and retained their integrity and stability over the time period of 60 days.

Table 4. Stability study of formulations at different temperatures for 60 days.

Parameters	Codes	Fresh	24 h	36 h	48 h	72 h	7 d	21 d	28 d	60 d
Color	C1	OW	OW	OW	OW	OW	OW	OW	OW	OW
	C2	OW	OW	OW	OW	OW	OW	OW	OW	OW
Odor	C1	-ve	-ve	-ve	-ve	-ve	-ve	-ve	-ve	-ve
	C2	-ve	-ve	-ve	-ve	-ve	-ve	-ve	-ve	-ve
Phase Separation	C1	-ve	-ve	-ve	-ve	-ve	-ve	-ve	-ve	-ve
	C2	-ve	-ve	-ve	-ve	-ve	-ve	-ve	-ve	-ve

Note: OW (off white), -ve (No change).

Figure 3. Photographs of formulations C1 and C2.

3.3. *In Vitro Drug Release*

It is claimed that effectiveness of any drug depends upon the drug release pattern across the cell membrane [35]. Several factors such as polymer, emulsifiers, surfactants and gelling agents greatly affect the spreadability and viscosity of a formulation, which in turn affects the drug release pattern from the topical preparation [36]. The release patterns of the drug from bacitracin-loaded cream without chitosan coating (C1) and bacitracin-loaded cream with chitosan coating (C2) are graphically presented in Figure 4. The percentage drug release of the C1 formulation was recorded as up to 74% at the end of 24 h, while the releases of the drug from the C2 formulation was found to be up to 57% for 24 h. The percentage drug release of C1 was significantly higher that of the C2 formulation (ANOVA; $p < 0.05$). The above-mentioned values show that the C2 formulation releases less of the drug as compared to C1; this is due to addition of polymer-chitosan. The presence of the gelling agent increases the integrity of the gel network, resulting in longer diffusion pathways of drug penetrating through the membrane and, hence, reduced drug release from C2 [35]. Jhaveri et al. found similar findings, claiming that release of the drug from any formulation depends upon polymer or gelling agent use, showing an inverse relationship [37].

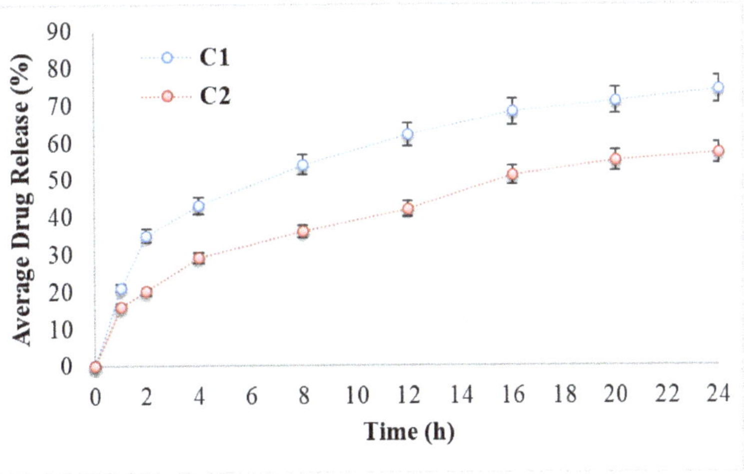

Figure 4. In vitro drug release profiles of C1 and C2.

3.4. *Antibacterial Activity Test*

The kill rate of several microbes was examined using the viable cell counting technique to determine antibacterial activity. In the present study, the disk diffusion technique was used to test the susceptibility of *S. aureus* (ATCC), *E. coli* (STCC), *P. aeruginosa* (ATCC) and *B. cereus* (ATCC) against C1 (bacitracin-loaded non-chitosan-decorated cream) and C2 (bacitracin-loaded chitosan-decorated cream). All the tests were based on the guidelines of the European Committee for Antimicrobial Susceptibility Testing [38]. Chitosan's antibacterial activity was tested against pathogenic clinical isolates and its antibiotic sensitivity was compared to a basic bacitracin-loaded non-chitosan-decorated cream. The inhibitory activity of chitosan was found to be greater against all the bacterial strains, in contrast with simple drug-loaded cream (ANOVA; $p < 0.05$). The zones of inhibition of C1 (bacitracin-loaded non-chitosan-decorated cream) were 2 ± 0.2, 28 ± 0.92, 15 ± 0.5 and 11 ± 1.25 mm, while the zones of inhibition of C2 (bacitracin-loaded chitosan-decorated cream) were 10 ± 0.6, 34 ± 1.5, 31 ± 0.76 and 21 ± 2.02 mm for different species, as shown in Table 5 and Figure 5. This was within the EUCAST-recommended quality control range [38]. Sukmark et al. explored the same phenomenon: that the antibacterial action of chitosan-decorated formulations from various sources show a larger zone of inhibition than basic antibiotics [39]. Similarly, Mauro et al. also showed that the antibacterial activity of a natural compound was enhanced in the presence of chitosan [40]. Chitosan, a cationic polymer, interacts with the anionic groups found on bacterial cell surfaces, which results in the alteration of the cell wall or outer membrane, followed by disturbances in the cytoplasmic membrane permeability and the death of the bacterial cell [41]. Chitosan also forms an impermeable layer on the surface of bacteria cells, affecting the transport of vital components into the cell [42]. The presence of polymer (chitosan) shows a synergistic effect in preventing the growth of microorganisms.

Table 5. Zone of inhibition for different bacterial strains.

	Zone of Inhibition (mm)			
Strains	*E. coli*	*S. aureus*	*P. aeruginosa*	*B. cereus*
C1	2 ± 0.2	28 ± 0.92	15 ± 0.5	11 ± 1.25
C2	10 ± 0.6	34 ± 1.5	31 ± 0.76	21 ± 2.02

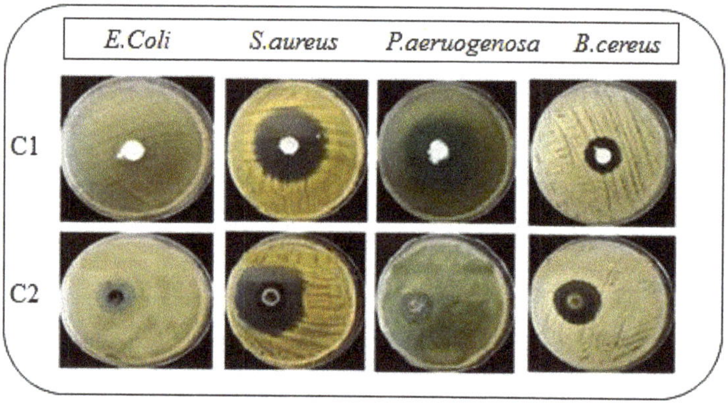

Figure 5. Antibacterial activity of the prepared formulations (C1 and C2) against various bacterial strains.

4. Conclusions

Topical antimicrobials have been used successfully to decrease bacterial infections in wounds for decades. Bacitracin is a broad-spectrum antibiotic with a wide range of biological intervention that may be used to prepare variety of formulations to treat inflammation, wounds and microbiological infections. In the present study, topical creams were prepared with and without chitosan coating. The prepared creams have optimum pH, viscosity, homogeneity, spreadability and drug content. The release of the drug from the cream was controlled in the presence of chitosan. The chitosan-decorated cream showed significantly larger zones of inhibition against different bacterial strains as compared to non-chitosan-decorated cream. This was attributed to the synergistic effect of chitosan, as chitosan acts as a strong antimicrobial. Based on these findings, the loading of chitosan and bacitracin into skin cream is a promising approach for further use in biomedical applications, predominantly in wound dressings.

Author Contributions: Writing—original draft and experimentation, R.Z.K. and M.S.L.; supervision, A.N.; writing—review and editing, S.A.R. and M.A.A.; formatting A.A.K. and S.A.A.; project administration A.N. All authors have read and agreed to the published version of the manuscript.

Funding: This research received no external funding.

Institutional Review Board Statement: Not applicable.

Informed Consent Statement: Not applicable.

Data Availability Statement: Not applicable.

Acknowledgments: All the authors are thankful to the Gomal Centre of Pharmaceutical Sciences (GCPS), Faculty of Pharmacy, Gomal University, D.I.Khan, KP, Pakistan for helping and providing facility and support. Princess Nourah bint Abdulrahman University Researchers Supporting Project number (PNURSP2022R186), Princess Nourah bint Abdulrahman University, Riyadh, Saudi Arabia.

Conflicts of Interest: The authors declare no conflict of interest.

References

1. Karamchandani, B.M.; Chakraborty, S.; Dalvi, S.G.; Satpute, S.K. Chitosan and its derivatives: Promising biomaterial in averting fungal diseases of sugarcane and other crops. *J. Basic Microbiol.* **2022**, *62*, 533–554. [CrossRef] [PubMed]
2. Del Valle, L.J.; Díaz, A.; Puiggalí, J. Hydrogels for biomedical applications: Cellulose, chitosan, and protein/peptide derivatives. *Gels* **2017**, *3*, 27. [CrossRef] [PubMed]
3. Shameli, K.; Saiful, S.A.; Yusefi, M. Cross-linked Chitosan-Based Hydrogels Nanocomposites for Treatment of Disease. *J. Res. Nanosci. Nanotechnol.* **2022**, *5*, 65–97. [CrossRef]
4. Morin-Crini, N.; Lichtfouse, E.; Torri, G.; Crini, G. Fundamentals and applications of chitosan. In *Sustainable Agriculture Reviews 35*; Springer International Publishing: Cham, Switzerland, 2019; pp. 49–123.
5. Mumtaz, S.; Ali, S.; Mumtaz, S.; Mughal, T.A.; Tahir, H.M.; Shakir, H.A. Chitosan conjugated silver nanoparticles: The versatile antibacterial agents. *Polym. Bull.* **2022**, *2022*, 1–18. [CrossRef]
6. Rashki, S.; Asgarpour, K.; Tarrahimofrad, H.; Hashemipour, M.; Ebrahimi, M.S.; Fathizadeh, H.; Khorshidi, A.; Khan, H.; Marzhoseyni, Z.; Salavati-Niasari, M.; et al. Chitosan-based nanoparticles against bacterial infections. *Carbohydr. Polym.* **2021**, *251*, 117108. [CrossRef] [PubMed]
7. Ravanfar, K.; Amniattalab, A.; Mohammadi, R. Curcumin-polyethylene glycol loaded on chitosan-gelatin nanoparticles enhances burn wound healing in rat. *J. Burn. Care. Res.* **2022**. [CrossRef] [PubMed]
8. Vanić, Ž.; Jøraholmen, M.W.; Škalko-Basnet, N. Nanomedicines for the topical treatment of vulvovaginal infections: Addressing the challenges of antimicrobial resistance. *Adv. Drug Deliv. Rev.* **2021**, *178*, 113855. [CrossRef] [PubMed]
9. Ma, J.; Wang, Y.; Lu, R. Mechanism and Application of Chitosan and Its Derivatives in Promoting Permeation in Transdermal Drug Delivery Systems: A Review. *Pharmaceuticals* **2022**, *15*, 459. [CrossRef]
10. King, A.L.; Finnin, M.S.; Kramer, C.M. *Significance of Open Wounds Potentially Caused by Non-Lethal Weapons*; Institute for Defense Analyses: Alexandria, VA, USA, 2019.
11. Meng, L.; Deresinski, S.; Holubar, M. Intraoperative bacitracin irrigations for the prevention of surgical site infections—Consider the alternatives. *Infect. Control Hosp. Epidemiol.* **2020**, *41*, 831–832. [CrossRef]
12. Sinha, P.; Dey, S.; Sen, A.; Akhter, K.; Kumar, A.; Singh, S. Bacteriological Profile of Organisms Isolated from Patients with Conjunctivitis in Katihar, Bihar. *J. Evol. Med. Dent. Sci.* **2021**, *10*, 1079–1083. [CrossRef]
13. Lade, H.; Kim, J.-S. Bacterial Targets of Antibiotics in Methicillin-Resistant Staphylococcus aureus. *Antibiotics* **2021**, *10*, 398. [CrossRef] [PubMed]
14. Negut, I.; Grumezescu, V.; Grumezescu, A.M. Treatment strategies for infected wounds. *Molecules* **2018**, *23*, 2392. [CrossRef]
15. Ijaz, N.; Durrani, A.I.; Rubab, S.; Bahadur, S. Formulation and characterization of Aloe vera gel and tomato powder containing cream. *Acta Ecol. Sin.* **2022**, *42*, 34–42. [CrossRef]
16. Limbraj, M.S.; Saher, Q.R.; Amol, J.; Prakash, P.M. FORMULATION AND EVALUATION OF HERBAL ANTISEPTIC BURN CREAM. *IJRAR-Int. J. Res. Anal. Rev.* **2020**, *7*, 195–202.
17. Chaiwong, N.; Phimolsiripol, Y.; Leelapornpisid, P.; Ruksiriwanich, W.; Jantanasakulwong, K.; Rachtanapun, P.; Seesuriyachan, P.; Sommano, S.R.; Leksawasdi, N.; Simirgiotis, M.J.; et al. Synergistics of Carboxymethyl Chitosan and Mangosteen Extract as Enhancing Moisturizing, Antioxidant, Antibacterial, and Deodorizing Properties in Emulsion Cream. *Polymers* **2022**, *14*, 178. [CrossRef] [PubMed]
18. Fernández-Campos, F.; Clares Naveros, B.; Lopez Serrano, O.; Alonso Merino, C.; Calpena Campmany, A. Evaluation of novel nystatin nanoemulsion for skin candidosis infections. *Mycoses* **2013**, *56*, 70–81. [CrossRef] [PubMed]
19. Ali Ishaq, M. Formulation and In Vitro Evaluation of Cream Containing *Vitis vinifera* Fruit Extract. Ph.D. Thesis, The Islamia University of Bahawalpur, Bahawalpur, Pakistan, 2020.
20. Katakam, L.N.R.; Katari, N.K. Development of in-vitro release testing method for permethrin cream formulation using Franz Vertical Diffusion Cell apparatus by HPLC. *Talanta* **2021**, *4*, 100056. [CrossRef]
21. Bdewe Bdewe, S.A. Partial Purification and Characterization of Protease Enzymes from *Bacillus cereus*. Master Thesis, Erciyes Üniversitesi, Kayseri, Turkey, 2020.
22. Joice, J.; Ramya, G.; Florence, J.F.; Kanmani, R.; Rakkini, A.M.; Rosaline, L.A.M.; Khalifa, A.S.; Elfasakhany, A.; Brindhadevi, K. Synthesis, characterization and photocatalytic activity of potassium Titanate nanocatalyst. *Appl. Nanosci.* **2022**, *2022*, 1–10. [CrossRef]
23. Rahmandari, F.; Swastawati, F.; Kurniasih, R. Quality Characteristics of Body Cream with the Addition of Gelatin from Tilapia (Oreochromis niloticus) Scales as an Emulsifier. *IOP Conf. Ser. Earth Environ. Sci.* **2021**, *750*, 012008. [CrossRef]
24. Chhabra, R.P. Non-Newtonian fluids: An introduction. In *Rheology of Complex Fluids*; Springer International Publishing: Cham, Switzerland, 2010; pp. 3–34.

25. Sadozai, S.K.; Zafar, A.; Sajjad, S. Topically Applied Products. In *Essentials of Industrial Pharmacy*; Springer International Publishing: Cham, Switzerland, 2022; pp. 151–175.
26. Yadav, A.; Mishra, D.K.; Paliwal, P.; Farooqui, N.; Gawshinde, A. Formulation and Evaluation of Polyherbal Antiaging Cream. *Asian J. Pharm. Technol.* **2021**, *11*, 284–288. [CrossRef]
27. Rudyak, V.Y.; Minakov, A.; Pryazhnikov, M. Preparation, characterization, and viscosity studding the single-walled carbon nanotube nanofluids. *J. Mol. Liq.* **2021**, *329*, 115517. [CrossRef]
28. Rai, P.; Poudyl, A.P.; Das, S. Pharmaceutical Creams and their use in wound healing: A Review. *J. Drug Deliv. Ther.* **2019**, *9*, 907–912.
29. Soriano-Ruiz, J.L.; Calpena-Capmany, A.C.; Cañadas-Enrich, C.; Bozal-de Febrer, N.; Suñer-Carbó, J.; Souto, E.B.; Clares-Naveros, B. Biopharmaceutical profile of a clotrimazole nanoemulsion: Evaluation on skin and mucosae as anticandidal agent. *Int. J. Pharm.* **2019**, *554*, 105–115.
30. Estanqueiro, M.; Amaral, M.; Sousa Lobo, J. Comparison between sensory and instrumental characterization of topical formulations: Impact of thickening agents. *Int. J. Cosmet. Sci.* **2016**, *378*, 389–398. [CrossRef] [PubMed]
31. Dantas, M.G.B.; Reis, S.A.G.B.; Damasceno, C.M.D.; Rolim, L.A.; Rolim-Neto, P.J.; Carvalho, F.O.; Quintans Junior, L.J.; da Silva Almeida, J.R.G. Development and evaluation of stability of a gel formulation containing the monoterpene borneol. *Sci. World J.* **2016**, *2016*, 7394685. [CrossRef]
32. Netto MPharm, G.; Jose, J. Development, characterization, and evaluation of sunscreen cream containing solid lipid nanoparticles of silymarin. *J. Cosmet. Dermatol.* **2018**, *17*, 1073–1083. [CrossRef]
33. Rodrigues Ueoka, A.; Pedriali Moraes, C.A. Development and stability evaluation of liquid crystal-based formulations containing glycolic plant extracts and nano-actives. *Cosmetics* **2018**, *5*, 25. [CrossRef]
34. Pakzad, Y.; Fathi, M.; Omidi, Y.; Mozafari, M.; Zamanian, A. Synthesis and characterization of timolol maleate-loaded quaternized chitosan-based thermosensitive hydrogel: A transparent topical ocular delivery system for the treatment of glaucoma. *Int. J. Biol. Macromol.* **2020**, *159*, 117–128. [CrossRef]
35. Murthy, S.N.; Shivakumar, H.N.; Suresh, S. Challenges in Design of Drug Delivery Systems. *Fund. Drug Deliv.* **2021**, 15–38.
36. Daood, N.M.; Jassim, Z.E.; Gareeb, M.M.; Zeki, H. Studying the effect of different gelling agent on the preparation and characterization of metronidazole as topical emulgel. *Asian J. Pharm. Clin. Res.* **2019**, *12*, 571–577.
37. Jhaveri, J.; Raichura, Z.; Khan, T.; Momin, M.; Omri, A. Chitosan nanoparticles-insight into properties, functionalization and applications in drug delivery and theranostics. *Molecules* **2021**, *26*, 272. [CrossRef] [PubMed]
38. Matuschek, E.; Brown, D.F.; Kahlmeter, G. Development of the EUCAST disk diffusion antimicrobial susceptibility testing method and its implementation in routine microbiology laboratories. *Clin. Microbiol. Infect.* **2014**, *20*, O255–O266. [CrossRef] [PubMed]
39. Matica, M.A.; Aachmann, F.L.; Tøndervik, A.; Sletta, H.; Ostafe, V. Chitosan as a wound dressing starting material: Antimicrobial properties and mode of action. *Int. J. Mol. Sci.* **2019**, *20*, 5889. [CrossRef] [PubMed]
40. Mauro, M.; Pinto, P.; Settanni, L.; Puccio, V.; Vazzana, M.; Hornsby, B.L.; Fabbrizio, A.; Di Stefano, V.; Barone, G.; Arizza, V. Chitosan Film Functionalized with Grape Seed Oil—Preliminary Evaluation of Antimicrobial Activity. *Sustainability* **2022**, *14*, 5410. [CrossRef]
41. Confederat, L.G.; Tuchilus, C.G.; Dragan, M.; Sha'at, M.; Dragostin, O.M. Preparation and antimicrobial activity of chitosan and its derivatives: A concise review. *Molecules* **2021**, *26*, 3694. [CrossRef] [PubMed]
42. Younes, I.; Rinaudo, M. Chitin and chitosan preparation from marine sources. Structure, properties and applications. *Mar. Drugs* **2015**, *16*, 1133–1174. [CrossRef] [PubMed]

antibiotics

Article

Antimicrobial Activities of Secondary Metabolites from Model Mosses

Lia R. Valeeva [1], Ashley L. Dague [2], Mitchell H. Hall [2], Anastasia E. Tikhonova [1], Margarita R. Sharipova [1], Monica A. Valentovic [3], Lydia M. Bogomolnaya [3,*] and Eugene V. Shakirov [2,3,*]

1 Institute of Fundamental Medicine and Biology, Kazan Federal University, Kazan 420008, Russia; lia2107@yandex.ru (L.R.V.); aetikhonova@mail.ru (A.E.T.); marsharipova@gmail.com (M.R.S.)
2 Department of Biological Sciences, College of Science, Marshall University, Huntington, WV 25701, USA; dague3@marshall.edu (A.L.D.); hall755@marshall.edu (M.H.H.)
3 Department of Biomedical Sciences, Joan C. Edwards School of Medicine, Marshall University, Huntington, WV 25755, USA; valentov@marshall.edu
* Correspondence: bogomolnaya@marshall.edu (L.M.B.); shakirov@marshall.edu (E.V.S.)

Abstract: Plants synthetize a large spectrum of secondary metabolites with substantial structural and functional diversity, making them a rich reservoir of new biologically active compounds. Among different plant lineages, the evolutionarily ancient branch of non-vascular plants (Bryophytes) is of particular interest as these organisms produce many unique biologically active compounds with highly promising antibacterial properties. Here, we characterized antibacterial activity of metabolites produced by different ecotypes (strains) of the model mosses *Physcomitrium patens* and *Sphagnum fallax*. Ethanol and hexane moss extracts harbor moderate but unstable antibacterial activity, representing polar and non-polar intracellular moss metabolites, respectively. In contrast, high antibacterial activity that was relatively stable was detected in soluble exudate fractions of *P. patens* moss. Antibacterial activity levels in *P. patens* exudates significantly increased over four weeks of moss cultivation in liquid culture. Interestingly, secreted moss metabolites are only active against a number of Gram-positive, but not Gram-negative, bacteria. Size fractionation, thermostability and sensitivity to proteinase K assays indicated that the secreted bioactive compounds are relatively small (less than <10 kDa). Further analysis and molecular identification of antibacterial exudate components, combined with bioinformatic analysis of model moss genomes, will be instrumental in the identification of specific genes involved in the bioactive metabolite biosynthesis.

Keywords: plant metabolite; Bryophytes; moss; *Physcomitrium patens*; *Sphagnum fallax*; antibacterial activity; exudate; extract

Citation: Valeeva, L.R.; Dague, A.L.; Hall, M.H.; Tikhonova, A.E.; Sharipova, M.R.; Valentovic, M.A.; Bogomolnaya, L.M.; Shakirov, E.V. Antimicrobial Activities of Secondary Metabolites from Model Mosses. *Antibiotics* **2022**, *11*, 1004. https://doi.org/10.3390/antibiotics11081004

Academic Editors: Fuhang Song and Yunjiang Feng

Received: 11 July 2022
Accepted: 21 July 2022
Published: 26 July 2022

1. Introduction

The emergence of bacterial drug resistance, especially in hospital settings, represents the next great frontier in healthcare. If no action is taken, diseases caused by antibiotic-resistant bacteria are predicted to kill up to 10 million people a year by 2050, similar to reported mortality rates for cancer [1]. Thus, in 2019, the World Health Organization listed antimicrobial resistance among the ten biggest health challenges [2]. Among different pathogens, Gram-positive bacteria represent particularly serious health concerns, as many key infections are caused by multidrug-resistant Gram-positive bacteria, including methicillin-resistant *Staphylococcus aureus* (MRSA), vancomycin-resistant *Enterococcus faecium* (VRE) and erythromycin-resistant group A Streptococcus (GAS) [3]. The rise of antibiotic-resistant microbial infections brings about further fears that the last remaining drugs to treat Gram-positive bacterial infections may become ineffective. Thus, the critical need to discover new potent antibiotics is becoming widely recognized.

Despite the clear demand for more antimicrobial agents, very few new antibiotics are reaching the market—the last entirely original class of antibiotic was discovered in the late 1980s. One of the critical barriers to progress in the field is that the search for natural antibiotics has been historically limited mostly to soil microorganisms (fungi, bacteria) that can be propagated in the laboratory setting, whereas other phyla remained relatively untapped. At the same time, plants produce a variety of bioactive secondary metabolites, peptides and various small molecules with unique biological functions, including defense from environmental threats and resistance to microbial pathogens [4,5]. Some well-known examples of herbal-based pharmaceuticals include acetylsalicylic acid (isolated from willow bark), artemisinin (an antimalaria drug from Artemisia annua) and Taxol (an anticancer drug from Pacific yew conifer).

There are at least five classes of known secondary metabolites produced by plants: terpenes, aromatics, glucosinolates, benzoxazinoids and green leaf volatiles [6]. In the model flowering plant *Arabidopsis thaliana*, various glucosinolates and benzoxazinoids play important roles in the defense against *Pseudomonas syringae*. Glycoalkaloids, glucosinolates and cannabinoids produced by *Solanum nigrum*, *Armoracia rusticana* and *Cannabis sativa* display antimicrobial activity against Gram-positive and Gram-negative bacteria [6]. Additionally, secreted peptides and small proteins are also involved in plant defense and adaptation to environmental stress [7–9]. Therefore, identification and characterization of new plant compounds with antibacterial activity can be a viable route to new antibiotic discovery.

Although some data exist on the presence of bioactive secondary metabolites from the flowering plants (Angiosperms), little is known about secondary metabolites from other groups of the plant kingdom, especially from the early diverging non-vascular plants, currently represented by the ancient Bryophyte division. With over 20 thousand extant species, Bryophytes (mosses, liverworts and hornworts) are the second most diverse group of land plants after flowering plants. Among all terrestrial plants, Bryophytes are considered by many scientists as a nearly unexplored natural reservoir of new biologically active secondary metabolites [10]. Well adapted to different environmental and stress conditions, Bryophytes evolved a number of defense mechanisms, including production of various antimicrobial compounds [11]. In a study of several mosses, high antimicrobial activity was found against fungi and both Gram-negative and Gram-positive bacteria [12]. Metabolites from several other mosses were also shown to effectively inhibit growth of Gram-negative or Gram-positive bacteria [10,13]. In addition to mosses, liverworts, such as *Marchantia polymorpha* and *Conocephalum conicum*, were also described as potent producers of bioactive compounds with antimicrobial activity [14].

Overall, Bryophytes produce a number of unique natural compounds with antimicrobial properties; however, very few of them have been biochemically isolated or characterized in detail [15]. The current inadequacy in their comprehensive analysis stems largely from the small physical size of Bryophytes, significant gaps in their classification [16] and numerous technical obstacles, such as the lack of powerful genomic and proteomic tools for their analysis [17,18]. However, these limitations can be circumvented by the use of model Bryophytes, such as *Physcomitrium patens* (formerly *Physcomitrella patens*) and *Sphagnum fallax*, whose genomes have been completely sequenced [19,20] and standard laboratory techniques for their maintenance, transformation and even growth in bioreactor conditions have been well established [21,22]. Though such model mosses are often used to investigate unique biological characteristics of Bryophytes, such as adaptations to life on land and to drought stress [23], they have previously not been exhaustively studied for antimicrobial potential.

Here, to extend the arsenal of available antimicrobials, we aim to characterize antibacterial activity of intracellular and extracellular metabolites produced by the model mosses *P. patens* and *S. fallax*. We show that both mosses synthetize polar and non-polar intracellular compounds with antimicrobial activity against Gram-negative *Pseudomonas syringae* bacteria. Interestingly, in contrast to intracellular metabolites, secreted water-soluble *P. patens* exudates display specific inhibitory activity against *S. aureus* and other Gram-positive bacteria, but not against Gram-negative bacterial species. Analysis of exudate stability under various physical conditions indicated that secreted metabolites are stable after multiple freezing/thawing cycles and in different light conditions, but their antimicrobial activity is substantially reduced following sample boiling or treatment with the proteinase K enzyme. Furthermore, size fractionation experiments indicated that the bioactive moss compounds are <10 kDa in size. Taken together, our data suggest that secreted antibacterial moss compounds are likely peptides or small proteins. Overall, our approach of using model mosses with completely sequenced genomes for the identification of novel antibacterial compounds will allow future bioinformatic analyses to identify specific genes involved in their biosynthetic pathways.

2. Results

2.1. Determination of Antibacterial Activity of Intracellular Moss Metabolites

We first tested model moss extracts for putative antibacterial activity displayed by intracellular metabolites. Metabolites were extracted from 10-day-old protonema tissues of the previously sequenced *Physcomitrium patens* ecotype Gransden (Gd) [19] and 30-day-old gametophores of *Sphagnum fallax* strain MW (Table 1). To determine optimal extraction conditions, we employed different combinations of solvents and extraction time: 80% methanol (for polar compounds) and hexane (for non-polar metabolites) extractions for 24 h and 45 h, followed by the qualitative assessment of extracted metabolites for antibacterial activity using the disk-diffusion (DDM) assay (clsi.org).

Table 1. Moss growth and intracellular metabolite extraction conditions.

Moss Species	Growth Time, Days	Extraction Solvent	Metabolite Extraction Time, h
P. patens Gd	10	80% methanol	24, 45
		hexane	24, 45
S. fallax MW	30	80% methanol	24, 45
		hexane	24, 45

The 24 h methanol extracts from both *P. patens* and *S. fallax* showed statistically significant (*t*-test, $p \leq 0.05$) inhibitory activity against the phytopathogenic Gram-negative *P. syringae* DC3000 strain, though activity levels dropped below significance levels for 45 h methanol extracts from both mosses (Figure 1 left panel, Table 2). Hexane extracts from both mosses also displayed antibacterial activity (Figure 1 right panel, Table 2). Interestingly, extending the metabolite extraction time with hexane had a positive effect on antibacterial activity: the strongest inhibition of *P. syringae* growth was observed with 45 h hexane extracts from both *P. patens* Gd and *S. fallax* MW (*t*-test, $p \leq 0.01$), whereas activity levels of 24 h hexane extracts were not significant (Figure 1 right panel and Table 2). This observation suggests that the ratio of bioactive polar and non-polar compounds changes with extraction time for both mosses.

We also tested a smaller subset of moss extracts against other Gram-negative and Gram-positive bacteria (Table 2). Neither extract showed antibacterial activity against Gram-negative bacterium *Serratia marcescens*, and *P. patens* extracts showed no activity towards Gram-negative *Escherichia coli* or Gram-positive *Staphylococcus aureus* bacteria. Overall, we conclude that the model moss extracts tested in qualitative DDM assays displayed statistically significant antibacterial activity levels against Gram-negative phytopathogenic *P. syringae* bacteria, though this activity appeared relatively unstable in follow-up quantitative assays and was not pursued further.

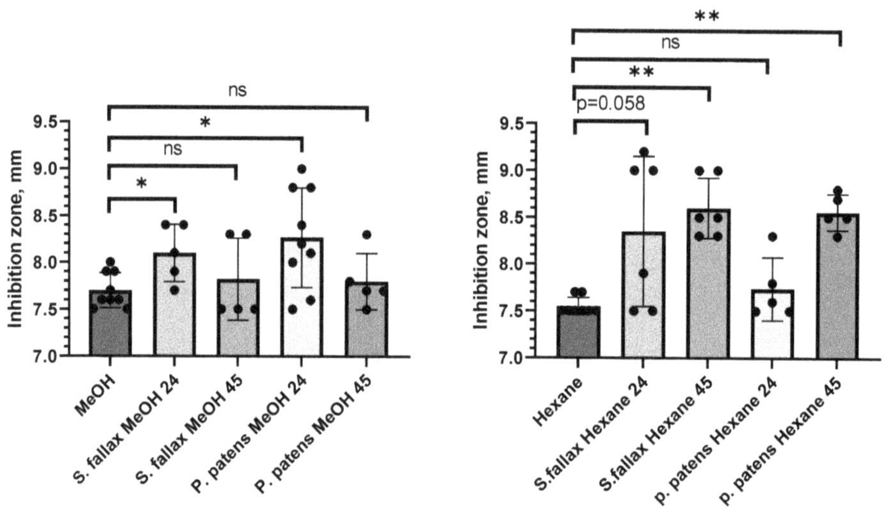

Figure 1. Polar and non-polar intracellular moss metabolites inhibit the growth of *Pseudomonas syringae* DC3000. The effect of intracellular fractions of *S. fallax* and *P. patens* mosses extracted with methanol (MeOH) (**left panel**) and hexane (**right panel**) on the growth of Gram-negative phytopathogenic *P. syringae* DC3000 was evaluated by the disk-diffusion assay (DDM); 80% methanol and hexane were used as negative controls. Moss metabolite extractions were performed for 24 or 45 h. Diameter of bacterial growth inhibition area (halo) around each cellulose disk containing moss metabolites was measured and plotted. Data represent the means from at least three independent experiments and a standard deviation. The asterisks indicate significance in an unpaired *t*-test; *—statistical significance $p \leq 0.05$; **—statistical significance $p \leq 0.01$; ns—not significant.

Table 2. Bacterial growth inhibition activity of moss extracts in DDM test.

Moss Line	Extraction Solvent	Extraction Time, h	Growth Inhibition Zone, mm				
			Negative Control [a]	*P. syringae* DC3000	*S. marcescens* SM6	*E. coli* TOP10	*S. aureus* ATCC25923
P. patens Gd	80% methanol	24	7.00 [b]	8.30 ± 0.53 *	7.00	8.03 ± 0.59	7.00
		45	7.00	7.80 ± 0.30	7.00	8.13 ± 0.47	7.00
	Hexane	24	7.00	7.74 ± 0.34	7.00	8.37 ± 0.40	NA
		45	7.00	8.56 ± 0.19 **	7.00	8.50 ± 0.50	NA
S. fallax MW	80% methanol	24	7.00	8.10 ± 0.31 *	7.00	NA	NA
		45	7.00	7.82 ± 0.44	7.00	NA	NA
	Hexane	24	7.00	8.35 ± 0.80	7.00	NA	NA
		45	7.00	8.60 ± 0.32 **	7.00	NA	NA

[a] Eighty percent methanol or hexane only; [b] cellulose disk diameter is 7 mm (no antibacterial activity). *—statistical significance $p \leq 0.05$; **—statistical significance $p \leq 0.01$; NA—not analyzed.

2.2. Identification of Antibacterial Activity in Moss Exudates

Several types of small extracellular peptides and metabolites from mosses and other plants are known to have antibacterial properties [5,9,24]. Thus, two *P. patens* ecotypes, Gransden (Gd) and Villersexel (Vx), were grown in liquid cultures for 1, 2 and 4 weeks and their exudates (water-soluble fractions secreted to the growth medium) were tested for the presence of antimicrobial activity. DDM assays detected substantial antimicrobial activity of *P. patens* exudates against Gram-positive *S. aureus* bacteria (Figure 2A), but not against Gram-negative *Salmonella* Typhimurium or *S. marcescens* (Figure 2B,C). Antibacterial activity of *P. patens* exudates against *S. aureus* was detectable even at the first time point of analysis (1 week of moss growth) and further increased with longer moss growth times (Table 3).

Interestingly, antibacterial activity against *S. aureus* in exudates from the Gd ecotype appeared to reach its maximum at 2 weeks of growth (Figure 2D), whereas in exudates from the Vx ecotype this activity continued to increase over time and peaked at 4 weeks of moss growth (Figure 2E). Collectively, our data indicate that extracellular exudates of two different ecotypes of the model moss *P. patens* display high growth inhibitory activity against Gram-positive bacterium *S. aureus* ATCC25293.

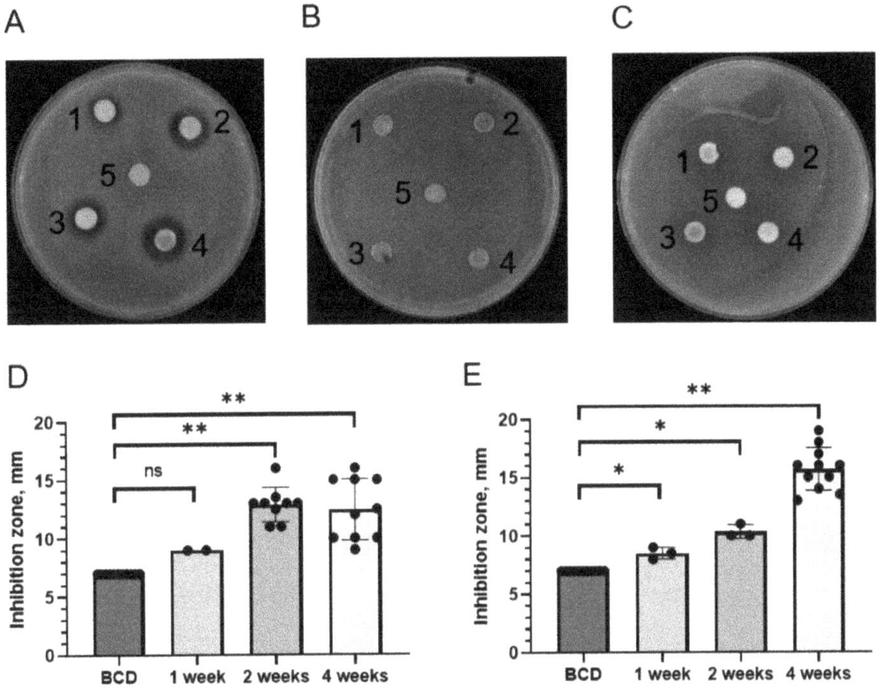

Figure 2. Bacterial growth inhibitory activity of extracellular metabolites from *P. patens* ecotypes. (**A–C**) Representative pictures of qualitative DDM assays with filter disks soaked with exudates from four-week-old *P. patens* Gd ecotype and placed on top of *S. aureus* ATCC25923 (**A**), *S. enterica ser.* Typhimurium ATCC14028s (**B**) and *S. marcescens* SM6 (**C**) bacterial lawns. 1–4, disks with *P. patens* exudates; 5, negative control disks (BCD medium only). (**D**,**E**) Diameter of *S. aureus* growth inhibition area (halo) around each cellulose disk containing secreted moss metabolites from one-week, two-week and four-week-old *P. patens* Gd (**D**) or Vx (**E**) ecotypes was measured and plotted. Data represent the means from at least three independent experiments and a standard deviation. The asterisks indicate significance in an unpaired *t*-test; *—statistical significance $p \leq 0.05$; **—statistical significance $p \leq 0.01$; ns—not significant.

Table 3. Bacterial growth inhibition activity of exudates from *P. patens* ecotypes in DDM assays against *S. aureus*, in mm.

P. patens Ecotype	Bacterial Growth Inhibition Zone, in mm			
	No Exudate Control	1-Week-Old Moss Exudate	2-Week-Old Moss Exudate	4-Week-Old Moss Exudate
Gd	7 [a]	9 ± 0.01	13.17 ± 1.27 **	12.97 ± 2.36 **
Vx	7	8.5 ± 0.71 *	10.33 ± 0.58 *	15.88 ± 1.65 **

[a] Cellulose disk diameter is 7 mm (no antibacterial activity). *—statistical significance $p \leq 0.05$; **—statistical significance $p \leq 0.01$.

2.3. Quantitative Analysis of Antibacterial Activities of P. patens Exudate

Although the DDM assay is a very powerful qualitative method for initial detection of antibacterial activity, it is not suitable for quantitative characterization of moss exudates because of the differences in the diffusion capacity of various exudate components in the solid medium. Thus, following CLSI guidelines, we employed a broth microdilution method to determine minimum inhibitory concentration (MIC) of metabolites present in *P. patens* exudates. Since the genome of the *P. patens* Gd ecotype was previously fully sequenced [19], we specifically focused on exudates from the two-week and four-week-old Gd ecotype in MIC assays. First, we performed a series of positive and negative control experiments. As expected for a negative control test, the addition of resuspended BCD medium concentrate alone did not inhibit *S. aureus* culture growth in the Mueller–Hinton medium (Figure 3A). In contrast, the addition of a range of carbenicillin and chloramphenicol antibiotic dilutions inhibited the growth of *S. aureus* (Figure 3B) and other Gram-positive bacteria (Figure 3C,D).

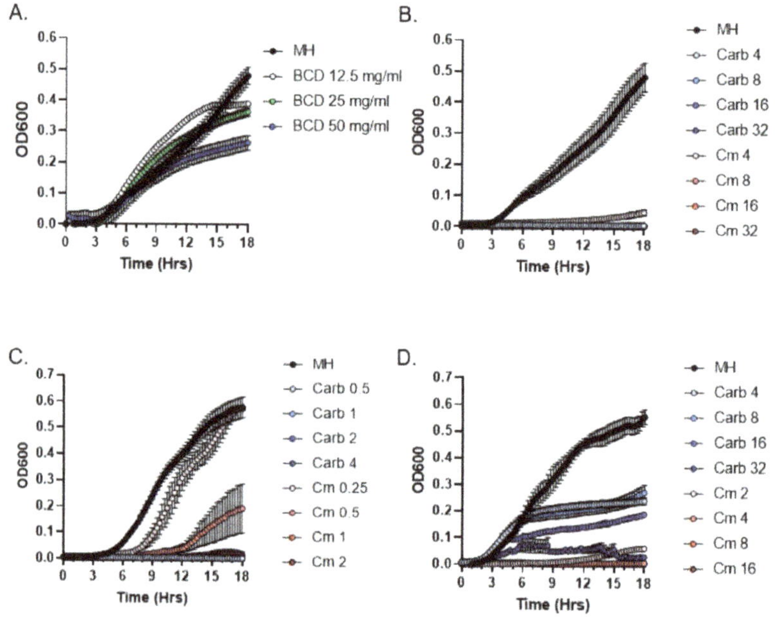

Figure 3. MIC analysis of the effects of BCD medium and antibiotics carbenicillin (Carb) and chloramphenicol (Cm) on bacterial growth. Growth curves of *S. aureus* (**A**,**B**), *Streptococcus pyogenes* (**C**) and *Enterococcus faecium* (**D**) in the presence of BCD medium (**A**) or in a range of Carb and Cm concentrations in µg/mL (**B**–**D**) are shown.

Next, we used lyophilized moss exudates (containing all secreted *P. patens* compounds and inorganic salts from the BCD medium) that were serially diluted to 50, 25, 12.5 and 6.25 mg/mL concentrations in 96-well microtiter plates containing *S. aureus* cultures to evaluate bacterial growth dynamics for 18 h (Figure 4). Interestingly, exudates from two-week-old *P. patens* Gd cultures completely inhibited the growth of *S. aureus* at a 25 mg/mL concentration (Figure 4A), whereas four-week-old *P. patens* Gd exudates displayed a 2-fold lower MIC value of 12.5 mg/mL (Figure 4B). These quantitative MIC data indicate that, unlike the prediction from qualitative DDM assays, antibacterial activity in *P. patens* Gd exudates actually continues to increase over time, supporting the notion that both Gd and Vx ecotypes show apparently similar patterns of antibacterial activity accumulation as their cultures get older.

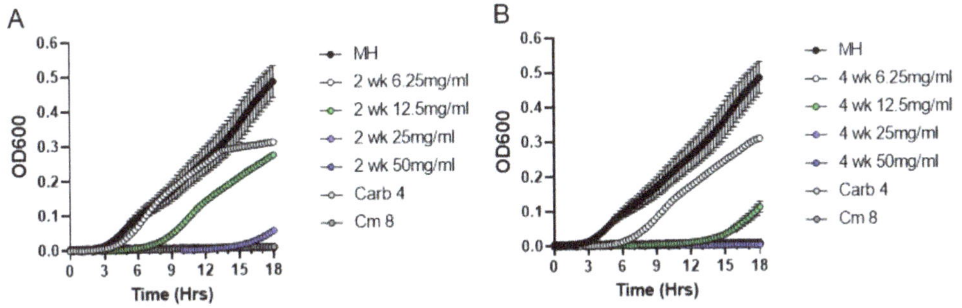

Figure 4. Broth microdilution method to determine minimum inhibitory concentration (MIC) of metabolites present in *P. patens* Gd exudates. Exudates from two-week-old (**A**) or four-week-old (**B**) *P. patens* Gd ecotype were tested in MIC assays against *S. aureus* ATCC25923. Growth curve of *S. aureus* cells was monitored in the presence of 6.25, 12.5, 25 and 50 mg/mL of exudate solution. MH—negative control, no exudate added. Carbenicillin (Carb, 4 µg/mL) and chloramphenicol (Cm, 8 µg/mL) treatments were used as positive controls for *S. aureus* growth inhibition.

We next asked if *P. patens* exudates are also active against other Gram-positive bacteria besides *S. aureus*. Specifically, we tested exudates against *Streptococcus pyogenes* and *Enterococcus faecium* strains, which represent close relatives of GAS and VRE bacteria from the "Biggest Threats" CDC list [3]. Indeed, exudates from the four-week-old *P. patens* Gd strain were able to inhibit growth of both pathogens (Figure 5). Interestingly, MIC values were 6.25 mg/mL for *S. pyogenes* (Figure 5A) and 25 mg/mL for *E. faecium* (Figure 5B), indicating that *S. pyogenes* bacteria are more sensitive to the effects of the Gd moss exudate than other tested bacteria and implying that sensitivity to the *P. patens* exudate varies among different Gram-positive bacteria. Finally, we used the MIC assays to confirm DDM data that *P. patens* exudates are not effective against Gram-negative bacteria. Indeed, four-week-old *P. patens* Gd exudates were not effective against *S.* Typhimurium (Figure 5C) or *S. marcescens* (Figure 5D). Taken together, our data indicate that *P. patens* moss secretes potent antimicrobial metabolites with high specificity against Gram-positive bacteria.

2.4. Stability of Antibacterial Compounds in Moss Exudates

Analysis of exudate stability under different physical conditions can aid in determining the likely chemical nature of the antibacterial exudate components. Thus, we subjected *P. patens* Gd exudates to different treatment regimens, such as heating, light exposure, proteinase K treatment, repeated freezing and thawing, and evaluated residual antibacterial activity using the MIC assays. As expected, neither treatment affected *P. patens* exudates diluted to a 6.25 mg/mL concentration, which showed no inhibition of *S. aureus* growth (Figure 6A). When used at the MIC concentration of 12.5 mg/mL, *P. patens* exudates fully retained antibacterial activity against *S. aureus* after 3 h exposure to room temperature (RT) with or without direct sunlight, and after repeated freezing and thawing cycles (Figure 6B), indicating that these treatments do not affect antimicrobial activity of the Gd exudate. In a sharp contrast, antibacterial activity was substantially reduced when *P. patens* exudates at MIC values (12.5 mg/mL) were either boiled or treated with proteinase K (Figure 6B). Furthermore, boiling substantially reduced antibacterial activity even at a higher exudate concentration of 25 mg/mL (Figure 6C). Though not fully sufficient to rule out other possibilities, these results are consistent with the peptide or protein nature of secreted antimicrobial *P. patens* metabolites.

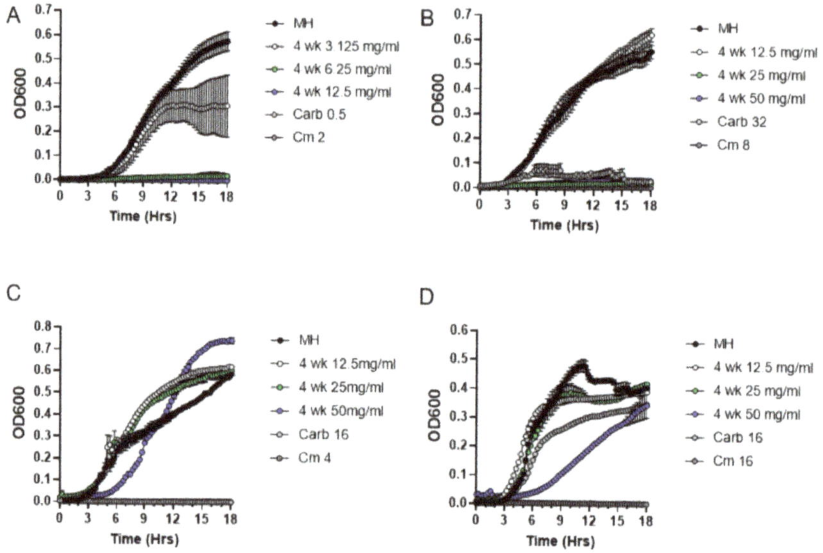

Figure 5. Antibacterial activity (MIC assay) of *P. patens* exudate from four-week-old Gd strain against Gram-positive and Gram-negative bacteria. Growth curve of *Streptococcus pyogenes* (**A**), *Enterococcus faecium* (**B**), *Salmonella* Typhimurium (**C**) and *Serratia marcescens* (**D**) cells was monitored in the presence of different concentrations of *P. patens* exudate solution. MH—negative control with untreated bacterial cultures grown in Mueller–Hinton broth medium. Different concentrations (in μg/mL) of carbenicillin (Carb) and chloramphenicol (Cm) were used as positive controls for bacterial growth inhibition.

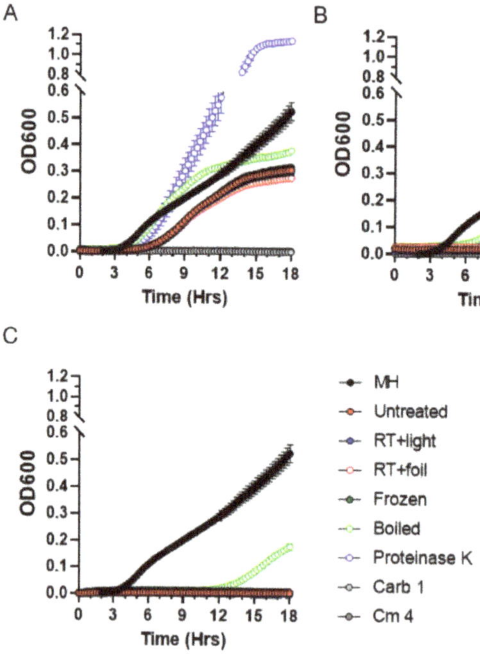

Figure 6. Residual antibacterial activity of *P. patens* exudate after different treatment regimens.

Residual activity of exudate from four-week-old *P. patens* Gd culture was tested by MIC assays against *S. aureus* following treatments for temperature (Frozen, Boiled) and light/dark (RT + light, RT + foil) sensitivity, as well as proteinase K treatment. Residual activity was tested at exudate concentrations 6.25 mg/mL (**A**), 12.5 mg/mL (**B**) and 25 mg/mL (**C**). Carbenicillin (Carb, 1 μg/mL) and chloramphenicol (Cm, 4 μg/mL) were used as positive controls. MH—*S. aureus* growth in liquid MH medium without exudate addition.

2.5. Size Fractionation of Bioactive Exudate Metabolites

To determine the approximate molecular weight range of secreted antimicrobial *P. patens* compounds, we performed a size fractionation experiment using the Amicon Ultra-15 ultrafiltration system (3 kDa and 10 kDa molecular weight cutoffs). Following size fractionation, MIC data indicated that antibacterial activity against *S. aureus* was completely lost in the largest >10 kDa fraction (Figure 7A). In contrast, antibacterial activity was fully retained in the smaller 3–10 kDa size fraction (Figure 7B) and also to some degree in the smallest <3 kDa fraction (Figure 7C). Specifically, the MIC value of the most active 3–10 kDa fraction (25 mg/mL) was only 2-fold less than that of the unfractionated Gd exudate (Figure 4B). Overall, these data indicate that the apparent molecular weight of the bioactive *P. patens* exudate compounds is less than 10 kDa, which correlates well with their presumed peptide or small protein nature determined through proteinase K and boiling sensitivity assays.

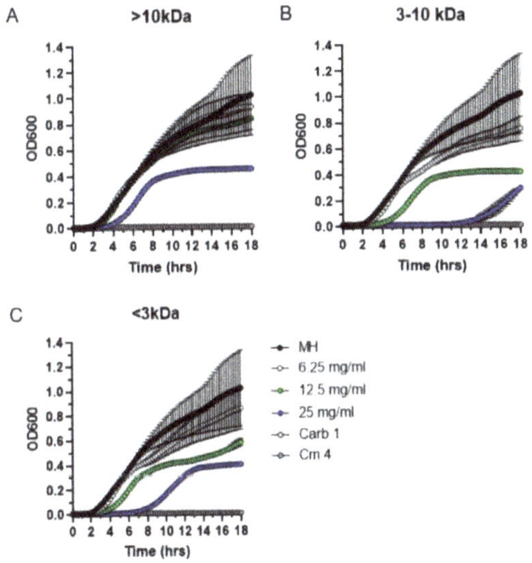

Figure 7. Size fractionation of *P. patens* exudate. Exudate from four-week-old *P. patens* Gd culture was fractionated into three molecular weight fractions, >10 kDa (**A**), 3–10 kDa (**B**), <3 kDa (**C**). Each exudate fraction was analyzed by MIC assay at three different concentrations (6.25, 12.5 and 25 mg/mL). MH—*S. aureus* growth in liquid MH medium without exudate addition. Carbenicillin (Carb, 1 μg/mL) and chloramphenicol (Cm, 4 μg/mL) were used as positive controls.

3. Discussion

Bryophytes produce a number of different compounds with unique biological activities [25], though less is known about antimicrobial metabolites from model mosses. Model Bryophytes provide a promising avenue for natural product discovery as they confer several important advantages over other plants, including completely sequenced genomes, well-developed laboratory techniques and transgenic manipulation methods for follow-up biotechnological and metabolomic applications [26–28]. Despite some recent advances in model moss proteomics, metabolomes and secretomes of model mosses are not yet extensively studied. Here, we evaluated intracellular and extracellular fractions from *P. patens* and *S. fallax* for the presence of potent antibacterial activity.

To initiate our analysis, we first evaluated intracellular compounds of *P. patens* and *S. fallax* for their antibacterial potential. Both polar (methanol-based) and non-polar (hexane-based) extracts inhibited the growth of Gram-negative *P. syringae* DC3000 bacteria. These findings are in the agreement with results from another study which evaluated extracts from 42 different Bryophytes and detected moderate antimicrobial activity against several bacterial species [29]. Overall, though the presence of antibacterial activity in the methanol and hexane extracts of *P. patens* and *S. fallax* mosses appeared promising, the bioactive metabolites were not stable during subsequent extract processing steps, and completely lost their activities after lyophilization (data not shown). Thus, we focused on the antibacterial potential of secreted exudate fractions, which remained stable throughout the experiments.

Unlike *S. fallax*, all *P. patens* strains can be easily propagated in liquid cultures where they secrete exudates containing a number of extracellular compounds [8,21]. First, we tested exudates from two laboratory *P. patens* strains, Gd and Vx, for their ability to inhibit bacterial growth in disk diffusion assays. Exudates from both *P. patens* strains had high antimicrobial activity against Gram-positive *S. aureus* bacteria, suggesting that the production of antibacterial metabolites is not specific to any one individual strain, but is likely a common feature of different *P. patens* isolates. To get a broader understanding of bioactive moss metabolite specificity, we assayed a panel of different Gram-positive and Gram-negative bacterial species. In addition to *S. aureus* bacteria, *P. patens* exudates showed growth inhibitory activity against two other Gram-positive bacteria, *E. faecium* and *S. pyogenes*. In a stark contrast, no antibacterial activity was observed with any exudates against Gram-negative bacteria *Salmonella*, *S. marcescens* or *E.coli*. Taken together, these data indicate that *P. patens* produces secreted metabolites with antibacterial activity specific only against Gram-positive but not Gram-negative bacteria.

The narrow specificity of moss metabolites only against Gram-positive bacteria is intriguing and may stem from the structural differences in bacterial cell walls or antibiotic resistance mechanisms. For instance, some features of Gram-negative bacteria may make it hard for moss antimicrobials to reach their targets: (a) the presence of the outer and inner membranes and the periplasmic space, which may represent a physical barrier to peptide antibiotics [30]; (b) very efficient proteases and powerful efflux systems which quickly remove antibiotics [31]; and (c) the presence of antimicrobial stress-responding proteins in periplasmic space, such as SipA in *Vibrio cholerae* and SapA in *Actinobacillus pleuropneumoniae* [32,33]. Other, currently unknown, mechanisms may also be involved.

Interestingly, time course experiments indicated that antibacterial activity in *P. patens* exudates increases over time: it is already detectable after 1 week of moss culture growth and continues to increase after 2 weeks and further after 4 weeks. Two possible scenarios can be envisioned that explain accumulation of antibacterial activity in moss exudates over time. First, the bioactive moss compounds may be relatively stable in the environment under the chosen *P. patens* growth conditions and accumulate in the exudate as the moss culture continues to grow. Alternatively, antimicrobial compounds may be relatively unstable and degrade quickly, but their higher abundance in the four-week-old exudate may simply reflect a moss biomass increase over the longer growth period.

To start addressing the question of bioactive moss metabolite stability and to obtain a glimpse into the potential molecular nature of antimicrobial moss compounds, we performed a series of thermo- and photostability assays. Our MIC experiments indicated that *P. patens* exudates fully retained antibacterial activity against *S. aureus* after various physical challenges, but activity was greatly diminished when *P. patens* exudates were either boiled or treated with proteinase K. These results are consistent with the peptide or small protein nature of antimicrobial *P. patens* metabolites. Furthermore, our size fractionation data also indicated that the antibacterial activity is associated with the smaller <10 kDa molecular weight fractions. Thus, the apparently small molecular weight of bioactive *P. patens* exudate compounds correlates well with their presumed peptide nature.

Several types of peptide antibiotics have been described in the literature. One type of such antibiotic represents mostly bacterial polypeptides not synthesized on ribosomes, including polymyxins, bacitracins and glycopeptides [34]. Another type of antimicrobial peptide is ribosomally synthesized by many diverse organisms as an inherent component of the natural host defense system. Though these compounds are often less well characterized, they represent a potentially very promising opportunity for natural antibiotic development. Specifically, *P. patens* is known to produce over 400 secreted proteins, most of which are pathogen defense-related and 72 are secreted only in the presence of chitosan, an elicitor of the plant pathogen response pathway [24]. In addition to known secreted proteins, *P. patens* also exudes over 500 cryptic peptides with presumed antibacterial and regulatory activities [8,9]. However, previous data indicated that the synthesis and secretion of such cryptic peptides are typically induced by phytopathogens, whereas our results indicate that *P. patens* cells constitutively secrete antibacterial compounds when grown under standard laboratory conditions. Thus, identification of such constitutively synthesized metabolites offers an important advantage over pathogen-induced antibacterial peptides, as moss cells in our experiments are not artificially manipulated, have no reduction in growth rate and experience no other unwanted physiological changes during growth in culture.

4. Materials and Methods

4.1. Moss Strains and Growth Conditions

S. fallax strain MW (a gift from Dr. David Weston and Dr. Megan Patel, Oak Ridge National Laboratory) and *P. patens* ecotypes Gransden (Gd) [19] and Villersexel (Vx) (gifts from Dr. Pierre-François Perroud and Dr. Stefan Rensing, Philipps-Universität Marburg) were propagated on Petri plates with BCD agar medium containing 1 mM $MgSO_4$, 1.84 mM KH_2PO_4 pH 6.5, 10 mM KNO_3, 0.045 mM $FeSO_4$, 1 mM $CaCl_2$, and the trace elements of 9.93 mM H_3BO_3, 2.2 mM $CuSO_4 \times 5H_2O$, 1.96 mM $MnCl_2 \times 4H_2O$, 0.231 mM $CoCl_2 \times 6H_2O$, 0.191 mM $ZnSO_4 \times 7H_2O$, 0.169 mM KI and 0.103 mM $Na_2MoO_4 \times 2H_2O$, supplemented with 5.5 mM ammonium tartrate and 0.7% agar [35]. Moss tissue was passaged weekly by homogenizing with the IKA Ultra-Turrax T10 basic tissue dispenser, followed by plating on cellophane disks placed on solid BCD medium in Petri dishes. Moss plates were grown in a plant growth chamber (Model 7300, Caron Products) at 22 °C, 65% humidity, 880 lux light intensity and 12/12 h light/dark conditions.

4.2. Intracellular Metabolite Extraction

Polar and non-polar moss metabolites were extracted with 80% methanol and hexane treatments, respectively. Mosses were grown for 10 days (*P. patens*) or 30 days (*S. fallax*) on Petri dishes with cellophane disks placed on solid BCD agar medium. Moss tissue (1.5 g) was collected with a spatula, excess moisture was removed by blotting with a paper towel and tissue was ground to a thin powder with mortar and pestle (model 29-151, Genesee Scientific) using liquid nitrogen. Tissue powder was transferred into a tube with 15 mL of the appropriate solvent, wrapped in aluminum foil and metabolites were extracted by maceration at RT for 24 or 45 h. Extracts were collected by centrifugation at 4300× *g* for 15 min (centrifuge model 5424R, Eppendorf) and filtered through a 0.45 µm syringe filter. Samples were first concentrated by drying under a stream of nitrogen, and subsequently, fully dried in a lyophilizer (model 7382021, Labconco). Dry pellets were weighted and stored at −80 °C before use. For analysis, samples were dissolved in 80% methanol or hexane at a final concentration of 100 µg/µL.

4.3. Preparation f Extracellular Metabolites from Moss Exudates

For the analysis of secreted metabolites, *P. patens* ecotypes were grown in 250 mL flasks containing 100 mL of liquid BCD medium on the orbital shaker (with rotation 150 rpm) at 22 °C, 65% humidity, 880 lux light intensity and 12/12 h light/dark conditions. Moss cultures were grown for 1–4 weeks depending on the experiment, and exudates were collected by filtering first through a 70 μm cell strainer and then through a 0.45 μm syringe filter, and were flash frozen in liquid nitrogen. Unused BCD culture medium was processed similarly as a negative control. Dry samples were weighted and stored at −80 °C until needed. For experimental analysis, dry samples were dissolved in sterile BCD medium with ammonium tartrate in the final concentration of 100 μg/μL.

4.4. Tests for Antibacterial Activity

Antimicrobial activity of crude moss extracts and exudates was analyzed against Gram-positive bacteria (*Staphylococcus aureus* ATCC 25923, *Streptococcus pyogenes* ATCC 12344 and *Enterococcus faecium* ATCC 35667) and Gram-negative bacteria (*Serratia marcescens SM6, Salmonella enterica ser.* Typhimurium *ATCC14028s, Escherichia coli TOP10* and *Pseudomonas syringae DC3000*). Strains were grown in LB medium (Gram-negative bacteria) or Tryptic soy medium (Gram-positive bacteria).

Disk-diffusion test (DDM). Inhibition of bacterial growth by moss metabolites was determined by the disk-diffusion method on LB agar according to CLSI guidelines (www.clsi.org, accessed on 1 July 2022). Bacterial cultures were grown overnight (ON) at 35 °C. Bacterial inoculum (CFU = 1×10^7/plate) was prepared by dilution of 25 μL of ON culture in 5 mL of TOP agar (LB broth powder 25 g/L, 0.7% agar), stirred and poured on the Petri dish containing 20 mL of regular LB agar. Then, 17.5 mg of moss metabolites were added to each sterile Whatman disk (disk diameter = 7 mm). Disks soaked with 80% methanol, hexane or liquid BCD medium were used as negative controls. Disks soaked with metabolites were placed on top of inoculated plates and incubated at 35 °C for 18 h. The diameter of the bacterial growth inhibition area (halo) around each cellulose disk containing moss metabolites was then measured in mm and plotted. The diameter of the cellulose disk itself was 7 mm. All experiments were carried out in duplicate on at least three separate occasions.

Broth microdilution method to determine minimum inhibitory concentration (MIC). A determination of minimum inhibitory concentration (MIC) of moss metabolites was performed using 96-well microtiter plates in a BioTek plate reader spectrophotometer, the Synergy HTX [36]. Bacterial cultures were grown ON at 35 °C in MH (Mueller–Hinton) broth. Overnight cultures were subcultured at a 1:100 ratio in fresh MH broth and incubated at 35 °C with shaking (200 rpm) until each bacterial suspension reached turbidity equal to a 0.5 McFarland standard. Each resulting culture was further diluted and used to inoculate a 96-well dish containing MH broth to a final concentration of approximately 1×10^7 CFU/mL. Metabolites were added using serial dilutions (50, 25, 12.5, 6.25 mg/mL or less, depending on the experiment), and the 96-well dish was sealed with a Breathe-Easy membrane (Diversified Biotech) to reduce evaporation and incubated for 18 h at 35 °C. The optical density at 600 nm (OD_{600}) was measured every 15 min using a spectrophotometer (BioTek Synergy HTX). Bacterial growth in the presence of BCD medium concentrate was processed the same way as the moss exudate was used as a negative control, whereas carbenicillin and chloramphenicol dilutions were used as positive controls. Bacterial cultures grown in MH medium in the absence of metabolites were used as the general control for bacterial growth. The MIC was defined as the lowest concentration of an antimicrobial agent that inhibited the visible growth of bacteria. All experiments were performed in four biological and three technical replicates.

4.5. Metabolite Stability Test

Extracellular metabolites were analyzed for their stability using different treatments: boiling for 10 min, freezing/thawing, sensitivity to light and proteinase K. Lyophilized moss exudates were dissolved in MH medium and used in MIC analysis following treatments. For the freezing/thawing experiment, exudate samples were frozen in liquid nitrogen and, subsequently, thawed in 37 °C water bath three times. For the thermostability assay, exudates were incubated for 3 h at RT in a microcentrifuge tube covered with foil.

For the light stability analysis, samples were exposed to white light in the transparent microcentrifuge tube for 3 h at RT. For the proteinase stability assay, samples were treated with 0.33 μg of proteinase K and incubated for 3 h at 37 °C. After treatments, all samples were subjected to MIC analysis.

4.6. Size Fractionation of Extracellular Moss Metabolites

The Amicon Ultra-15 ultrafiltration system (3 kDa and 10 kDa cutoff columns) (Millipore) was used to separate exudate components by their molecular weight. Fractionation was conducted following manufacturer instructions to obtain fractions <3 kDa, 3–10 kDa and >10 kDa. All procedures were performed at 4 °C.

4.7. Data analysis

Data were reported as mean ± standard deviation of 3 independent experiments with 2 biological replicates. Statistical significance was determined using the unpaired t-test with Welch's correction; $p < 0.05$. Analyses were performed using GraphPad Prism v.9.3.1, San Diego, CA, USA.

5. Conclusions

This study provided new data on the antibacterial activity of extracellular and intracellular metabolites from different species of model mosses. These results highlight the applicability of mosses as a source of new bioactive compounds, as well as their biotechnological potential in medicine and agriculture. Specifically, the unique combination of *P. patens* facile genetics and genomic tools with advanced biochemical and proteomic analysis methods will make it possible to not only discover and characterize novel secondary metabolites or bioactive peptides from this previously underexplored Bryophyte model, but also to perform bioinformatics analyses to identify genes for their biosynthetic pathways.

Author Contributions: All authors contributed significantly to this work. L.R.V., L.M.B. and E.V.S. designed the experiments. L.R.V., A.L.D., M.H.H., A.E.T., M.A.V. and L.M.B. performed the experiments. L.R.V., M.R.S., L.M.B. and E.V.S. analyzed data and prepared figures. L.R.V., L.M.B. and E.V.S. wrote the paper with contributions from all other authors. Correspondence and requests for materials should be addressed to L.M.B. or E.V.S. All authors have read and agreed to the published version of the manuscript.

Funding: This work was supported in part by the NASA West Virginia Space Grant Consortium Training Grant #NNX15AI01H to D.A.L., by the WV Higher Education Policy Commission, Division of Science and Research Grant number dsr.20.1698-001 to M.H.H., and by the Kazan Federal University Strategic Academic Leadership Program (Priority-2030). Research in the Valentovic Lab was supported by NIH grants P20GM103434, 2R15CA161491-03, R15AI15197-01 and R15HL145573-01.

Data Availability Statement: The authors confirm that the data supporting the findings of this study are available within the article.

Acknowledgments: We thank David Weston and Megan Patel (Oak Ridge National Laboratory) for sharing the *S. fallax* MW strain, and Pierre-François Perroud and Stefan Rensing (Philipps-Universität Marburg) for sharing *P. patens* ecotypes Gd and Vx.

Conflicts of Interest: The authors declare no conflict of interest.

References

1. Murray, C.J.L.; Ikuta, K.S.; Sharara, F.; Swetschinski, L.; Aguilar, G.R.; Gray, A.; Han, C.; Bisignano, C.; Rao, P.; Wool, E.; et al. Global burden of bacterial antimicrobial resistance in 2019: A systematic analysis. *Lancet* **2022**, *399*, 629–655. [CrossRef]
2. The SDG Knowledge Hub. WHO Identifies Top Health Challenges, Begins Five-Year Health Plan. 2019. Available online: http://sdg.iisd.org/news/who-identifies-top-health-challenges-begins-five-year-health-plan/ (accessed on 11 July 2022).
3. Centers of Disease Control and Prevention. 2019 AR Threats Report. Available online: https://www.cdc.gov/drugresistance/biggest-threats.html (accessed on 11 July 2022).
4. Cragg, G.M.; Kingston, D.G.I.; Newman, D.J. *Anticancer Agents from Natural Products*; CRC Press: Boca Raton, FL, USA, 2011.

5. Erb, M.; Kliebensteinb, D.J. Plant Secondary Metabolites as Defenses, Regulators, Primary Metabolites: The Blurred Functional Trichotomy. *Plant Physiol.* **2020**, *184*, 39–52. [CrossRef]
6. Lelario, F.; Scrano, L.; De Franchi, S.; Bonomo, M.G.; Salzano, G.; Milan, S.; Milella, L.; Bufo, S.A. Identification and antimicrobial activity of most representative secondary metabolites from different plant species. *Chem. Biol. Technol. Agric.* **2018**, *5*, 13. [CrossRef]
7. Alexandersson, E.; Ali, A.; Resjö, S.; Andreasson, E. Plant secretome proteomics. *Front. Plant Sci.* **2013**, *4*, 9. [CrossRef] [PubMed]
8. Fesenko, I.; Azarkina, R.; Kirov, I.; Kniazev, A.; Filippova, A.; Grafskaia, E.; Lazarev, V.; Zgoda, V.; Butenko, I.; Bukato, O.; et al. Phytohormone treatment induces generation of cryptic peptides with antimicrobial activity in the Moss *Physcomitrella patens*. *BMC Plant Biol.* **2019**, *19*, 9. [CrossRef] [PubMed]
9. Lyapina, I.; Filippova, A.; Kovalchuk, S.; Ziganshin, R.; Mamaeva, A.; Lazarev, V.; Latsis, I.; Mikhalchik, E.; Panasenko, O.; Ivanov, O.; et al. Possible role of small secreted peptides (SSPs) in immune signaling in bryophytes. *Plant Mol. Biol.* **2021**, *106*, 123–143. [CrossRef]
10. Frahm, J.P. Recent developments of commercial products from Bryophytes. *Bryologist* **2004**, *107*, 277–283. [CrossRef]
11. Xie, C.-F.; Lou, H.-X. Secondary Metabolites in Bryophytes: An Ecological Aspect. *Chem. Biodivers.* **2009**, *6*, 303–312. [CrossRef]
12. Veljić, M.; Tarbuk, M.; Marin, P.D.; Ćirić, A.; Soković, M.; Marin, M. Antimicrobial Activity of Methanol Extracts of Mosses from Serbia. *Pharm. Biol.* **2008**, *46*, 871–875. [CrossRef]
13. Olofin, T.A.; Akande, A.O.; Oyetayo, V.O. Assessment of the antimicrobial properties of fractions obtained from bryophytes. *J. Microbiol. Antimicrob.* **2013**, *5*, 50–54. [CrossRef]
14. Asakawa, Y. Biologically active compounds from bryophytes. *Pure Appl. Chem.* **2007**, *79*, 557–580. [CrossRef]
15. Mishra, R.; Pandey, V.K.; Chandra, R. Potential of Bryophytes as therapeutics. *Int. J. Pharm. Sci. Res.* **2014**, *5*, 3584–3593.
16. Medina, R.; Johnson, M.G.; Liu, Y.; Wickett, N.J.; Shaw, A.J.; Goffinet, B. Phylogenomic delineation of *Physcomitrium* (Bryophyta: Funariaceae) based on targeted sequencing of nuclear exons and their flanking regions rejects the retention of *Physcomitrella*, *Physcomitridium* and *Aphanorrhegma*. *J. Syst. Evol.* **2019**, *57*, 404–417. [CrossRef]
17. Erxleben, A.; Gessler, A.; Vervliet-Scheebaum, M.; Reski, R. Metabolite profiling of the moss *Physcomitrella patens* reveals evolutionary conservation of osmoprotective substances. *Plant Cell Rep.* **2012**, *31*, 427–436. [CrossRef]
18. Asakawa, Y.; Ludwiczuk, A. Chemical Constituents of Bryophytes: Structures and Biological Activity. *J. Nat. Prod.* **2018**, *81*, 641–660. [CrossRef] [PubMed]
19. Rensing, S.A.; Lang, D.; Zimmer, A.D.; Terry, A.; Salamov, A.; Shapiro, H.; Nishiyama, T.; Perroud, P.-F.; Lindquist, E.A.; Kamisugi, Y.; et al. The *Physcomitrella* genome reveals evolutionary insights into the conquest of land by plants. *Science* **2008**, *319*, 64–69. [CrossRef] [PubMed]
20. Weston, D.J.; Turetsky, M.R.; Johnson, M.G.; Granath, G.; Lindo, Z.; Belyea, L.R.; Rice, S.K.; Hanson, D.T.; Engelhardt, K.A.M.; Schmutz, J.; et al. The Sphagnome Project: Enabling ecological and evolutionary insights through a genus-level se-quencing project. *New Phytol.* **2018**, *217*, 16–25. [CrossRef]
21. Decker, E.L.; Reski, R. Mosses in biotechnology. *Curr. Opin. Biotechnol.* **2020**, *61*, 21–27. [CrossRef]
22. Rensing, S.A.; Goffinet, B.; Meyberg, R.; Wu, S.-Z.; Bezanilla, M. The *Moss Physcomitrium* (*Physcomitrella*) *patens*: A Model Organism for Non-Seed Plants. *Plant Cell* **2020**, *32*, 1361–1376. [CrossRef]
23. Frank, W.; Ratnadewi, D.; Reski, R. *Physcomitrella patens* is highly tolerant against drought, salt and osmotic stress. *Planta* **2005**, *220*, 384–394. [CrossRef]
24. Lehtonen, M.; Takikawa, Y.; Rönnholm, G.; Akita, M.; Kalkkinen, N.; Ahola-Iivarinen, E.; Somervuo, P.; Varjosalo, M.; Valkonen, J.P.T. Protein Secretome of Moss Plants (*Physcomitrella patens*) with Emphasis on Changes Induced by a Fungal Elicitor. *J. Proteome Res.* **2014**, *13*, 447–459. [CrossRef] [PubMed]
25. Klavina, L.; Springe, G.; Nikolajeva, V.; Martsinkevich, I.; Nakurte, I.; Dzabijeva, D.; Steinberga, I. Chemical Composition Analysis, Antimicrobial Activity and Cytotoxicity Screening of Moss Extracts (Moss Phytochemistry). *Molecules* **2015**, *20*, 17221–17243. [CrossRef] [PubMed]
26. Cove, D.; Bezanilla, M.; Harries, P.; Quatrano, R. Mosses as Model Systems for the Study of Metabolism and Development. *Annu. Rev. Plant Biol.* **2006**, *57*, 497–520. [CrossRef] [PubMed]
27. Bowman, J.L.; Kohchi, T.; Yamato, K.T.; Jenkins, J.; Shu, S.; Ishizaki, K.; Yamaoka, S.; Nishihama, R.; Nakamura, Y.; Berger, F.; et al. Insights into Land Plant Evolution Garnered from the *Marchantia polymorpha* Genome. *Cell* **2017**, *171*, 287–304. [CrossRef] [PubMed]
28. Heck, M.A.; Lüth, V.M.; Gessel, N.; Krebs, M.; Kohl, M.; Prager, A.; Joosten, H.; Decker, E.L.; Reski, R. Axenic in vitro cultivation of 19 peat moss (*Sphagnum* L.) species as a resource for basic biology, biotechnology, and paludiculture. *New Phytol.* **2021**, *229*, 861–876. [CrossRef]
29. Vollár, M.; Gyovai, A.; Szűcs, P.; Zupkó, I.; Marschall, M.; Csupor-Löffler, B.; Bérdi, P.; Vecsernyés, A.; Csorba, A.; Liktor-Busa, E.; et al. Antiproliferative and Antimicrobial Activities of Selected Bryophytes. *Molecules* **2018**, *23*, 1520. [CrossRef]
30. Choi, U.; Lee, C.-R. Distinct Roles of Outer Membrane Porins in Antibiotic Resistance and Membrane Integrity in Escherichia coli. *Front. Microbiol.* **2019**, *10*, 953. [CrossRef]
31. Gruenheid, S.; Le Moual, H. Resistance to antimicrobial peptides in Gram-negative bacteria. *FEMS Microbiol. Lett.* **2012**, *330*, 81–89. [CrossRef]

32. Saul-McBeth, J.; Matson, J.S. A Periplasmic Antimicrobial Peptide-Binding Protein Is Required for Stress Survival in Vibrio cholerae. *Front. Microbiol.* **2019**, *10*, 161. [CrossRef]
33. Xie, F.; Wang, Y.; Li, G.; Liu, S.; Cui, N.; Liu, S.; Langford, P.R.; Wang, C. The SapA Protein Is Involved in Resistance to Antimicrobial Peptide PR-39 and Virulence of *Actinobacillus pleuropneumoniae*. *Front. Microbiol.* **2017**, *8*, 811. [CrossRef]
34. Bogomolnaya, L.M.; Tilvawala, R.; Elfenbein, J.R.; Cirillo, J.D.; Andrews-Polymenis, H.L. Linearized Siderophore Products Secreted via MacAB Efflux Pump Protect *Salmonella enterica* Serovar Typhimurium from Oxidative Stress. *mBio* **2020**, *11*, e00528-20. [CrossRef] [PubMed]
35. Ashton, N.V.; Cove, D.J. The Isolation and Preliminary Characterisation of Auxotrophic and Analogue Resistant Mutants of the Moss, *Physcomitreila patens*. *Molec. Gen. Genet.* **1977**, *154*, 87–95. [CrossRef]
36. Shirshikova, T.V.; Sierra-Bakhshi, C.G.; Kamaletdinova, L.K.; Matrosova, L.E.; Khabipova, N.N.; Evtugyn, V.G.; Khilyas, I.V.; Danilova, I.V.; Mardanova, A.M.; Sharipova, M.R.; et al. The ABC-Type Efflux Pump MacAB Is Involved in Protection of Serratia marcescens against Aminoglycoside Antibiotics, Polymyxins, and Oxidative Stress. *mSphere* **2021**, *6*, e00033-21. [CrossRef] [PubMed]

antibiotics

Article

Secondary Metabolite Variation and Bioactivities of Two Marine *Aspergillus* Strains in Static Co-Culture Investigated by Molecular Network Analysis and Multiple Database Mining Based on LC-PDA-MS/MS

Yuan Wang [1,2], Evgenia Glukhov [3], Yifan He [3], Yayue Liu [1,2], Longjian Zhou [1,2], Xiaoxiang Ma [1,2], Xueqiong Hu [1], Pengzhi Hong [1,2], William H. Gerwick [3] and Yi Zhang [1,2,3,*]

[1] College of Food Science and Technology, Guangdong Ocean University, Guangdong Provincial Key Laboratory of Aquatic Product Processing and Safety, Guangdong Province Engineering Laboratory for Marine Biological Products, Guangdong Provincial Engineering Technology Research Center of Seafood, Shenzhen Institute of Guangdong Ocean University, Zhanjiang Municipal Key Laboratory of Marine Drugs and Nutrition for Brain Health, Research Institute for Marine Drugs and Nutrition, Guangdong Ocean University, Zhanjiang 524088, China; wangyuan5614@163.com (Y.W.); liuyayue@gdou.edu.cn (Y.L.); zhoulongjian@gdou.edu.cn (L.Z.); maxiaoxiang97@outlook.com (X.M.); hxwz247@163.com (X.H.); hpzgd13902501729@126.com (P.H.)

[2] Collaborative Innovation Center of Seafood Deep Processing, Dalian Polytechnic University, Dalian 116034, China

[3] Center for Marine Biotechnology and Biomedicine, Scripps Institution of Oceanography, and the Skaggs School of Pharmacy And Pharmaceutical Sciences, University of California, San Diego, La Jolla, CA 92093, USA; eglukhov@ucsd.edu (E.G.); roderick98hyf@gmail.com (Y.H.); wgerwick@health.ucsd.edu (W.H.G.)

* Correspondence: hubeizhangyi@163.com or zhangyi@gdou.edu.cn; Tel.: +86-759-2396046

Citation: Wang, Y.; Glukhov, E.; He, Y.; Liu, Y.; Zhou, L.; Ma, X.; Hu, X.; Hong, P.; Gerwick, W.H.; Zhang, Y. Secondary Metabolite Variation and Bioactivities of Two Marine *Aspergillus* Strains in Static Co-Culture Investigated by Molecular Network Analysis and Multiple Database Mining Based on LC-PDA-MS/MS. *Antibiotics* **2022**, *11*, 513. https://doi.org/10.3390/tantibiotics11040513

Academic Editors: Fuhang Song and Yunjiang Feng

Received: 19 February 2022
Accepted: 6 April 2022
Published: 12 April 2022

Abstract: Co-culture is known as an efficient way to explore the metabolic potential of fungal strains for new antibiotics and other therapeutic agents that could counter emerging health issues. To study the effect of co-culture on the secondary metabolites and bioactivities of two marine strains, *Aspergillus terreus* C23-3 and *Aspergillus. unguis* DLEP2008001, they were co-cultured in live or inactivated forms successively or simultaneously. The mycelial morphology and high-performance thin layer chromatography (HPTLC) including bioautography of the fermentation extracts were recorded. Furthermore, the agar cup-plate method was used to compare the antimicrobial activity of the extracts. Based on the above, liquid chromatography-photodiode array-tandem mass spectrometry (LC-PDA-MS/MS) together with Global Natural Products Social molecular networking (GNPS) and multiple natural products database mining were used to further analyze their secondary metabolite variations. The comprehensive results showed the following trends: (1) The strain first inoculated will strongly inhibit the growth and metabolism of the latter inoculated one; (2) Autoclaved *A. unguis* exerted a strong inducing effect on later inoculated *A. terreus*, while the autoclaved *A. terreus* showed high stability of its metabolites and still potently suppressed the growth and metabolism of *A. unguis*; (3) When the two strains are inoculated simultaneously, they both grow and produce metabolites; however, the *A. terreus* seemed to be more strongly induced by live *A. unguis* and this inducing effect surpassed that of the autoclaved *A. unguis*. Under some of the conditions, the extracts showed higher antimicrobial activity than the axenic cultures. Totally, *A. unguis* was negative in response but potent in stimulating its rival while *A. terreus* had the opposite effect. Fifteen MS detectable and/or UV active peaks showed different yields in co-cultures vs. the corresponding axenic culture. GNPS analysis assisted by multiple natural products databases mining (PubChem, Dictionary of Natural Products, NPASS, etc.) gave reasonable annotations for some of these peaks, including antimicrobial compounds such as unguisin A, lovastatin, and nidulin. However, some of the peaks were correlated with antagonistic properties and remain as possible novel compounds without mass or UV matching hits from any database. It is intriguing that the two strains both synthesize chemical 'weapons' for antagonism, and that these are upregulated when needed in competitive co-culture environment. At the same time, compounds not useful in this antagonistic setting are downregulated in their expression. Some of the natural products produced during antagonism are unknown chlorinated metabolites and deserve further study for their antimicrobial properties. In summary, this study disclosed the different responses of two *Aspergillus* strains in co-culture, revealed their metabolic variation, and displayed new opportunities for antibiotic discovery.

Keywords: *Aspergillus terreus; Aspergillus unguis*; co-culture; antimicrobial activity; LC-PDA-MS/MS; molecular network; database mining

---

1. Introduction

The ocean supports an amazing variety of marine life and is a crucial part of the biosphere. Marine organisms, including microbes, have developed complex metabolic mechanisms to adapt themselves to the unique environment of high salinity, high pressure, low oxygen, and oligotrophy. As a result, secondary metabolites (SMs) with novel structures and rich activities are produced and provide a rich source of drug lead compounds [1]. Although medical sciences have made significant progress, infectious diseases caused by bacteria, fungi, and viruses still pose a substantial threat to public health. Due to the development of antibiotic resistance, finding new antibiotics remains an essential task for scientists worldwide. Nevertheless, most of the current antibacterial agents derived from natural products were isolated from terrestrial sources, while marine organisms are still primarily untapped resources for new biologically active natural products, and especially antibiotics [2].

Previous studies have shown that fungi contain many diverse biosynthetic gene clusters that encode for secondary metabolites, but under artificial culture conditions in the laboratory, most fungal functional gene clusters are silent (i.e., not expressed). The methods for enriching the diversity of expressed fungal metabolites include changing the composition of the medium, changing the environmental conditions, adding epigenetic modifiers, and co-cultivating with other fungi or bacteria [3–5]. Specifically, the co-cultivation of microorganisms from different sources often creates competition and antagonism. To compete for the limited natural resources in such an environment, or for living space or to maintain information transmission between species, the microbes often produce secondary metabolites that are not produced when cultured separately [6].

Mass spectrometry (MS)-based metabolomics is increasingly playing an important role in efficient natural products studies. These approaches enable the accurate offline and online comparison of constituent differences among samples including big data samples. However, these data sets are often too large for manual analysis as more than 1000 MS/MS spectra can be collected from just one extract sample. Global Natural Products Social molecular networking (GNPS) is a data-driven open platform for the storage, analysis, and dissemination of MS/MS spectra. It provides the ability to visualize data sets from different users and compare these with all publicly available reference spectra to annotate known molecules and discover putative analogs [7]. For example, Oppong-Danquah. et al. used a GNPS molecular networking-based screening method to annotate metabolites with crop protection activity in co-cultures of several marine fungi, significantly improving the efficiency of discovery and identification of trace novel natural products [8]. GNPS is a continuously developing platform for accurate dereplication and annotation tasks, and thus is best complemented with other approaches at the present time.

In preliminary studies, our laboratory obtained a marine *Aspergillus terreus* strain C23-3 from a coral collected in Xuwen Natural Reserve of South China Sea and a marine *A. unguis* strain DLEP2008001 from a seaweed collected at the intertidal zone of Dalian City by the Yellow Sea of Northern China. Intriguingly, our previous research and the reports from other groups indicated that both these two species/strains can produce potent antibiotics as well as other bioactive compounds. For example, *A. terreus* produces butyrolactones showing antibacterial, antitumor, antioxidant, antiviral, enzymes (glucosidase, glucuronidase, and cyclin-dependent protein kinase 5) inhibitory, neuroprotective, anti(-neuro-)inflammatory, and axonal growth promoting activities [9–14], as well as lipid lowering lovastatins and acetylcholinesterase (AChE) inhibitory territrems [15,16]. As for *A. unguis*, it was known to produce halogenated and non-halogenated depsidones which were reported to possess antibacterial, antifungal, brine shrimp larvacidal, enzyme inhibitory (AChE and aromatase), diphenyl-picryl hydrazyl (DPPH) free radical scavenging, and neuroprotective activities [17–25].

Because both strains are producers of antibiotics and neuroactive agents, it was appealing to investigate the effect of co-culture conditions on expression of their secondary metabolites, as this might result in the discovery of new antibiotics or anti-neurodegenerative agents. In this paper, we investigated the high-performance thin layer chromatographic (HPTLC) profiles, bioactivities, LC-MS/MS based GNPS molecular networking, and multiple natural product database mining of secondary metabolites deriving from co-culturing of these two marine fungi.

2. Results

2.1. Morphological Comparison

The two strains *A. unguis* and *A. terreus* were statically cultivated for a total of 28 days (in one stage or two stages) in seawater potato sucrose broth under 7 experimental condition groups, including: G1) *A. unguis* axenically for 28 days (abbreviated as axU); G2) *A. terreus* axenically for 28 days (axT); G3) *A. unguis* 7-day culture-inactivation + live *A. terreus* for the following 21 days (iacU-livT); G4) Live *A. unguis* 7-day culture + live *A. terreus* for the following 21 days (livU-livT); G5) live *A. unguis* and live *A. terreus* inoculated simultaneously (livU/livT) and co-cultivated for 28 days; G6) *A. terreus* 7-day culture-inactivation + live *A. unguis* for the following 21 days (iacT-livU); G7) Live *A. terreus* 7-day culture + live *A. unguis* for the following 21 days (livT-livU) (see details of the culture experiments in Section 4.2.1).

When *A. unguis* grows axenically, it initially forms scattered bright yellow colonies and then merges into a dark brown mycoderm (Figure 1G1). When *A. terreus* grows alone, it initially forms white colonies and then expands to form an off-white to brown mycoderm (Figure 1G2). If *A. terreus* is inoculated on the autoclaved 7-day culture of *A. unguis*, it still grows but just as small scattered off-white colonies on the dead mycoderm of *A. terreus* without forming its own continuous mycoderm (Figure 1G3). If *A. terreus* is inoculated on the live 7-day culture of *A. unguis*, no obvious growth of *A. terreus* is observed (Figure 1G4). When *A. unguis* and *A. terreus* are inoculated simultaneously, two types of mycoderms are observed: dark brown and off-white (Figure 1G5). When *A. unguis* is inoculated on the autoclaved 7-day culture of *A. terreus*, it does not grow as well as the axenic *A. unguis* culture but still forms its own complete mycoderm on the surface of the dead mycoderm of *A. terreus* (Figure 1G6). However, if *A. unguis* is inoculated on the live 7-day culture of *A. terreus*, the newcomers' growth is not obvious (Figure 1G7). Comparing the different culture experiments, it was found that the early inoculated fungus (even if autoclaved) will inhibit the growth of the late inoculated one, and the live fungus showed stronger inhibition than the inactivated one.

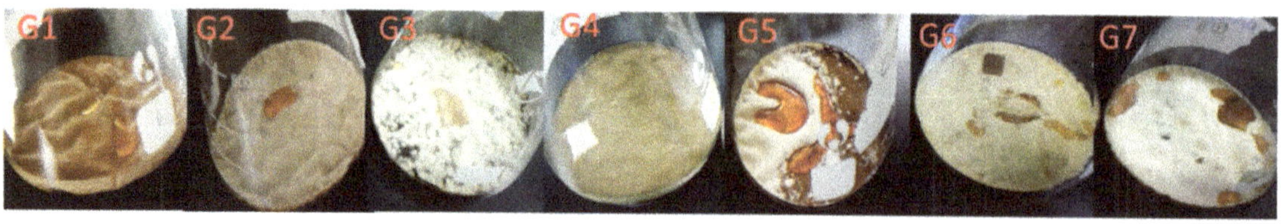

Figure 1. The morphology of axenic cultures and co-cultures in different experiments (28 days in total). (**G1–G7**), respectively, represent: axU (**G1**), axT (**G2**), iacU-livT (**G3**), livU-livT (**G4**), livU/livT (**G5**), iacT-livU (**G6**), livT-livU (**G7**).

2.2. Comparison of HPTLC Fingerprints

The profiles of secondary metabolites including their antioxidant as well as anti-AChE constituents were demonstrated using HPTLC images that were observed under 254 nm and 365 nm, colored by anisaldehyde reagent and potassium ferricyanide-ferric chloride (PFFC) reagent, and revealed by DPPH free radical scavenging and AChE inhibitory bioautographies, respectively. These HPTLC images revealed the variation between axenic cultures and co-cultures in different ways, including some dramatical changes.

The UV images (under 254 nm and 365 nm) clearly showed rich secondary metabolic profiles from the cultures. In the profile of axenic *A. unguis* (lane 1 in Figure 2A under 254 nm), the big dark spot with Rf value of 0.70 was judged to be comprised of depsidones according to our previous study on this strain [18]. For axenic *A. terreus*, the dark spot with Rf value of 0.28 was recognized from previous work as butyrolactone I [11]. These annotations were also supported by the LC-PDA-MS/MS analysis as described below in Section 2.4.

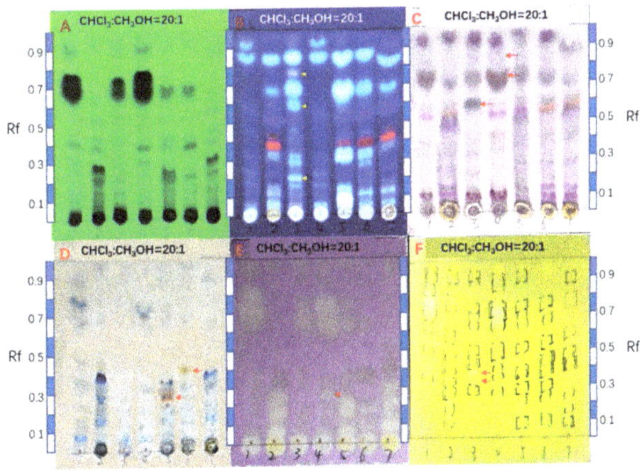

Figure 2. HPTLC fingerprints of the axenic and co-cultural extracts. (**A**) is the UV images of experiments G1–G7 under 254 nm (the sample numbers were marked with pencil below the starting line). (**B**) is the UV images of G1–G7 under 365 nm. (**C**) is the image of sulfuric acid-anisaldehyde colorized plate of G1–G7. (**D**) is the image of potassium ferricyanide-ferric chloride (PFFC) colorized plate of G1–G7. (**E**) is the DPPH free radical scavenging autographic image of G1–G7. (**F**) is acetylcholinesterase inhibitory bioautographic image of G1–G7. The developing agent was chloroform:methanol = 20:1 (*v*/*v*). The rulers beside the TLC plate are taken as references for Rf value calculation. The yellow or red arrows mark the new metabolites produced only under co-cultural conditions.

In the experiment G3 (iacU-livT), some *A. terreus* metabolites disappeared including the spots at Rf 0.43 (orange fluorescence), Rf 0.38 (red fluorescence), and Rf 0.17–0.30 (dark blue, including butyrolactone I). However, other putative *A. terreus* metabolites were enhanced in their production, including white fluorescent spots at Rf 0.84 and Rf 0.68. Moreover, several new constituents appeared in this co-culture including fluorescent spots at Rf 0.76 (light orange), Rf 0.58 (white), and Rf 0.20 (white) as marked with the yellow arrows (Figure 2B under 365 nm). Remarkably, the typical *A. unguis* depsidone metabolites still appeared in this co-culture (Figure 2A under 254 nm), indicating their thermostability against autoclaving.

In the experiment G4 (livU-livT, inoculated in tandem), the UV images were basically the same as those of the axenic *A. unguis*, indicating that the later inoculated *A. terreus* was extremely suppressed in its growth and production of metabolites. This is consistent with the morphological observation described above.

In experiment G5 (livU/livT, inoculated simultaneously), the UV images greatly differed from those of G3 and G4, but closely resembled those of G2 (the axenic *A. terreus*). However, some *A. terreus* products were produced in lower yields such as the spot with Rf 0.38 (red fluorescence), or even vanished such as the one with Rf 0.43 (orange fluorescence). In contrast, some spots, like the white fluorescent spots with Rf 0.84, 0.67, 0.32, and 0.08, were significantly enhanced to a much higher extent than in co-culture G3. As for *A. unguis* metabolites, only a small quantity of depsidone metabolites (Rf 0.67) was observed under 254 nm. This situation agreed with the growth advantage of *A. terreus* vs. *A. unguis* in Figure 1G5.

In the experiment G6 (iacT-livU), the UV image under 365 nm was similar to that of axenic *A. terreus*, suggesting that the *A. terreus* metabolites were quite thermotolerant against autoclaving and were stable over a three-week period. The depsidones of *A. unguis* were present but with much lower yield compared to the axenic *A. unguis* culture. Therefore, even the autoclaved *A. terreus* can remarkably inhibit the growth or metabolism of *A. unguis*.

Likewise, in experiment G7 (livT-livU, inoculated in tandem), the UV image displayed almost identical features with the axenic *A. terreus* cultures; however, the depsidones from *A. unguis* could barely be observed.

The results of anisaldehyde and PFFC colorization (Figure 2C,D) were consistent with the UV findings. Additionally, they revealed the production of new metabolite during the co-cultivation experiments. For example, the following were new compounds: blue gray spot at Rf 0.58 in experiment G3, gray spot at Rf 0.82 in G4 (both with anisaldehyde detection), brown spot at Rf 0.27 and brownish spot at Rf 0.40 (both with PFFC detection).

The two bioautographies revealed the antioxidant and anti-AChE constituents in the different culture experiments (Figure 2E,F). The depsidones showed antioxidant and anti-AChE activities, while butyrolactone I showed antioxidant activity. Their variation in amounts (Figure 2A) were partially reflected in the changes of the bioactive spots in these bioautography experiments. Nevertheless, some minute new anti-AChE spots (Rf 0.32 and 0.36) were also observed in co-culture G3, and differences in highly polar constituents remaining at the point of application cannot be excluded because the mobile phase (chloroform:methanol = 20:1 (v/v)) was not polar enough to mobilize them in the chromatogram.

Generally, the HPTLC suggested the following trends: (1) the strain firstly inoculated will strongly inhibit the growth and metabolism of the later inoculated one, (2) the autoclaved *A. unguis* exerted a strong inducing effect on the later inoculated *A. terreus*, while the autoclaved *A. terreus* showed high stability of its metabolites and still potently suppressed the growth and metabolism of *A. unguis*, (3) when the two strain were inoculated simultaneously, they both grew and produced metabolites. However, the *A. terreus* seemed to be more strongly induced by live *A. unguis* and this inducing effect surpassed that of the autoclaved *A. unguis*. Finally, *A. unguis* was negative in response and agonism but potent in stimulating its rival while *A. terreus* had the opposite effect.

2.3. Antimicrobial Activity

Based on the above preliminary co-cultural product HPTLC analysis, the antimicrobial activities of the extracts were further tested against several indicator strains, including Methicillin-resistant *Staphylococcus aureus* (MRSA), *Bacillus subtilis*, *Pseudomonas aeruginosa*, *Vibro parahemolyticus*, *V. alginolyticus*, *Shewanella putrefaciens*, *Yersinia pseudotuberculosis*, and *Candida albicans*. The results are shown in Table 1 and the representative photos below in Figure 3.

Table 1. The total extract amounts and antimicrobial activities from experiments G1–G7, which were measured using the Oxford Cup method (dosage: 200 mL/well, concentration = 1 mg/mL, concentration for both controls = 0.1 mg/mL, n = 4).

Sample Number	Total Sample Amount (Yield: mg/flask)	Diameters of Inhibition Zones Against Indicator Microbes (mm) [#]							
		MRSA	*Bacillus subtilis*	*Pseudomonus aeruginosa*	*Vibro parahemolyticus*	*Vibro alginolyticus*	*Shewanella putrefaciens*	*Yersinia pseudotuberculosis*	*Candida albicans*
axU (G1)	345 ± 40	15.7 ± 0.6	13.2 ± 0.5	14.0 ± 0.9	18.6 ± 0.9	17.8 ± 0.8	17.1 ± 1.2	-	13 ± 0.5
axT (G2)	624 ± 10	-	-	-	7.3 ± 0.4	8.1 ± 0.6	9.3 ± 0.5	-	-
iacU-livT (G3)	560 ± 20	-	7.5 ± 0.8	8.9 ± 0.9	11.8 ± 0.4	7.3 ± 0.9	7.1 ± 0.5	-	14.2 ± 0.7
livU-livT (G4)	309 ± 30	13.0 ± 0.4	15.7 ± 0.5	15.6 ± 1.0	17.8 ± 0.6	16.2 ± 0.6	17.4 ± 0.5	-	11.7 ± 0.5
livU/livT (G5)	420 ± 80	10.2 ± 0.6	14.3 ± 0.7	14.2 ± 0.7	-	14.3 ± 0.5	18 ± 0.8	-	13.0 ± 0.8
iacT-livU (G6)	440 ± 20	-	7.0 ± 0.4	-	-	9.6 ± 0.6	10.4 ± 0.8	-	12.3 ± 0.5
livT-livU (G7)	638 ± 40	8.1 ± 0.3	-	-	8. 1 ± 0.9	-	-	-	9.2 ± 0.7
Ampicillin	-	14.1 ± 0.4	17.5 ± 0.2	19.7 ± 0.5	24.1 ± 0.9	18.6 ± 0.4	16.1 ± 0.6	11.3 ± 0.2	-
Ketoconazole	-	-	-	-	-	-	-	-	16.3 ± 0.8

[#]: three times average ± standard deviation; -: no activity or very weak activity.

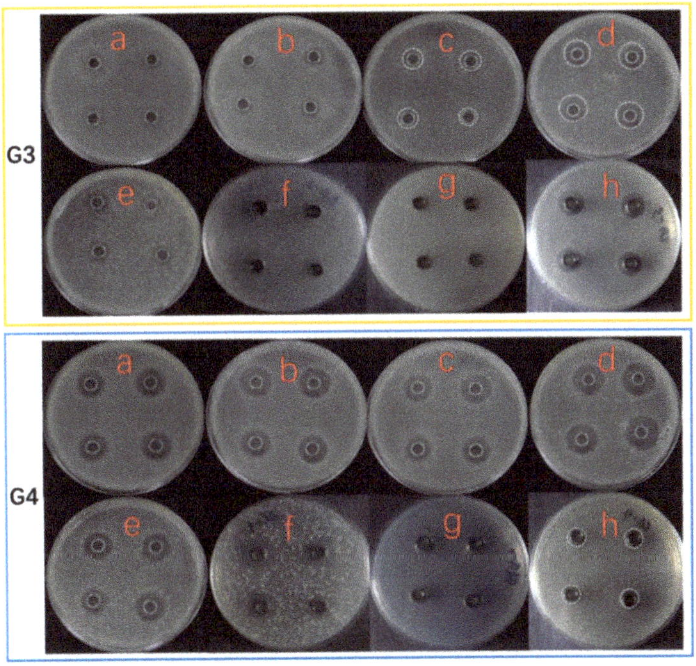

Figure 3. The antimicrobial activities of representative co-cultures (G3 & G4). The figures (**a–h**) for G3 & G4 are the antimicrobial results against MRSA, *Bacillus subtilis*, *Pseudomonas aeruginosa*, *Vibro parahemolyticus*, *V. alginolyticus*, *Shewanella putrefaciens*, *Yersinia pseudotuberculosis*, and *Candida albicans*, sequentially. For some relatively weak inhibition zones, circles in dash lines were used to mark them.

In these experiments, axenic *A. unguis* extracts exhibited much stronger broad-spectrum antimicrobial activity against the selected indicator strains than axenic *A. terreus* extracts. However, as revealed by morphological evaluation and HPTLC fingerprints, the *A. terreus* possessed strong inhibition against *A. unguis*, even by its autoclaved medium. This antifungal activity, though, may be specific to *A. unguis*, since *C. albicans* showed low sensitivity to the axenic *A. terreus* products.

When *A. unguis* grew well and fully, such as in the livU-livT (G4) experiment, the overall activity of the co-culture reached comparable levels of axU (G1). However, it was noticed that its activities against *B. subtilis* and *P. aeruginosa* were remarkably higher than those of axU (G1), which was possibly related to the higher depsidone metabolites yield under this condition as shown by HPTLC.

While *A. unguis* was inactivated by autoclaving after the first week or grew weakly when co-inoculated or late inoculated, with low yields of depsidones, the antimicrobial spectra and the inhibition potency of the co-cultural products would be generally weakened, as depicted in the results of iacU-livT(G3), livU/livT(G5), iacT-livU(G6), and livT-livU(G7). Nevertheless, the anti-*B. subtilis* and anti-*S. putrefaciens* activities of livU/livT(G5) were enhanced by co-culture, and the anti-*C. albican* activity of iacU-livT(G3) also became higher than axU (G1), suggesting that new antimicrobial substances may be produced in co-cultures. Taking into account, too, the remarkable increase in total extract amounts, the extent of the activity enhancement is even more dramatic because the activities reported in Table 1 are the results from identical sample concentrations. Considering that the strong inducing effect of *A. unguis* towards *A. terreus* and the antifungal potential of *A. terreus* (against *A. unguis* in co-cultures), the antifungal activity of iacU-livT(G3) was possibly from the induced products of *A. terreus*.

2.4. *Metabolits Profile Comparison by LC-PDA-MS/MS and Multiple Database Mining*

To further investigate the metabolic profile variation occurring as a result of co-cultivation, and to annotate putative antimicrobial compounds and their derivatives, LC-PDA-MS/MS analysis and GNPS molecular network-based analyses were performed for the extracts of the seven experimental setups. As a supplement to GNPS automatic metabolite annotation, manual searching was also performed to find possible candidates for the compounds showing dramatic changes in yield. This latter analysis used several open accessible natural product databases including PubChem, Dictionary of Natural Products (DNP), NPASS, Natural Product Atlas, and Nmrdata (WeiPu). The deduced molecular weights, isotopic patterns (for chlorinated metabolites), UV features, and taxon information (mainly within the Genus of *Aspergillus* and expanded to the kingdom of fungi when necessary) were queried in these database searches.

In general, the LC-UV profiles (under 280 and 360–370 nm) together with the LC-MS BPC profiles (under both positive and negative MS modes) (Figures 4–7) show the similar trends in fungal metabolite production in co-cultures as those observed by HPTLC. Especially, the 360–370 nm UV monitoring (although without fluorescent detection) showed rich upregulated peaks from *A. terreus* (G5 vs. G3) as displayed by the fluorescent components in HPTLC results but not by MS monitoring. These observations suggest that live *A. unguis* had a stronger inducing ability to *A. terreus* than the autoclaved one. These results also suggestion that the fluorescent substances may be not readily ionizable, or that they are present in very low quantity.

Totally, 15 MS detectable main peaks were chosen based on their significant variation in yield compared with the corresponding axenic cultures (Table 2) and submitted to annotation by GNPS and the multiple database mining approach described above. Their complete information is summarized in Table 3; the peaks are marked in Figures 4–7 and the selected annotated structures are presented in Figure 8 and partially by GNPS in Figures 9 and 10. By comparison of the axenic and co-culture LC profiles, peaks 1, 2, and 5–13 were assigned to be *A. terreus* metabolites, and peaks 2–4, 14, and 15 were assigned to *A. unguis* metabolites. Specially, peak 2 was produced by both strains.

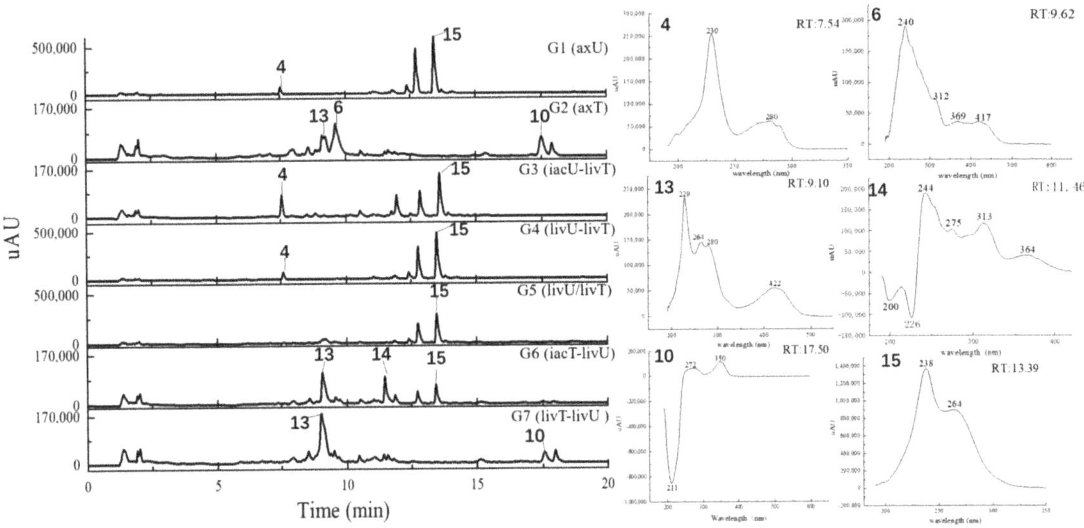

Figure 4. The HPLC traces of the culture extracts detected under the UV wavelength of 280 nm and the UV spectra for the featured peaks. The samples include axU (G1), axT (G2), iacU-livT (G3), livU-livT (G4), livU/livT (G5), iacT-livU (G6), and livT-livU (G7). The numbers marked on the peaks or in the UV spectra are numbers for the peaks with remarkable yield changes detected by mass spectrometry and are consistent with the peak numbers in Table 2.

Table 2. The changing folds for the yields of the differential peaks.

	Feature Peak Number	G3(iacU-livT)	G4(livU-livT)	G5(livU/livT)	G6(iacT-livU)	G7(livT-livU)
G2(axT)	1	↓0.14	↓<0.01	↓0.02	↓0.13	↑1.58
	2	↑1.59	↑1.96	↑2.11	↑2.24	↑2.46
	5	↓0.23	↓0.02	↓0.08	↑1.49	↑3.49
	6	↓0.12	↓<0.01	↓0.11	↓0.7	↓0.72
	7	↓0.21	↓0.08	↑3.57	↑3.15	↑4.42
	8	↓0.52	↓0.19	↑8.41	↑7.31	↑10.47
	9	↓0.02	↓0.06	↓<0.01	↓0.08	↓0.05
	10	↓<0.01	↓<0.01	↓0.09	↓0.64	↑1.36
	11	↓0.02	↓0.04	↓0.52	↑1.33	↑3.03
	12	↓0.2	↓0.06	↓0.08	↓0.02	↓0.57
	13	↓0.03	↓0.02	↓0.14	↓0.27	↓0.91
G1(axU)	2	↑4.57	↑5.66	↑6.07	↑6.45	↑7.09
	3	↓0.05	↑3.46	↑3.08	↓<0.01	↓0.01
	4	↑3.65	↑3.48	↓0.78	↓0.09	↓0.14
	15	↓0.23	↓0.56	↓0.5	↓0.14	↓0
	14	↓0	↓0	↑1	↑13.39	↑1.89

Note: ↓: Production decreased fivefold and more; ↓: Production decreased less than fivefold; ↑: Production increased fivefold and more; ↑: Production increased less than fivefold; ↑: New metabolite (in large quantities) in co-culture; ↑: New metabolite (in small amounts) in co-culture.

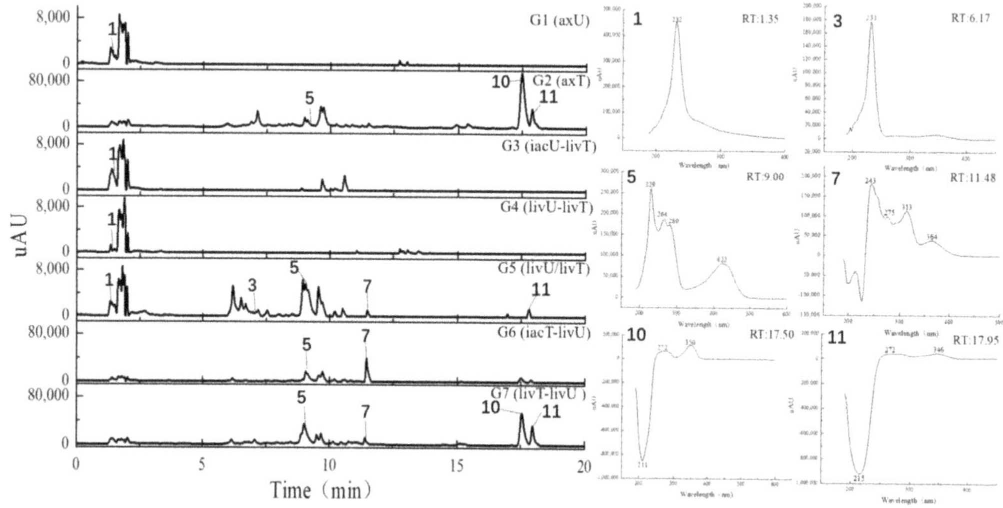

Figure 5. The HPLC traces of the culture extracts detected under the UV wavelength of 360–370 nm and the UV spectra for the featured peaks. The samples include axU (G1), axT (G2), iacU-livT (G3), livU-livT (G4), livU/livT (G5), iacT-livU (G6), and livT-livU (G7). The numbers marked on the peaks or in the UV spectra are numbers for the peaks with remarkable yield changes detected by mass spectrometry and are consistent with the peak numbers in Table 2.

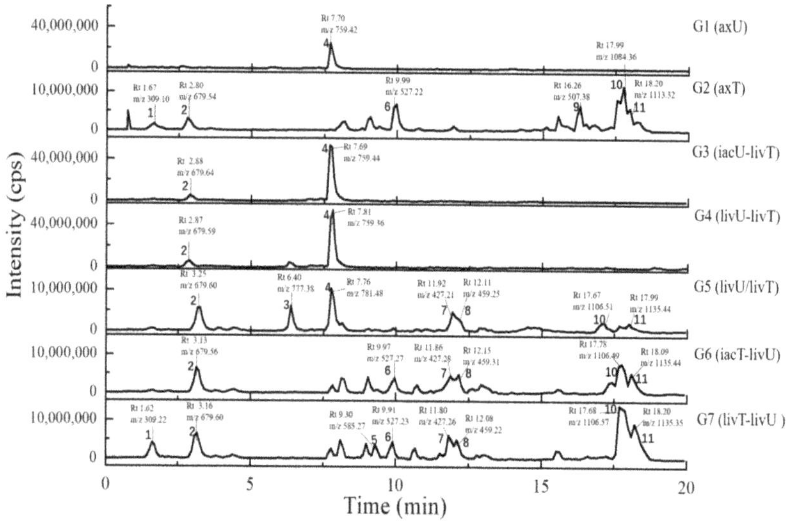

Figure 6. The LC-MS traces (base peak chromatographies, BPC) under positive ion mode of the culture extracts. The samples include axU (G1), axT (G2), iacU-livT (G3), livU-livT (G4), livU/livT (G5), iacT-livU (G6), and livT-livU (G7). The numbers marked on the peaks or in the UV spectra are numbers for the peaks with remarkable yield changes detected by mass spectrometry and are consistent with the peak numbers in Table 2.

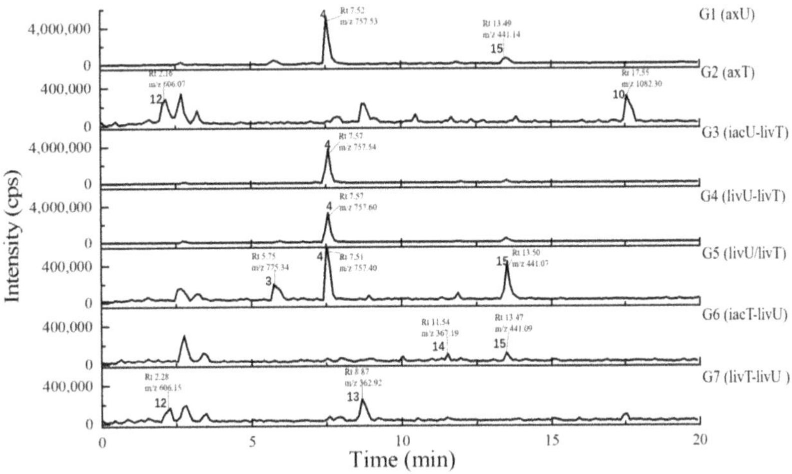

Figure 7. The LC-MS traces (base peak chromatographies, BPC) under negative ion mode of the culture extracts. The samples include axU (G1), axT (G2), iacU-livT (G3), livU-livT (G4), livU/livT (G5), iacT-livU (G6), and livT-livU (G7). The numbers marked on the peaks or in the UV spectra are numbers for the peaks with remarkable yield changes detected by mass spectrometry and are consistent with the peak numbers in Table 2.

Table 3. Multiple database mining of the main peaks that show remarkably different yields in base peak chromatographies (BPCs) of their LC-MS profiles using positive and negative modes of ionization.

Peak Number	Presence in Sample	*m/z* Value Measured	Retention Time (min)	UV Maximum Measured (nm)	Compound Hits in Library	Molecular Weight in Libraries	Libraries & IDs	MS^2 Similarity (Cosine)	Molecular Formula	UV Maximum Absportive Peaks in Libraries/Literature (nm)	Bioresource	DOI	Biological Activity	Structures Code of the Compound Hits
1	G2, G7	309.10 $[M + H]^+$ (presumed)	1.67	232	1,4-Bis(piperidin-1-ylmethyl)piperazine-2,5-dione	308.33	Dictionary of Natural Products	N/A	$C_{16}H_{28}N_4O_2$	N/A (isolate amide)	*A. terreus*	N/A	N/A	1-1 *
2	G1–G7	679.59 $[M + H]^+$ 677.88 $[M - H]^-$	2.87	231	3β-(β-D-glucopyranosyloxy)olean-12-ene-23,28,30-trioic acid	678.81	NMRDATA, 1331571; Natural Product Atlas, NPA026397; Pubchem, 146682840	N/A	$C_{36}H_{54}O_{12}$	N/A (isolate double bonds)	*A. amstelodami*	10.1002/cbdv.2019/00237	anti-melanogenic and anti-allergic activity	2-1 *
3	G1, G4, G5	777.38 $[M + H]^+$ 775.34 $[M - H]^-$	6.40	231	Aspergilasine B	775.84	NMRDATA, 999923; Dictionary of Natural Products	N/A	$C_{42}H_{49}NO_{13}$	202, 240	*A. flavipes QCS12*	10.1021//acs.orglett./7b02146	no inhibitory activities against seven cancer cell lines up to a concentration of 40 μM.	3-1
4	G1, G3–G7	759.36 $[M + H]^+$ 757.60 $[M - H]^-$	7.81	230, 280	Unguisin A	758.92	NMRDATA, 29553	N/A	$C_{40}H_{54}N_8O_7$	290, 281, 274, 219	*A. unguis*	10.1021//np980539z; 10.1039//C7OB00316A	moderately inhibited *Staphylococcus aureus*;as an anion receptor with high affinity for phosphate and pyrophosphate	4-1 *
5	G2, G7	585.27 $[M + H]^+$ 583.24 $[M - H]^-$	9.30	229, 264, 280, 422	Aspergilol A	584.62	NMRDATA, 895659; Pubchem, 132915662; Natural Product Atlas, NPA009011; Dictionary of Natural Products	N/A	$C_{34}H_{32}O_9$	196, 293, 452	*A. versicolor*	10.1016//j.tet.2015.10.038	possessing antioxidant activities	5-1
					Aspergilol B	584.61	Dictionary of Natural Products	N/A	$C_{34}H_{32}O_9$	194, 293, 462	*Aspergillus*	10.1016//j.tet.2015.10.038	possessing antioxidant activities	5-2
					Aspergilol G	584.61	Dictionary of Natural Products	N/A	$C_{34}H_{32}O_9$	206, 265, 295, 458	*Aspergillus*	10.1016/J.BMCL.2017.01.032	N/A	5-3 *
6	G2, G6, G7	527.22 $[M + H]^+$	9.99	240, 312, 369, 417	Territrem B	526.57	GNPS, CCMSLIB00005436075	0.70	$C_{29}H_{34}O_9$	195, 220, 236, 330, 284	*A. terreus*	10.3390/md12126113	strong inhibitory activity against acetylcholinesterase, potent antifouling activity	6-1 *
						526.57	Dictionary of Natural Products	N/A						
7	G2, G5–G7	427.26 $[M + Na]^+$	11.80	243, 275, 313, 364	Lovastatin	404.54	GNPS, CCMSLIB00000852214	0.76	$C_{24}H_{36}O_5$	231, 238, 247	*A. terreus*	10.1080/10826068.2020.1805624	the competitive inhibitors of the enzyme hydroxy-methyl-glutaryl coenzyme A (HMG-CoA) reductase	7-1 *
						404.54	Dictionary of Natural Products	N/A						
8	G5–G7	459.31 $[M + H]^+$ (presumed)	12.15	246, 285	Unannotated statin	N/A	N/A	N/A	N/A	N/A	N/A	N/A	N/A	N/A
					Aspernolide D	458.46	NMRDATA, 152713; Pubchem, 46930025; Natural Product Atlas, NPA003511	N/A	$C_{24}H_{26}O_9$	290	*A. terreus RCBC1002*	10.1248/cpb.58.1221	Inactive against all bacterial strains	8-1
9	G2, G7	507.38 $[M + H]^+$ (presumed)	16.26	N/A (no obvious absorption)	Terretonin G	506.58	NMRDATA, 809567; Dictionary of Natural Products	N/A	$C_{27}H_{38}O_9$	End absorption	*Aspergillus* sp. *OPMF00272*	10.1038/ja.2014.46	Moderate antimicrobial activity against Gram-positive bacteria	9-1 *

Table 3. *Cont.*

Peak Number	Presence in Sample	*m/z* Value Measured	Retention Time (min)	UV Maximum Measured (nm)	Compound Hits in Library	Molecular Weight in Libraries	Libraries & IDs	MS^2 Similarity (Cosine)	Molecular Formula	UV Maximum Absportive Peaks in Libraries/Literature (nm)	Bioresource	DOI	Biological Activity	Structures Code of the Compound Hits
10	G2, G5–G7	1106.49 [M + Na]$^+$ 1082.38 [M − H]$^-$	17.78	272, 350	Epichloenin A	1083.15	Dictionary of Natural Products	N/A	$C_{46}H_{74}N_{12}O_{18}$	N/A (containing a,b-unsaturated amides)	*Epichloe¨ festucae*	10.1371/journal.ppat.1003332	as an important molecular/cellular signal for controlling fungal growth and hence the symbiotic interaction.	**10-1 ***
					Astalluminoxide	1084.22	Natural Product Atlas, NPA032177	N/A	$C_{60}H_{66}AlN_9O_9$	201, 222, 272, 349	*A. terreus BCC51799*	10.1016/j.tet.2020.131496	moderate to weak cytotoxicity against both cancerous and non-cancerous cells.	**10-2 ***
11	G2, G5–G7	1135.35 [M + Na]$^+$	18.20	273, 346	N/A	N/A	N/A	N/A	N/A	N/A	N/A	N/A	N/A	N/A
12	G2, G7	606.07 [M − H]$^-$ (presumed)	2.16	231	N/A	N/A	N/A	N/A	N/A	N/A	N/A	N/A	N/A	N/A
13	G2, G7	362.92 [M − H]$^-$ (presumed)	8.87	226, 264, 278, 420	Austamide	363.41	Dictionary of Natural Products	N/A	$C_{21}H_{21}N_3O_3$	234, 256, 282, 392	*A.ustus*	10.1016/s0040-4039(01)97170-9	toxic to ducklings	**13-1 ***
					Circumdatin B	363.37	Dictionary of Natural Products	N/A	$C_{20}H_{17}N_3O_4$	284, 358	*A. ochraceus*	10.1021/jo981536u	Inactive in the assay against NCI's 60 cancer cell line panel	**13-2**
					Asperimide A	363.41	Natural Product Atlas, NPA028229	N/A	$C_{22}H_{21}NO_4$	229, 278, 360	*A.terreus*	10.1016/j.fitote.2018.10.011	not found exhibited cytotoxicity	**13-3 ***
14	G6, G7	367.19 [M − H]$^-$ (presumed) (367.19:369.11:371.19 = 9:6:1, in intensity) revealing to be dichlorinated compound	11.54	244, 275, 313, 361	Cosmochlorin A	369.24	Dictionary of Natural Products; Natural Product Atlas, NPA030107	N/A	$C_{18}H_{18}Cl_2O_4$	323	*Cosmospora vilior* IM2-155	10.1016/j.phytol.2016.09.007	moderate antimicrobial activity against gram-positive bacteria and fungi; partially restored the growth inhibition caused by hyperactivated Ca^{2+}-signaling in mutant yeast and showed GSK-3b inhibition	**14-1 ***
					Cosmochlorin B	369.24	Dictionary of Natural Products; Natural Product Atlas, NPA030108	N/A	$C_{18}H_{18}Cl_2O_4$	230, 290	*Cosmospora vilior* IM2-155	10.1016/j.phytol.2016.09.007	Inactive against microbes; similar restoring the growth inhibition activity to cosmochlorin A, promoting osteoclast formation	**14-2 ***
					Cosmochlorin C	369.24	Dictionary of Natural Products; Natural Product Atlas, NPA030109	N/A	$C_{18}H_{18}Cl_2O_4$	323	*Cosmospora vilior* IM2-155	10.1016/j.phytol.2016.09.007	Similar antimicrobial activity to cosmochlorin A	**14-3 ***
					Penicillixanthone	369.15	Dictionary of Natural Products; Natural Product Atlas, NPA008373	N/A	$C_{16}H_{10}Cl_2O_6$	230, 294, 369	*Penicillium* sp. PSU-RSPG99	10.1016/j.tet.2014.05.105	No antimycobacterial and cytotoxic activities	**14-4 ***

Table 3. *Cont.*

Peak Number	Presence in Sample	*m/z* Value Measured	Retention Time (min)	UV Maximum Measured (nm)	Compound Hits in Library	Molecular Weight in Libraries	Libraries & IDs	MS^2 Similarity (Cosine)	Molecular Formula	UV Maximum Absportive Peaks in Libraries/Literature (nm)	Bioresource	DOI	Biological Activity	Structures Code of the Compound Hits
15	G1, G3–G6	441.07 $[M-H]^-$ (441.07:443.02:444.99:446.12 = 27:27:9:1, in intensity)	13.50	237, 264	Nidulin	443.70	GNPS, CCMSLIB00005436077	0.70	$C_{20}H_{17}Cl_3O_5$	267	*A. unguis*	10.1055/s-0031-1298228 10.1080/14786419.2013.879305	aromatase inhibitory and antimicrobial and DNA damaging activities	15-1 *
							Pubchem, 6450195; Dictionary of Natural Products	N/A						

N/A: indicates not applicable or with no record or having limited accessibility. Compounds can be found by name in the Dictionary of Natural Product online database with more details and ID numbers. The asterisks (*) mark hits from databases with relatively higher reliability based on their similarities, including not only molecular weights, but also at least one of the following characteristics, like GNPS MS^2 similarity, UV features, isotopic patterns (for chlorinated compounds), and taxonomy, to the featured peaks.

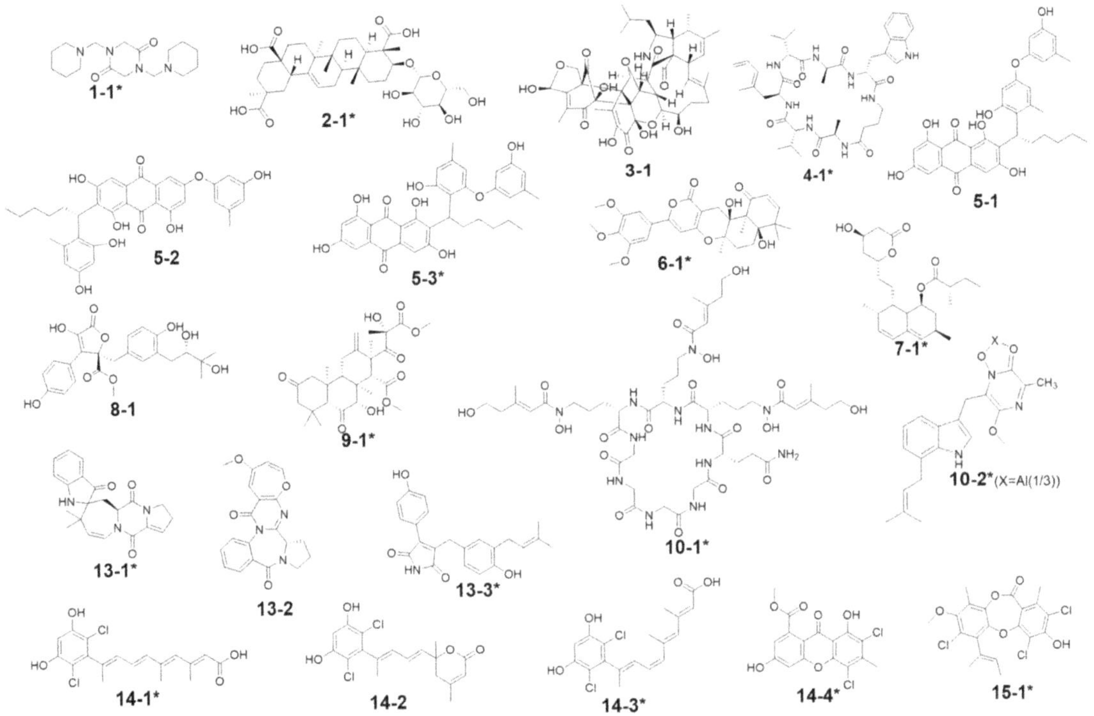

Figure 8. Annotated compound structures for the peaks 1–15. The asterisks (*) mark hits from databases with relatively higher reliability based on their similarities, including not only molecular weights, but also at least one of the following characteristics like GNPS MS^2 similarity, UV features, isotopic patterns (for chlorinated compounds), and taxonomy, to the featured peaks. Compounds **10-1** and **10-2** represent two alternative annotations for peak 10 (see Table 3 for further data on each of these annotations). This terminology also applies for the other base compound numbers.

Compared with the axenic *A. terreus*, most of its metabolites except peak 2 were remarkably decreased in yield when *A. terreus* was inoculated one week after *A. unguis* (i.e., groups G3: iacU-livT and G4: livU-livT). Moreover, live *A. unguis* showed a stronger suppressive effect than the autoclaved material, indicating the inhibition of *A. unguis* metabolites towards *A. terreus* (Table 2).

Peak 7 was annotated to be the typical *A. terreus* metabolite lovastatin by MS/MS similarity to GNPS records. Its UV features showed a difference to related literature values [15,16] which was possibly caused by unknown impurities in the peak. Peak 8 appeared in the same cluster in the GNPS molecular network (the node with average *m/z* 427.256 for peak 7 and the node with average *m/z* 459.273 for peak 8 in Figure 9A) and was therefore also proposed to be a statins. Peaks 7 and 8 were dramatically upregulated in G5 and G7 (when *A. terreus* was inoculated simultaneously with or prior to *A. unguis*, respectively), indicating that they were employed in antagonistic responses to *A. unguis*. This phenomenon is also consistent with the antifungal activity reported for statin natural products [26,27]. In G6 (iacT-livU), these two compounds remained in high concentration even though the 1-week *A. terreus* culture was autoclaved, indicating that they were produced in the early stage of growth and possessed good stability to persist in their inhibition to *A. unguis*. This was indicated by the weakened mycelial growth and downregulated metabolites observed in HPTLC and LC profiles of group G6 *A. unguis*.

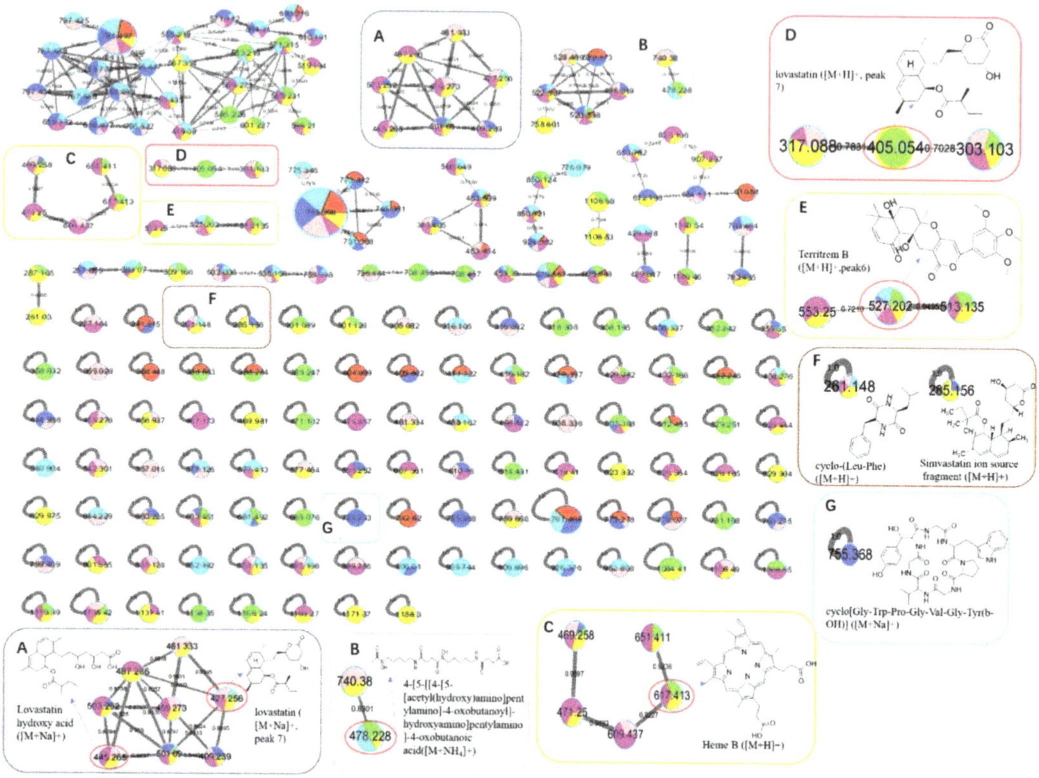

Figure 9. The GNPS molecular network based on positive ion MS/MS spectral similarity, showing a selection of amplified clusters. The nodes display the measured average masses of the molecular ions with identical MS/MS spectra. The sizes of the nodes reflect the relative amount of the corresponding compounds. The different colors of sections in the nodes represent different samples, i.e., : axU (G1), axT (G2), iacU-livT (G3), livU-livT (G4), livU/livT (G5), iacT-livU (G6), and livT-livU (G7), respectively. (**A**) is an enlarged cluster for statins including the sodiated ion of peak 7 (lovastatin). (**B**) is an enlarged cluster containing a possible fusarine-like siderophore. (**C**) is an enlarged cluster containing heme B. (**D**) is an enlarged cluster containing protonated ion of peak 7. (**E**) is an enlarged cluster containing peak 6 (territrem B). (**F**) conclude two nodes annotated as a diketopiperazine and a simvastatin fragment. (**G**) is a node annotated as a cyclopeptide.

Four peaks (1, 5, 10, and 11) all increased in G7 (livT-livU) but were reduced in the simultaneously inoculated G5 experiment (livT/livU). Among them, peaks 5 and 11 were also increased in group G6 (iacT-livU) but to a lesser extent than G7, but peaks 1 and 10 were decreased in G6. Their varied production levels suggest that all four compounds may contribute to the competitive success of pre-inoculated *A. terreus* growth in co-cultures; however, their potency may be too low to be effective in the spontaneous inoculation co-culture G5. Peaks 5 and 11 were deduced to be produced in the early stage of growth and stable, whereas peaks 1 and 10 may be not stable or are synthesized late in the growth cycle. Although GNPS did not provide any annotations for these latter metabolites, query of multiple databases provided some clues as to their identities. For peak 1 there was 1 hit, a diketopiperazine with end absorption similar to what was observed. There were 3 possible hits (anthraquinones aspergilols A, B, and G) for peak 5 which included UV absorptions most similar to aspergilol G [28]; however, only by UV comparison, aspergilols A and B cannot be definitely excluded since they were also reported to have similar UV absorptions to peak 5. Two hits were found for peak 10 with similar UV profiles as the cyclopeptide epichloenin A and diketopiperazine-aluminum salt astalluminoxid (the former with reported antifungal activity) [29,30]. Peak 11 has a similar molecular weight to the antifungal Tyr-containing cyclic peptide KK-1 from the database Natural Product Atlas (record NPA028479), but their distinct UV properties rule out the possibility of such a match. [31].

Figure 10. The GNPS molecular network based on negative ion MS/MS spectral similarity showing a selection of amplified clusters. The nodes display the measured average masses of the molecular ions with identical MS/MS spectra. The sizes of the nodes reflect the relative amount of the corresponding compounds. The different colors of sections in the nodes represent different samples, i.e., : axU (G1), axT (G2), iacU-livT (G3), livU-livT (G4), livU/livT (G5), iacT-livU (G6), and livT-livU (G7), respectively. (**A**) is an enlarged cluster containing peak 15 (nidulin). (**B**) is an enlarged cluster containing nornidulin.

Another four peaks (6, 9, 12, and 13) were all downregulated in experimental groups G5, G6, and G7. Peak 9 was nearly eliminated in all the groups, indicating that it may be not important to this fungal strains' arsenal of allelochemicals. The down regulation of these four compounds is consistent with their not being involved in the antagonistic behavior of this fungus, and is consistent with economizing unused metabolic pathways during stress conditions. GNPS and database mining gave reliable annotations for peak 6 as the known AChE inhibitor territrem B [15,32,33] and for related peak 9 the antibacterial sesterterpene terretonin G. The latter annotation had similar UV characteristics [34]. For peak 12, five compounds with similar molecular weights and high polarity were found, i.e., Aspergillussanone H, Nigerasperone B, Aurasperone B, and Fumigatosides C-D from *Aspergillus* spp. [35–40], but their rich UV absorptive peaks ruled them out since peak 12 essentially only showed end absorption. Nevertheless, two possibilities, austamide (diketopiperizine) and asperimide A (containing a maleimide ring), had the most similar UV spectra to peak 13 out of the 3 hits [41,42]. Likewise, most of the *A. unguis* metabolites (peaks 3, 4, and 15) all were reduced in groups G6 and G7 when *A. unguis* was inoculated 1 week afterwards, reflecting the inhibition of *A. terreus* metabolites to *A. unguis* (Table 2). However, these three compounds changed in quite different ways in experimental groups G3, G4, and G5. For instance, peak 3 was produced in substantially higher yields in G5 (livU/livT, spontaneous inoculation) and G4 (livU-livT, *A. unguis* inoculated first) in contrast to its very low production level in G3 (iacU-livT). These results suggest that it was probably produced by *A. unguis* to oppose *A. terreus* and it was possibly a non-thermostable metabolite synthesized in its early growth stage. GNPS analysis along with a query of multiple databases did not provide a reasonable annotation based on molecular weight (776 Da), (MS^2 features), and UV (end absorption) (Table 3).

Peak 4 was annotated as the cyclopeptide unguisin A by combining the multiple database mining [43], our previous work on this strain [17], and manual interpretation of its MS/MS spectrum (Figure 11). This compound was increased by more than 3 folds in *A. unguis* first inoculated groups G4 (livU-livT) and G3 (iacU-livT), whereas it was moderately downregulated in G5 (livU/livT). This profile suggests that it was useful in maintaining the predominance of *A. unguis* in co-culture, but is not likely a major component of its antagonistic arsenal.

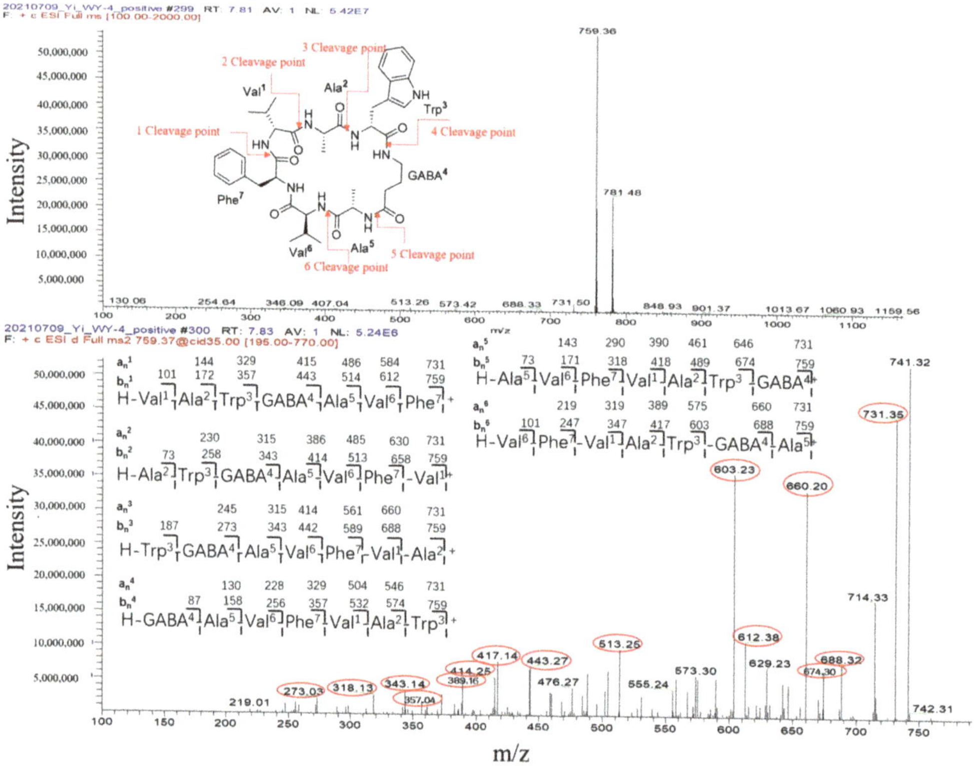

Figure 11. The interpretation of MS/MS spectrum of peak 4 (cyclopeptide unguisin A).

Peak 15 showed a typical isotopic pattern for a tri-chlorinated compound by mass spectrometry (Table 3 and Figure S20), and was thus annotated as nidulin based on GNPS matching, multiple database mining [23], and our previous studies [17,18]. Its production in groups G3 (iacU-livT), G4 (livU-livT), and G5 (livU/livT) was reduced to different extents but not eliminated. In co-cultures G4 and G5 with live *A. unguis*, its yield still reached about 50% of the level observed in G1, suggesting that it played a positive yet minor role in this antagonistic response.

Noteworthily, peak 14 had an isotopic cluster typical of a dichlorinated metabolite (Table 3 and Figure S19) with four dichlorinated but UV distinct 'hits' in the kingdom of fungi, but no hits from the genus *Aspergillus* by GNPS matching or query of multiple databases [44,45]. Interestingly, it was not detected in monocultures of *A. unguis* and *A. terreus*. However, considering the observed halogenation ability of *A. unguis* [17,18,46], it is likely a metabolite of this latter fungus. It is intriguing that it was not produced in monoculture nor in G4 when it had the advantage of earlier inoculation, but only when the two strains were inoculated at the same time. As the competition was more intense as in G7 (livT-liv U) and *A. terreus* preemptively produced inhibitory factors as in G6 (iacT-liv U), its production was increased. This profile suggests that it might be a metabolite that provides resistance to the stress imposed by *A. terreus* metabolites. Furthermore, it is interesting that autoclaved *A. terreus* had a much stronger ability to induce production of peak 14 than live cultures, inferring that some potent and thermostable inducers are produced in the early stage of *A. terreus* growth.

Peak 2 is able to be produced by both *Aspergillus* species; however, the yield in *A. terreus* was about triple that of *A. unguis*; and under different co-culture conditions, its yields were higher than the monocultures; but it was not possible to discern which strain contributed more under the co-culture conditions. GNPS and multiple database mining did not give a clear indication of its identity with only one hit suggested, an anti-melanogenic triterpenoid saponin from *Aspergillus*. It had an identical molecular weight and end UV absorption [47]; however, more studies are necessary to confirm its identity and function as a potential allelochemical.

Briefly, the variation in features in different co-culture experiments revealed that both strains have allelochemicals that are produced in antagonistic conditions. Some metabolites appear to be involved in maintaining the predominance of first inoculated strain, others may be involved in anti-stress responses, and still others are down regulated in these conditions, presumably because they are not involved in these competitive interactions.

In addition to these 15 peaks observed in the LC-MS profiles, GNPS also annotated several other metabolites by MS/MS similarity comparisons as shown in Figures 9 and 10. Some of these also varied in production levels in the different culture experiments, such as a siderophore-like desferrioxamine with *m/z* 478.228 together with its congener showing up in axenic *A. terreus* (G2) and other cultures in cluster F. Additionally, an open-ring di-chlorinated depsidone with *m/z* value 426.979 was produced in extremely low yield (Figure 10). However, classical GNPS networking can potentially sum the intensity of ions into the same node that possess 'identical' m/z values and MS/MS profiles but quite different retention times (e.g., isomers). Therefore, this quantification is not as reliable as EIC integration of the original LC-MS profiles.

3. Discussion

Based on the comparison of monocultures (axenic cultures) and co-cultures in different configurations for their mycelial morphology, HPTLC-bioautography analysis of secondary metabolic profiles, antimicrobial tests, and LC-PDA-MS/MS analysis, the first inoculated *Aspergillus* strains were commonly observed to predominate in co-cultures. This was true even when the culture was autoclaved one week into the experiment, as it still suppressed the growth and production of metabolites in the latter inoculated strain. When both strains of *Aspergillus* were inoculated simultaneously, they each grew in reasonable yield and produced their secondary metabolites (SMs). When *A. terreus* was inoculated before *A. unguis*, it produced its characteristic metabolites, including statins, undetermined peaks (e.g., 1, 5, 11), and desferrioxamines. Similarly, when *A. unguis* was first inoculated in the co-culture, it produced its typical metabolites, including unguisin A, nidulin, and undetermined peak 3. Meanwhile, the strains dramatically downregulated several metabolites that possessed no or only weak antimicrobial activity, and therefore, these latter compounds seem unrelated to the antagonistic phenotype of each fungus. This was especially the case when they were the latter inoculated strain in the co-cultural experiment. Nevertheless, *A. unguis* produced new dichlorinated metabolites only when it was extremely stressed by the simultaneous or second in the sequence inoculation, but not when it grew alone or was inoculated prior to *A. terreus*. On the other hand, *A. terreus* was induced to synthesize unknown compounds that did not ionize well in MS, but had strong white fluorescence on TLC analysis. This was especially the situation when simultaneously inoculated with *A. unguis*, which may be responsible for the enhanced antibacterial activity of extract of G5. In this case, the known antimicrobial agent nidulin was downregulated by *A. unguis* in this co-culture condition. Given this profile of expression, these unknown compounds may have anti-stress activities in the two *Aspergillus* strains. The overall results suggest that *A. unguis* was relatively ineffective in mounting an agonistic phenotype in culture, but was quite potent in stimulating its rival *A. terreus* to mount a strong response. The opposite was true for *A. terreus* in that it produced strongly antagonistic natural products but elicited little response from the competing fungus *A. unguis*.

Previous studies have revealed that co-cultivated microbes, such as fungi, often secrete extracellular diffusible SMs like phenols and quinones as well as enzymes such as phenoloxidases, peroxidases, and lignin-degrading enzymes, to suppress rivals or compete for new resources [48]. These competitive interactions occur even between different intraspecies strains; for example, the non-aflatoxigenic *Aspergillus flavus* can inhibit the aflatoxigenic *A. flavus* via antifungal SMs and antioxidants [49]. Numerous induced microbial SMs are involved in microbial antagonism as manifested by the fact that 37% have been reported with antibacterial, 7% with antifungal, and 35% with cytotoxic activities, together with 9% having other related activities including 3% as siderophores, 2% as α-glucosidase inhibitors, 1% as ATP synthesis inhibitors, and 3% as pesticides [50]. It has also been reported that the first inoculated strain in co-culture experiments usually predominates in the production of SMs, and that their yields can be even higher than in the corresponding mono-culture conditions. This concept was demonstrated by the co-cultivation of *Streptomyces rimosus* and *A. terreus* in a stirred tank bioreactor [51], and is reinforced by the findings of this present study.

Co-culture has increasingly been recognized as an efficient approach to activate silent biosynthetic gene clusters (BGC). This has been partially attributed to chromatin remodeling via epigenetic modifications [50,52]. For example, intimate physical contact of *Streptomyces rapamycinicus* to hyphae of *Aspergillus nidulans* has been shown to trigger the latter's multi-subunit transcriptional co-activator complex SAGA/ADA, which in turns leads to histone H3K9 acetylation of transcription factor *basR* and *ors* BGC and the final production of SMs such as orsellinic acid [52,53]. Similar mechanisms are also known to occur in fungus-fungus co-cultures. The up-regulation of *O*-methylmellein was observed in the co-culture of two plant pathogens, *Eutypa lata* and *Botryosphaeria obtuse*. Interestingly, its production was also upregulated in the fungus *Stagonospora nodorum* by application of the epigenetic modifiers SAHA and nicotinamide [48,54,55]. Moreover, in a previous study by our group, a higher yield of unguisin A was observed in a *A. unguis* monoculture supplemented with both chemical epigenetic modifier procaine chloride and NaBr [17]. This previous finding is consistent with its significantly higher yields in the co-culture experiments G3 and G4 with *A. terreus*, and suggests that regulation of production of unguisin A may involve epigenetic mechanisms.

A literature-based search of MS-based features encountered in this study further supported their potential chemecological roles in these co-culture experiments. For example, unguisin A was initially reported from marine fungus *Emericella unguis* (the teleomorph of *A. unguis*); and while our previous investigations as well as a series of other studies did not detect potent antibacterial or antifungal activity for unguisin A [17,43,56], it was reported to be a promiscuous binder to various anions with particularly high affinity for phosphate and pyrophosphate [56]. Phosphorus is critical to fundamental life processes, and competition for phosphate is known to exist in microalgae-bacteria and microalgae-microalgae co-culture systems. Moreover, arbuscular mycorrhizal fungi have an important role in phosphorus absorption for their plant hosts [57–59]. Therefore, it is reasonable that in the *A. unguis*-*A. terreus* co-culture system of this present study, unguisin A may help *A. unguis* to acquire more phosphate and thus maintain its predominance over *A. terreus* when it is the first inoculated strain. It was surprising that in the G3 experiment (iacU-livT), this compound was maintained at high levels; this may be attributed to its relatively good stability. Additionally, it is conceivable *A. terreus* may use this *A. unguis*-derived compound to enhance its phosphate assimilation, as 'borrowing' SMs from rival microbes is not uncommon [60].

Statin molecules such as lovastatin and simvastatin are typical *A. terreus* metabolites. They are used as lipid lowering drugs by inhibiting HMG-CoA in cholesterol biosynthesis, and have been shown to inhibit the growth of various *Aspergillus* spp. (*A. fumigatus, A. flavus, A.niger, A. terreus,* etc.) and yeasts like *Candida albicans*. This growth inhibition occurs via multiple mechanisms, including HMG-CoA inhibition, iron starvation, induction of DNA fragmentation, fungal cell morphogenesis disruption, and others [27,61,62]. Variation in the production levels of statins was also investigated when *A. terreus* was co-cultivated with *Penicillium rubens, Chaetomium globosum,* and *Mucor racemosus*, respectively [63,64]. These experiments showed that the yields of lovastatin and its derivatives did not increase or were even lower than the yields in *A. terreus* monoculture, due partially to transformation into monacolin J. Nevertheless, in the present study the yields of lovastatin and its analog peak 8 were significantly larger compared to the monoculture; this may have been caused by the strong inducing effect of *A. unguis* towards *A. terreus*.

Halogenated compounds are an important class of marine fungal natural products (NPs). In 1994–2019, a total of 217 halogenated compounds were discovered from marine fungi, among which 88% were chlorinated compounds. Moreover, 18.9% were reported with antimicrobial activities, 35 were from the genus *Aspergillus* including 4 from *A. unguis* (all as depsidones), and none from *A. terreus* [65]. The PubChem database mainly records natural products from two original databases, the Natural Product Atlas (NPA) and the Natural Product Occurrence Database (NPOD). Duplicate records occurring in both databases are not excluded. Our search of PubChem returned 141 (from NPA) and 377 (from NPOD) *A. terreus* natural products, with 12 chlorinated ones (3 from NPA and 9 from NPOD) (total proportion 2.3%). From *A. unguis* we found that there were 19 NPs from NPA plus 58 NPs from NPOD with totally 24 (3 from NPA and 21 from NPOD) chlorinated ones (total proportion 31.2%). These data indicate that *A. unguis* possesses a much higher potential for producing chlorinated NPs than *A. terreus*. The tri-chlorinated NP nidulin that was found in this study is also known from *A. unguis* and *A. nidulans*, and has been characterized to possess antibacterial and moderate antifungal activities [17,18,64–68]. In co-cultural experiments G3-G6 in this study, the yield of nidulin was reduced although it remained at considerable levels of 14% to 56% of the yield from monocultures. We noticed that the extracts (1 mg/mL) of all the groups containing *A. unguis* that were growing well, such as axU (G1), showed comparable antimicrobial activity to the positive control ampicillin (0.1 mg/mL). This potent activity was possibly derived from nidulin, the main metabolite of *A. unguis*, and may also come from synergetic effects from other components. Furthermore, in experiments G5–G7, an unknown dichlorinated metabolite was upregulated under stress conditions. These results suggest that chlorinated NPs have chemical defensive functions in this fungus.

Siderophores are unique molecules that various organisms have evolved for the capture and assimilation of iron that is necessary for live. As such, they are intimately involved in iron resource competition during microbial co-cultures [51]. Hydroxamines (also as desferrioxamines, DFOs) are the most common siderophore family in nature and the main siderophore type produced by fungi. Fungi of the genus of *Aspergillus* are reported to produce DFOs consisting of ferrichromes and linear/cyclic fusarines [69,70]. The node with *m/z* value of 478.228 (Figure 9) was annotated by GNPS as a linear fusarine, although being different from any of the known *Aspergillus* fusarines. GNPS networking revealed that in the two monocultures, it was only produced by *A. terreus*. It was also present in the co-culture experiments G3, G5, and G7 where the live *A. terreus* was relatively easy to grow. These results are consistent with the use of this fusarine analog by *A. terreus* for iron uptake. These features along with the others discussed above may play important roles in the co-culture system.

In this current study, MS^2-based molecular networking was employed based on data from an ion trap MS/MS instrument operating at low mass resolution. Using the GNPS platform, the combination of MS^1 and MS^2 data provided good putative annotations for metabolites observed in these co-culture experiments. However, for compounds unmatched by the GNPS analysis, a tedious manual searching across multiple NP databases was performed using likely molecular weights, taxonomical information, and UV spectral features. Nevertheless, solid verification of annotated or unknown metabolites requires confirmation by scaled-up fermentation, isolation, and structural elucidation by comprehensive spectroscopic methods. This confirmatory work is necessary and common to all mass spectrometry-based metabolomics studies, and is made especially challenging in co-culture studies which are inherently more difficult to reproducibly control.

In the current study, predominance of one fungus over the other in these co-culture experiments mainly relied on macroscopic morphological observations and comparisons of metabolite yields, neither of which method is direct or completely accurate. In future studies, microscopic observation, real-time quantitative PCR, microbiota amplifier high-throughput sequencing for ITS rDNA or metagenome sequencing, or transcriptomics approaches could be used to understand variations in the fungal community and gene expression more fully and accurately.

In summary, this study is the first report on the co-culture between *Aspergillus terreus* and *A. unguis* and reveals that *A. terreus* is more aggressive and responds to its rival's presence, while *A. unguis* is less robust in its response but potent in stimulating its rival's allelochemicals. It reveals their tendency to maintain dominance by synthesizing secondary metabolites and discloses the production of an unknown dichlorinated compound by *A. unguis* and strongly fluorescent products by *A. terreus* under stress. These results provide a deeper understanding of fungal co-culture mechanisms as well as the discovery of new natural products. This study also revealed a couple of shortcomings in the use of GNPS for compound annotation. The limited number of library records impacts the extent of automatic matching that can be achieved by GNPS. Additionally, an automatic analysis of isotopic peak clusters for recognizing the occurrence of halogenated compounds is current lacking. These aspects make necessary a complementary manual analysis of the data as well as querying of multiple databases via spectral features and taxonomic information.

4. Materials and Methods

4.1. Materials

The marine fungus A. *terreus* C23-3 was collected from the Xuwen Coral Reserve in Zhanjiang and is now preserved in the Guangdong Provincial Microbial Culture Collection, with the deposit number GDMCC No. 60316. The marine fungus A. *unguis* DLEP2008001 was derived from a red alga from the seaside of Fujiazhuang, Dalian, Liaoning Province, China and deposited in China General Microbiological Culture Collection Center with number CGMCC 3372. *Bacillus subtilis* MCCC 1A03710 was purchased from China Marine Microbial Culture Collection; *Pseudomonas aeruginosa* and *Candida albicans*, were from American Typical Culture Collection with numbers ATCC 9027 and ATCC 10231, respectively; *Vibrio parahaemolyticus* was donated by Professor Wen Chongqing, Fisheries College, Guangdong Ocean University; *V. alginolyticus* and *Shewanella putrefaciens* were presented by Professor Liu Ying from the School of Food Science and Technology, Guangdong Ocean University; *Methicillin-resistant Staphylococcus aureus* A7983 was gifted by Professor Yu Zhijun from Dalian Friendship Hospital. The acetylcholinesterase (AChE, from electric eels) and DPPH were purchased from Sigma-Aldrich (St. Louis, MO, USA). All the organic mobile phase solvents for LC-MS were from Merck (Darmstadt, Germany). All the other reagents were with analytical purity.

4.2. Methods

4.2.1. Co-cultivation and Extraction of Strains

The strain A. *terreus* C23-3 and A. *unguis* DLEP2008001 were activated at 28 °C overnight and then inoculated, respectively, into conical culture flasks pre-filled with 200 mL of sterilized seawater potato solid medium for cultivation for 3–4 days until rich spores grew on the colonies. Then spore suspensions were prepared as seeds by washing the spores with sterile saline. Afterwards, the seed suspensions were inoculated into

500 mL Erlenmeyer flask each filled with 100 mL seawater potato sucrose liquid medium (containing 20 g sea salt, 20 g sucrose, and 500 mL potato juice per liter) according to the following group design. Each group contained three replicates.

Experiment grouping:

G1: *A. unguis* cultivated separately for 4 weeks (axenic *A. unguis*, abbreviated as axU);
G2: *A. terreus* cultured separately for 4 weeks (axenic *A. terreus*, abbreviated as axT);
G3: Inactivated *A. unguis* + live *A. terreus* (abbreviated as iacU-livT. In detail, *A. unguis* was inoculated first, cultured for one week and then iactivated by autoclaving; Afterwards, *A. terreus* was inoculated into the same flask and cultured for the next three weeks);
G4: Live *A. unguis* + live *A. terreus* (abbreviated as livU-livT; Similar to G3, but the first inoculated *A. unguis* was not autoclaved);
G5: Live *A. unguis*/Live *A. terreus* (abbreviated as livU/livT; The two strains were simultaneously inoculated into the same flask and cultivated for 4 weeks);
G6: Inactivated *A. terreus* + live *A. unguis* (abbreviated as iacT/livU; Similar to G3, but the inoculation order was opposite);
G7: Live *A. terreus* + *A. unguis* (abbreviated as livT-livU; Similar to G4, but the inoculation order was opposite).

During the 4-week cultivation, daily observation and photography were made to record the morphology of the cultures. After the four weeks, the fermentation broth was extracted three times with equal volumes of ethyl acetate. The mycelium was extracted three times with methanol assisted by ultrasonication; then the two extracts were concentrated, combined, and evaporated to dryness, weighed, and kept at 4 °C for further use.

4.2.2. Thin Layer Chromatography (TLC) Analysis and Bioautography

All samples were prepared as methanol solution with concentration of 10 mg/mL of TLC analysis. The mobile phase was chloroform: methanol (volume ratio 20:1), the solid phase was Silica gel 60 F254 plate produced by Merck, and the volume was 10 μL. The plates were observed under 254 and 365 nm UV lights, stained by anisaldehyde-sulfuric acid reagent and potassium ferricyanide-ferric chloride (PFFC) reagent, respectively, or displayed for bioactive spots by DPPH free radical scavenging and AChE inhibitory bioautographies [71]. All the parallel samples were preliminarily checked for repeatability by TLC before formal experiments. All experimental results were recorded by photography.

4.2.3. Antimicrobial Assay

The screening was performed using the bilayer agar plate-Oxford cup method [72,73]. The upper agar (5 mL) containing 0.5 mL of bacterial suspension with a concentration of 1×10^8 CFU/mL. h = The Oxford cups (with inner diameter of 6 mm and outer diameter of 8 mm) were filled with 200 μL of sample (concentration = 1 mg/mL). The plates were incubated at 37 °C for 16–18 h. The results were expressed in average diameters of inhibition zones and standard deviations. The standard Muller–Hinton agar and Sabouraud agar were used for antibacterial and antifungal test, respectively (the medium of Vibrio parahaemolyticus contains 1% sodium chloride).

4.2.4. LC-MS/MS Analysis

The nominal mass resolution LC–MS/MS analyses were run on a Thermo Finnigan LC-PDA-MS/MS system equipped with PDA Plus detector and a LCQ Advantage Plus ion-trap mass spectrometer (ThermoFinnigan, San Jose, CA, USA).

All the extract samples were prepared as methanol solutions in which the dryness contents were in proportion to the total crude extract amounts for each culture by weighting, dissolving with LC-MS pure methanol, and pretreating with Agilent SPE column. The DAD detector signal collection wavelength was 190–600 nm, and the monitoring wavelength was 210 nm, 254 nm, and 280 nm. The chromatographic column is a Phenomenex Kinetex C18 100A reverse-phase chromatographic column (100 × 4.60 mm, 5 μm). The detail of the mobile phase is listed in Table 4.

Table 4. Chromatographic analysis conditions.

Injection Volume (μL)	Elution Conditions		Flow Rate (mL/min)
	Time (min)	Proportion	
25	0.00–1.00	30% ACN-H_2O	0.6
	1.00–10.00	30–99% ACN-H_2O	
	10.00–16.00	99% ACN-H_2O	
	16.00–16.20	99–30% ACN-H_2O	
	16.20–20.00	30% ACN-H_2O	

Note: The Mobile phase contained 0.1% formic acid.

Mass spectrometry detection conditions: the mass scan range set to m/z 100–2000 Da, electrospray ionization; ion source: ion source voltage: 4 kV, capillary temperature: 325 °C, normalized collision energy: 35 eV, ion transfer tube voltage: 10 V. Trigger signal intensity threshold of the secondary mass spectrum: 1×10^5 CPS (Count Per Second) for positive ion mode, and 1×10^4 CPS for negative ion mode.

4.2.5. GNPS Molecular Network Analysis

Standard pipeline for GNPS molecular networking was performed by referring to the previous reports [6,70]. The parameters for clustering and compound matching were set as: minimal matching fragments to be 4; minimal cluster size to be 2; cosine threshold to be 0.7; searching database scope to be the whole GNPS library. The data visualization is carried out with Cytoscape 3.7.2 software.

4.2.6. Multiple Natural Product Databases Mining

The multiple natural product databases mining were performed on several open accessible online databases including the PubChem (https://pubchem.ncbi.nlm.nih.gov/#, 19 January 2022), the Dictionary of Natural Products (DNP) (http://dnp.chemnetbase.com/, 19 January 2022), the NPASS (http://bidd.group/NPASS/, 19 January 2022), the Natural Product Atlas (https://www.npatlas.org/, 19 January 2022), and the Nmrdata (WeiPu) (http://www.nmrdata.com/, 19 January 2022) using deduced molecular weights with error range of targeted MW ± 1 Da (or 2 Da for chlorinated metabolites), isotopic pattern (for chlorinated metabolites), UV features, and taxon information (mainly within the genus of *Aspergillus* and expanded to the kingdom of fungi when necessary).

Supplementary Materials: The following are available online at https://www.mdpi.com/article/10.3390/antibiotics11040513/s1, Table S1. The integrated areas for the differential compounds in co-cultures compared to the axenic cultures in extracted ion chromatographies. Figures S1–S20. The MS and MS/MS spectra of featured peaks under positive or negative mode. Table S2. Multiple database mining for the peaks without remarkable yield changes in LC-MS traces between axenic and co-cultures while annotated by GNPS matching.

Author Contributions: Conceptualization, Y.Z.; Funding acquisition, Y.Z. and P.H.; Investigation, Y.W., Y.Z., E.G., Y.H., Y.L., L.Z., X.M. and X.H.; Writing—first draft by Y.W., final draft by Y.Z., review & editing by W.H.G., E.G., Y.Z., Y.L., L.Z. and Y.H. All authors have read and agreed to the published version of the manuscript.

Funding: Special project in key fields of Guangdong provincial higher education institutions (Biomedicine and health care)(2021ZDZX2064); Guangdong Provincial Special Project in Science and Technology (2021A50114); the Yangfan Talent Project of Guangdong Province (201433009); the Basic Research Project of Shenzhen Science and Technology Innovation Commission (JCYJ20190813105005619); Shenzhen Dapeng New District Scientific and Technological Research and Development Fund (KJYF202001-07); the Innovation and Development Project about Marine Economy Demonstration of Zhanjiang City (Zhan-Hai-Chuang XM-202008-01B1); the Innovative Team Program of High Education of Guangdong Province (2021KCXTD021).

Conflicts of Interest: The authors declare no conflict of interest.

References

1. Newman, D.J.; Cragg, G.M. Natural products as sources of new drugs over the 30 years from 1981 to 2010. *J. Nat. Prod.* **2012**, *75*, 311–335. [CrossRef] [PubMed]
2. Mayer, A.M.S.; Guerrero, A.J.; Rodriguez, A.D.; Taglialatela-Scafati, O.; Nakamura, F.; Fusetani, N. Marine pharmacology in 2016–2017: Marine compounds with antibacterial, antidiabetic, antifungal, anti-inflammatory, antiprotozoal, antituberculosis and antiviral activities; affecting the immune and nervous systems, and other miscellaneous mechanisms of action. *Mar. Drugs* **2021**, *19*, 49. [CrossRef] [PubMed]
3. Marmann, A.; Aly, A.H.; Lin, W.H.; Wang, B.G.; Proksch, P. Co-cultivation—A powerful emerging tool for enhancing the chemical diversity of microorganisms. *Mar. Drugs.* **2014**, *12*, 1043–1065. [CrossRef]
4. Williams, R.B.; Henrikson, J.C.; Hoover, A.R.; Lee, A.E.; Cichewicz, R.H. Epigenetic remodeling of the fungal secondary metabolome. *Org. Biomol. Chem.* **2008**, *6*, 1895–1897. [CrossRef]
5. Peng, X.Y.; Wu, J.T.; Shao, C.L.; Li, Z.Y.; Chen, M.; Wang, C.Y. Co-culture: Stimulate the metabolic potential and explore the molecular diversity of natural products from microorganisms. *Mar. Life Sci. Technol.* **2020**, *3*, 363–374. [CrossRef]
6. Rateb, M.E.; Hallyburton, I.; Houssen, W.E.; Bull, A.T.; Goodfellow, M.; Santhanam, R. Induction of diverse secondary metabolites in *Aspergillus fumigatus* by microbial co-culture. *RSC Adv.* **2013**, *3*, 14444–14450. [CrossRef]
7. Wang, M.X.; Carver, J.J.; Phelan, V.V.; Sanchez, L.M.; Garg, N.; Peng, Y.; Nguyen, D.D.; Watrous, J.; Kapono, C.A.; Luzzatto-Knaan, T.; et al. Sharing and community curation of mass spectrometry data with Global Natural Products Social Molecular Networking. *Nat. Biotechnol.* **2016**, *34*, 828–837. [CrossRef]
8. Oppong-Danquah, E.; Parrot, D.; Blumel, M.; Labes, A.; Tasdemir, D. Molecular networking-based metabolome and bioactivity analyses of marine-adapted fungi co-cultivated with phytopathogens. *Front. Microbiol.* **2018**, *9*, 2072–2108. [CrossRef]
9. Ibrahim, S.R.M.; Mohamed, G.A.; Khedr, A.I.M. γ-Butyrolactones from *Aspergillus* species: Structures, biosynthesis, and biological activities. *Nat. Prod. Commun.* **2017**, *12*, 791–800. [CrossRef]
10. Chen, S.W.; Zhang, Y.; Niu, X.T.; Mohyuddin, S.G.; Wen, J.; Bao, M.L.; Yu, T.Y.; Wu, L.Y.; Hu, C.Y.; Yong, Y.H.; et al. Coral-derived endophytic fungal product, butyrolactone-I, alleviates LPS induced intestinal epithelial cell inflammatory response through TLR4/NF-κB and MAPK signaling pathways: An in vitro and in vivo studies. *Front. Nutr.* **2021**, *8*, 748118. [CrossRef]
11. Zhang, Y.Y.; Zhang, Y.; Yao, Y.B.; Lei, X.L.; Qian, Z.J. Butyrolactone-I from coral-derived fungus *Aspergillus terreus* attenuates neuro-inflammatory response via suppression of NF-κB pathway in BV-2 cells. *Mar. Drugs* **2018**, *16*, 202. [CrossRef]
12. Hur, J.; Jang, J.; Sim, J. A review of the pharmacological activities and recent synthetic advances of γ-butyrolactones. *Int. J. Mol. Sci.* **2021**, *22*, 2769. [CrossRef]
13. Ibrahim, S.R.M.; Mohamed, G.A.; Ross, S.A. Aspernolides L and M, new butyrolactones from the endophytic fungus *Aspergillus versicolor*. *Z. Nat.* **2016**, *72*, 155–160. [CrossRef] [PubMed]
14. Luo, Y.; Ma, H.M.; Zhang, X.Y.; He, X.; Wang, W.D.; Gang, C.; Wang, H.F.; Pei, Y.H. A new butyrolactone from *Aspergillus* sp. *Chem. Nat. Compd.* **2018**, *54*, 1035–1037. [CrossRef]
15. Ma, X.X.; Liu, Y.Y.; Nie, Y.Y.; Li, Y.M.; Wang, Y.; Xue, X.Y.; Hong, P.Z.; Zhang, Y. LC-MS/MS based molecular network analysis of the effects of chemical regulation on the secondary metabolites and biological activities of a marine-derived fungal strain *Aspergillus terreus* C23-3. *Biotechnol. Bull. Shengwu Jishu Tongbao* **2021**, *37*, 27–41. [CrossRef]
16. Alberts, A.W.; Chen, J.; Kuron, G.; Hunt, V.; Huff, J.; Hoffman, C.; Rothrock, J.; Lopez, M.; Joshua, H.; Harris, E.; et al. Mevinolin: A highiy potent competitive inhibitor of hydroxymethylglutaryl-coenzyme A reductase and a cholestesterol-lowering agent. *Biochemistry* **1980**, *77*, 3957–3961. [CrossRef] [PubMed]
17. Yang, W.C.; Bao, H.Y.; Liu, Y.Y.; Nie, Y.Y.; Yang, J.M.; Hong, P.Z.; Zhang, Y. Depsidone derivatives and a cyclopeptide produced by marine Fungus *Aspergillus unguis* under chemical induction and by its plasma induced mutant. *Molecules* **2018**, *23*, 2245. [CrossRef] [PubMed]
18. Zhang, Y.; Mu, J.; Feng, Y.; Wen, L.X.; Han, J. Four chlorinated depsidones from a seaweed-derived strain of *Aspergillus unguis* and their new biological activities. *Nat. Prod. Res.* **2014**, *28*, 503–506. [CrossRef]
19. Zhang, Y.; Yang, W.C.; Nei, Y.Y.; Yang, Z.Y.; Liu, Y.Y.; Song, C.; Hong, P.Z. Application of Aspergillusidone G in the Preparation of Neuroprotective Drugs. Patent Application CN11060473A, 24 December 2019.
20. Zhang, Y.; Yang, W.C.; Nei, Y.Y.; Yang, Z.Y.; Song, C.; Hong, P.Z. Application of Depsidone Compounds in the Preparation of Neuroprotective Drugs. Patent Application CN110559290A, 13 December 2019.
21. Sureram, S.; Kesornpun, C.; Mahidol, C.; Ruchirawat, S.; Kittakoop, P. Directed biosynthesis through biohalogenation of secondary metabolites of the marine-derived fungus *Aspergillus unguis*. *RSC Adv.* **2013**, *3*, 1781–1788. [CrossRef]
22. Ibrahim, S.R.M.; Mohamed, G.A.; Al Haidari, R.A.; El-Kholy, A.A.; Zayed, M.F.; Khayat, M.T. Biologically active fungal depsidones: Chemistry, biosynthesis, structural characterization, and bioactivities. *Fitoterapia* **2018**, *129*, 317–365. [CrossRef]
23. Sureram, S.; Wiyakrutta, S.; Ngamrojanavanich, N.; Mahidol, C.; Ruchirawat, S.; Kittakoop, P. Depsidones, aromatase inhibitors and radical scavenging agents from the marine-derived fungus *Aspergillus unguis* CRI282-03. *Planta Med.* **2012**, *78*, 582–588. [CrossRef] [PubMed]
24. Sadorn, K.; Saepua, S.; Bunbamrung, N.; Boonyuen, N.; Komwijit, S.; Rachtawee, P.; Pittayakhajonwut, P. Diphenyl ethers and depsidones from the endophytic fungus *Aspergillus unguis* BCC54176. *Tetrahedron* **2022**, *105*, 132612. [CrossRef]
25. Saetang, P.; Rukachaisirikul, V.; Phongpaichit, S.; Preedanon, S.; Sakayaroj, J.; Hadsadee, S.; Jungsuttiwong, S. Antibacterial and antifungal polyketides from the fungus *Aspergillus unguis* PSU-MF16. *J. Nat. Prod.* **2021**, *84*, 1498–1506. [CrossRef] [PubMed]

26. Gyetvai, A.; Emri, T.; Takacs, K.; Dergez, T.; Fekete, A.; Pesti, M.; Pocsi, I.; Lenkey, B. Lovastatin possesses a fungistatic effect against *Candida albicans*, but does not trigger apoptosis in this opportunistic human pathogen. *FEMS. Yeast Res.* **2006**, *6*, 1140–1148. [CrossRef] [PubMed]
27. Qiao, J.; Kontoyiannis, D.P.; Wan, Z.; Li, R.; Liu, W. Antifungal activity of statins against *Aspergillus* species. *Med. Mycol.* **2007**, *45*, 589–593. [CrossRef]
28. Huang, Z.; Nong, X.; Zhe, R.; Wang, J.; Zhang, X.; Qi, S. Anti-HSV-1, antioxidant and antifouling phenolic compounds from the deep-sea-derived fungus *Aspergillus versicolor* SCSIO 41502. *Bioorg. Med. Chem. Lett.* **2017**, *27*, 787–791. [CrossRef]
29. Johnson, L.J.; Koulman, A.; Christensen, M.; Lane, G.A.; Fraser, K.; Forester, N.; Johnson, R.D.; Bryan, G.T.; Rasmussen, S. An extracellular siderophore is required to maintain the mutualistic interaction of *Epichloe festucae* with *Lolium perenne*. *PLoS Pathog.* **2013**, *9*, 332–351. [CrossRef]
30. Bunbamrung, N.; Intaraudom, C.; Dramae, A.; Komwijit, S.; Laorob, T.; Khamsaeng, S.; Pittayakhajonwut, P. Antimicrobial, antimalarial and anticholinesterase substances from the marine-derived fungus *Aspergillus terreus* BCC51799. *Tetrahedron* **2020**, *76*, 131496. [CrossRef]
31. Yoshimi, A.; Yamaguchi, S.; Fujioka, T.; Kawai, K.; Gomi, K.; Machida, M.; Abe, K. Heterologous production of a novel cyclic peptide compound, KK-1, in *Aspergillus oryzae*. *Front. Microbiol.* **2018**, *9*, 690–702. [CrossRef]
32. Nong, X.H.; Wang, Y.F.; Zhang, X.Y.; Zhou, M.P.; Xu, X.Y.; Qi, S.H. Territrem and butyrolactone derivatives from a marine-derived fungus *Aspergillus terreus*. *Mar. Drugs.* **2014**, *12*, 6113–6124. [CrossRef]
33. Nie, Y.Y. Screening and Secondary Metabolite Research of Seaweeds and Marine Fungi with Anti-Alzheimer's Disease Related Activity. Master's Thesis, Guangdong Ocean University, Zhanjiang, China, 3 December 2017.
34. Fukuda, T.; Kurihara, Y.; Kanamoto, A.; Tomoda, H. Terretonin G, a new sesterterpenoid antibiotic from marine-derived *Aspergillus* sp. OPMF00272. *J. Antibiot.* **2014**, *67*, 593–595. [CrossRef] [PubMed]
35. Liu, J.; Wei, X.Y.; Kim, E.L.; Lin, X.P.; Yang, X.W.; Zhou, X.F.; Yang, B.; Jung, J.H.; Liu, Y.H. New glucosidated pyrazinoquinazoline indole alkaloids from fungus *Aspergillus fumigatus* derived of a jellyfish. *Tetrahedron* **2015**, *71*, 271–275. [CrossRef]
36. Gombodorj, S.; Yang, M.H.; Shang, Z.C.; Liu, R.H.; Li, T.X.; Yin, G.P.; Kong, L.Y. New phenalenone derivatives from *Pinellia ternata* tubers derived *Aspergillus* sp. *Fitoterapia* **2017**, *120*, 72–78. [CrossRef] [PubMed]
37. Katsuhiro, S.A.K.; Hiroshi, Y.; Shinji, F.; Takaaki, N.; Taizo, N. NF 00659A_1, A_2, A_3, B_1 and B_2, novel antitumor antibiotics produced by *Aspergillus* sp. NF 00659. *J. Antibiot.* **1997**, *4*, 318–324. [CrossRef]
38. Zhang, Y.; Li, X.M.; Wang, B.G. Nigerasperones A~C, new monomeric and dimeric naphtho-γ-pyrones from a marine alga-derived endophytic fungus *Aspergillus niger* EN-13. *J. Antibiot.* **2007**, *60*, 204–210. [CrossRef]
39. Shaaban, M.; Shaaban, K.A.; Abdel-Aziz, M.S. Seven naphtho-γ-pyrones from the marine derived fungus *Alternaria alternata*: Structure elucidation and biological properties. *Org. Med. Chem. Lett.* **2012**, *2*, 6–14. [CrossRef]
40. Priestap, H.A. New naphththopyrones from *Aspergillus fonsecaeus*. *Tetrahedron* **1984**, *40*, 3617–3624. [CrossRef]
41. Liao, G.F.; Wu, P.; Xue, J.H.; Liu, L.; Li, H.X.; Wei, X.Y. Asperimides A-D, anti-inflammatory aromatic butenolides from a tropical endophytic fungus *Aspergillus terreus*. *Fitoterapia* **2018**, *131*, 50–54. [CrossRef]
42. Steyn, P.S. Austamide, a new toxic metabolite from *Aspergillus ustus*. *Tetrahedron Lett.* **1971**, *12*, 3331–3334. [CrossRef]
43. Malmstrøm, J. Unguisins A and B: New cyclic peptides from the marine-derived fungus *Emericella unguis*. *J. Nat. Prod.* **1999**, *62*, 787–789. [CrossRef]
44. Shiono, Y.; Miyazaki, N.; Murayama, T.; Koseki, T.; Harizon; Katia, D.G.; Supratman, U.; Nakata, J.; Kakihara, Y.; Saeki, M.; et al. GSK-3β inhibitory activities of novel dichroloresorcinol derivatives from *Cosmospora vilior* isolated from a mangrove plant. *Phytochem. Lett.* **2016**, *18*, 122–127. [CrossRef]
45. Rukachaisirikul, V.; Satpradit, S.; Klaiklay, S.; Phongpaichit, S.; Borwornwiriyapan, K.; Sakayaroj, J. Polyketide anthraquinone, diphenyl ether, and xanthone derivatives from the soil fungus *Penicillium* sp. PSU-RSPG99. *Tetrahedron* **2014**, *70*, 5148–5152. [CrossRef]
46. Phainuphong, P.; Rukachaisirikul, V.; Phongpaichit, S.; Sakayaroj, J.; Kanjanasirirat, P.; Borwornpinyo, S.; Akrimajirachoote, N.; Yimnual, C.; Muanprasat, C. Depsides and depsidones from the soil-derived fungus *Aspergillus unguis* PSU-RSPG204. *Tetrahedron* **2018**, *74*, 5691–5699. [CrossRef]
47. Elsbaey, M.; Sallam, A.; El-Metwally, M.; Nagata, M.; Tanaka, C.; Shimizu, K.; Miyamoto, T. Melanogenesis inhibitors from the endophytic fungus *Aspergillus amstelodami*. *Chem. Biodivers.* **2019**, *16*, e1900237. [CrossRef]
48. Bertrand, S.; Bohni, N.; Schnee, S.; Schumpp, O.; Gindro, K.; Wolfender, J.L. Metabolite induction via microorganism co-culture: A potential way to enhance chemical diversity for drug discovery. *Biotechnol. Adv.* **2014**, *32*, 1180–1204. [CrossRef]
49. Sweany, R.R.; Mack, B.M.; Moore, G.G.; Gilbert, M.K.; Cary, J.W.; Lebar, M.D.; Rajasekaran, K.; Daman, K.E., Jr. Genetic responses and aflatoxin inhibition during co-culture of aflatoxigenic and non-aflatoxigenic *Aspergillus flavus*. *Toxins* **2021**, *13*, 794. [CrossRef]
50. Arora, D.; Gupta, P.; Jaglan, S.; Roullier, C.; Grovel, O.; Bertrand, S. Expanding the chemical diversity through microorganisms co-culture: Current status and outlook. *Biotechnol. Adv.* **2020**, *40*, 734–750. [CrossRef]
51. Boruta, T.; Scigaczewska, A.; Bizukojc, M. "Microbial Wars" in a stirred tank bioreactor: Investigating the co-cultures of *Streptomyces rimosus* and *Aspergillus terreus*, filamentous microorganisms equipped with a rich arsenal of secondary metabolites. *Front. Bioeng. Biotechnol.* **2021**, *9*, 713639. [CrossRef]
52. Liu, C.; Kakeya, H. Cryptic chemical communication: Secondary metabolic responses revealed by microbial co-culture. *Chem. Asian J.* **2020**, *15*, 327–337. [CrossRef]

53. Fischer, J.; Muller, S.Y.; Netzker, T.; Jager, N.; Gacek-Mattews, A.; Pezzini, F.; Schoeler, H.; Reichelt, M.; Gershenzon, J.; Krespach, M.K.; et al. Chromatin mapping identifies BasR, a key regulator of bacteria-triggered production of fungal secondary metabolites. *eLife* **2018**, *7*, 69–104. [CrossRef]
54. Glauser, G.; Gindro, K.; Fringeli, J.; Joffrey, D.J.P.; Rudaz, S.; Wolfender, J.L. Differential analysis of mycoalexins in confrontation zones of grapevine fungal pathogens by ultrahigh pressure liquid chromatography/time-of-flight mass spectrometry and capillary nuclear magnetic resonance. *J. Agric. Food Chem.* **2009**, *57*, 1127–1134. [CrossRef] [PubMed]
55. Yang, X.L.; Awakawa, T.; Wakimoto, T.; Abe, I. Induced biosyntheses of a novel butyrophenone and two aromatic polyketides in the plant pathogen *Stagonospora nodorum*. *Nat. Prod. Bioprospect.* **2013**, *3*, 141–144. [CrossRef]
56. Ariawan, A.D.; Webb, J.E.A.; Howe, E.N.W.; Gale, P.A.; Thordarson, P.; Hunter, L. Cyclic peptide unguisin A is an anion receptor with high affinity for phosphate and pyrophosphate. *Org. Biomol. Chem.* **2017**, *15*, 2962–2967. [CrossRef] [PubMed]
57. Lian, Z.R.; Wang, J.T. Competition between prorocentrum donghaiense and heterotrophic bacteria for phosphate. *Acta Hydrobiol. Sin. Shuisheng Shengwu Xuebao* **2010**, *36*, 663–668. [CrossRef]
58. Chiu, C.H.; Paszkowski, U. Mechanisms and impact of symbiotic phosphate acquisition. *Cold Spring Harbor Perspect. Biol.* **2019**, *11*, 603–636. [CrossRef]
59. Yang, Y.; Li, B.Q.; Cheng, C.P.; Liang, J.R.; Yang, Q.L.; Gao, Y.H. Interspecies competition between *Woloszynskia* sp. and *Alexandrium tamarense* based on phosphate concentration and initial cell density. *J. Xiamen Univ. Xiamen Daxue Xuebao* **2008**, *47*, 163–172.
60. Yamanaka, K.; Oikawa, H.; Ogawa, H.; Hosono, K.; Shinmachi, F.; Takano, H.; Sakuda, S.; Beppu, T.; Ueda, K. Desferrioxamine E produced by *Streptomyces griseus* stimulates growth and development of *Streptomyces tanashiensis*. *Microbiology* **2005**, *151*, 2899–2905. [CrossRef]
61. Macreadie, I.G.; Johnson, G.; Schlosser, T.; Macreadie, P.I. Growth inhibition of *Candida* species and *Aspergillus fumigatus* by statins. *FEMS Microbiol. Lett.* **2006**, *262*, 9–13. [CrossRef]
62. Tavakkoli, A.; Johnston, T.P.; Sahebkar, A. Antifungal effects of statins. *J. Pharmacol. Ther.* **2020**, *208*, 83–123. [CrossRef]
63. Boruta, T.; Milczarek, I.; Bizukojc, M. Evaluating the outcomes of submerged co-cultivation: Production of lovastatin and other secondary metabolites by *Aspergillus terreus* in fungal co-cultures. *Appl. Microbiol. Biotechnol.* **2019**, *103*, 5593–5605. [CrossRef]
64. Boruta, T.; Marczyk, A.; Rychta, K.; Przydacz, K.; Bizukojc, M. Confrontation between *Penicillium rubens* and *Aspergillus terreus*: Investigating the production of fungal secondary metabolites in submerged co-cultures. *J. Biosci. Bioeng.* **2020**, *130*, 503–513. [CrossRef] [PubMed]
65. Wang, C.; Lu, H.; Lan, J.; Zaman, K.H.A.; Cao, S. A review: Halogenated compounds from marine fungi. *Molecules* **2021**, *26*, 458. [CrossRef] [PubMed]
66. Dean, F.M.; Robertson, A.; Roberts, J.C.; Rapeb, K.B. Nidulin and 'ustin': Two chlorine-containing metabolic products of *Aspergillus nidulans*. *Nature* **1953**, *172*, 344. [CrossRef] [PubMed]
67. Kawahara, N.; Nakajima, S.; Satoh, Y.; Yamazaki, M.; Kawai, K. Studies on fungal products. XVIII. Isolation and structures of a new fungal depsidone related to nidulin and a new phthalide from *Emericella unguis*. *Chem. Pharm. Bull.* **1988**, *36*, 1970–1975. [CrossRef]
68. Isaka, M.; Yangchum, A.; Supothina, S.; Veeranondha, S.; Komwijit, S.; Phongpaichit, S. Semisynthesis and antibacterial activities of nidulin derivatives. *J. Antibiot.* **2019**, *72*, 181–184. [CrossRef]
69. Vala, A.K.; Dave, B.P.; Dube, H.C. Chemical characterization and quantification of siderophores produced by marine and terrestrial *Aspergilli*. *Can. J. Microbiol.* **2006**, *52*, 603–607. [CrossRef]
70. Renshaw, J.C.; Robson, G.D.; Trinci, A.P.J.; Wiebe, M.G.; Livens, F.R.; Collison, D.; Taylor, R.J. Fungal siderophores: Structures, functions and applications. *Mycol. Res.* **2002**, *106*, 1123–1142. [CrossRef]
71. Nie, Y.Y.; Yang, W.C.; Liu, Y.Y.; Yang, J.M.; Lei, X.L.; Gerwick, W.H.; Zhang, Y. Acetylcholinesterase inhibitors and antioxidants mining from marine fungi: Bioassays, bioactivity coupled LC–MS/MS analyses and molecular networking. *Mar. Life Sci. Technol.* **2020**, *2*, 386–397. [CrossRef]
72. Magaldi, S.; Mata-Essayag, S.; Hartung de Capriles, C.; Perez, C.; Colella, M.T.; Olaizola, C.; Ontiveros, Y. Well diffusion for antifungal susceptibility testing. *Int. J. Infect. Dis.* **2004**, *8*, 39–45. [CrossRef]
73. Panda, S.K. Ethno-medicinal uses and screening of plants for antibacterial activity from Similipal Biosphere Reserve, Odisha, India. *J. Ethnopharmacol.* **2014**, *151*, 158–175. [CrossRef]

antibiotics

MDPI

Article

New Antibacterial Secondary Metabolites from a Marine-Derived *Talaromyces* sp. Strain BTBU20213036

Fuhang Song [1], Yifei Dong [1], Shangzhu Wei [2], Xinwan Zhang [2], Kai Zhang [1] and Xiuli Xu [2,*]

[1] School of Light Industry, Beijing Technology and Business University, Beijing 100048, China; songfuhang@btbu.edu.cn (F.S.); dongyifei1930201016@st.btbu.edu.cn (Y.D.); zhangkai2030302071@st.btbu.edu.cn (K.Z.)
[2] School of Ocean Sciences, China University of Geosciences, Beijing 100083, China; 2111180016@cugb.edu.cn (S.W.); zhangxinwan@cugb.edu.cn (X.Z.)
* Correspondence: xuxl@cugb.edu.cn

Abstract: New polyketide-derived oligophenalenone dimers, bacillisporins K and L (**1** and **2**) and xanthoradone dimer rugulosin D (**3**), together with four known compounds, bacillisporin B (**4**), macrosporusone D (**5**), rugulosin A and penicillide (**6** and **7**), were isolated from the marine-derived fungus *Talaromyces* sp. BTBU20213036. Their structures were determined by detailed analysis of HRESIMS, 1D and 2D NMR data, and the absolute configurations were determined on the basis of calculated and experimental electronic circular dichroism (ECD). The antibacterial and antifungal activities of these compounds were tested against Gram-positive—*Staphylococcus aureus*, Gram-negative—*Escherichia coli*, and fungal strain—*Candida albicans*. These compounds showed potential inhibitory effects against *S. aureus* with minimum inhibitory concentrations ranging from 0.195 to 100 μg/mL.

Keywords: marine-derived fungus; *Talaromyces* sp.; antibacterial; *Staphylococcus aureus*; polyketide

Citation: Song, F.; Dong, Y.; Wei, S.; Zhang, X.; Zhang, K.; Xu, X. New Antibacterial Secondary Metabolites from a Marine-Derived *Talaromyces* sp. Strain BTBU20213036. *Antibiotics* **2022**, *11*, 222. https://doi.org/10.3390/antibiotics11020222

Academic Editors: William N. Setzer and Jesus Simal-Gandara

Received: 14 January 2022
Accepted: 8 February 2022
Published: 10 February 2022

1. Introduction

The discovery and introduction into clinics of antibiotics have made great contributions to human health. The most widely applied antibiotics in clinics were discovered in the period between the 1950s and 1960s, while the overuse of antibiotics lead to the emergence of drug-resistant bacteria, which is recognized globally by scientists nowadays [1,2]. The spread of multi-drug resistant microorganisms is getting more and more serious to humans [3–5]. It is estimated that 700,000 people died as a result of infections caused by antimicrobial resistant bacteria annually [6]. There is an urgent demand to screen new antibiotics in order to combat the infective diseases caused by drug-resistant bacteria.

Fungi from marine environments have proven to be important pools for structurally unique and biologically diverse natural products [7]. *Talaromyces* fungi belong to ascomycetous. A number of *Talaromyces* species have been isolated from marine environments, such as, *Talaromyces albobiverticillius*, *Talaromyces assiutensis*, *Talaromyces purpureogenus* [8–12]. Fungi of *Talaromyces* genus isolated from marine environments produced a series of bioactive natural products, such as oligophenalenones [11–15], terpenoids [16,17], naphthoquinones [18], spolyene and isocoumarin [19], diphenyl ether derivatives, sesquiterpene-conjugated amino acids [20,21], lactones [22], and ergosterol analog and bisanthraquinone [23]. In the course of our continuing investigation of bioactive natural products from marine-derived fungi [24–26], the fungal strain *Talaromyces* sp. BTBU20213036, which was obtained from a mud sample collected from the coastal region of Qingdao, Shandong Province, exhibited antimicrobial activity against *Staphylococcus aureus*. Fermentation scale-up of this strain was conducted in rice solid media. The study of the chemical constituents of the fermentation materials resulted in the isolation and characterization of three new secondary metabolites, including bacillisporins K and L (**1** and **2**) and rugulosin D (**3**), together with four known compounds, bacillisporin B [15], macrosporusone D [27], rugulosin A [28] and penicillide [29] (Figure 1). **1**, **2** and **4**–**6** showed potential antibacterial activities against *S. aureus*. Herein we report the details of isolation, structure elucidation, and antimicrobial activities evaluation of these compounds.

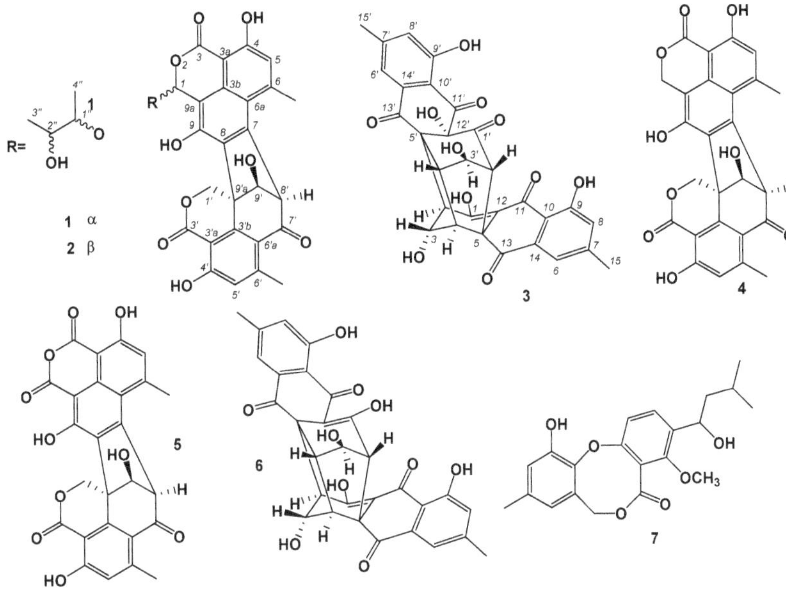

Figure 1. Chemical structures of **1–7**.

2. Results

Compound **1** was isolated as a light yellow amorphous powder. The molecular formula of **1** was determined to be $C_{30}H_{26}O_{11}$ based on high resolution electrospray ionization mass spectrum (HRESIMS) (*m/z* $[M + H]^+$ 585.1374, calcd for $C_{30}H_{27}O_{11}$, 585.1367), accounting for eighteen degrees of unsaturation (Figure S1). Figure S2 showed the High Performance Liquid Chromatography (HPLC) profile and ultraviolet (UV) spectrum of **1**. The ^1H, ^{13}C and Heteronuclear Single Quantum Correlation (HSQC) spectra of **1** (Figures S3–S5, Table 1) showed the presence of four methyl groups [δ_H 2.98/δ_C 24.4 (6-Me), δ_H 2.48/δ_C 23.2 (6′-Me), δ_H 0.75/δ_C 15.6 (C-3″), δ_H 0.99/δ_C 17.2 (C-4″)], one oxygenated methylene group [δ_H 5.12 and 4.95/δ_C 69.9 (C-1′)], five sp^3 methine groups [δ_H 6.86/δ_C 98.8 (C-1), δ_H 4.83/δ_C 64.9 (C-8′), δ_H 4.77/δ_C 85.1 (C-9′), δ_H 4.12/δ_C 78.6 (C-1″), δ_H 3.72/δ_C 68.9 (C-2″)], two aromatic methines [δ_H 6.96/δ_C 118.4 (C-5), δ_H 6.83/δ_C 119.6 (C-5′)], one sp^3 quaternary carbon [δ_C 49.5 (C-9′a)], as well as seventeen sp^2 quaternary carbons including one ketone carbonyl [δ_C 192.5 (C-7′)] and two lactone carboxyls [δ_C 168.3 (C-3), 167.8 (C-3′)]. ^1H-^1H Correlation Spectroscopy (COSY) spectrum (Figure 2 and Figures S6) indicated the side chain of C-3″/C-2″/C-1″/C-4″. By comparing the NMR data with those of **5**, one of the lactones was replaced by the acetal methine [δ_H 6.86 (s)/δ_C 98.8 (C-1)] and the linkage between C-1 and C-1″ through the oxygen atom was confirmed by Heteronuclear Multiple Bond Correlation (HMBC) correlations (Figure 2 and Figure S7, Table S1) from H-1 to C-3 and C-1″ and from H-1″ to C-1. The structure of **1** was further established by ^1H-^1H COSY (Figure 2 and Figure S6) and HMBC experiments. The relative configurations were deduced by the singlet peaks for H-8′ [δ_H 4.83 (s)] and H-9′ [δ_H 4.77 (brs)] and Rotating Frame Overhauser Spectroscopy (ROESY) correlation between H-8′ and H-9′ (Figure S8).

Table 1. ^{1}H (500 MHz) and ^{13}C NMR (125 MHz) data of **1**, **2** and **5** (in DMSO).

Position	1		2		5	
	δ_C	δ_H (J in Hz)	δ_C	δ_H (J in Hz)	δ_C	δ_H (J in Hz)
1	98.8	6.86, s	99.5	6.81, s	165.1	
3	168.3		168.1		164.4	
3a	96.4		96.2		98.4	
3b	130.3		130.2		134.6	
4	162.5		162.4		162.5	
5	118.4	6.96, s	118.8	6.97, s	120.1	7.14, s
6	146.7		146.6		148.1	
6a	118.4		118.4		118.0	
7	139.6		139.8		149.0	
8	134.3		133.8		132.5	
9	152.2		152.4		160.8	
9a	109.0		109.4		100.3	
1′	69.9	5.12, d (12.0) 4.95, d (12.0)	69.7	5.12, d (12.0) 5.00, d (12.0)	69.4	5.18, d (12.5) 5.00, d (12.5)
3′	167.8		167.9		167.6	
3′a	103.7		103.7		104.0	
3′b	147.6		147.8		147.1	
4′	163.1		163.3		163.5	
5′	119.6	6.83, s	119.7	6.83, s	120.1	6.87, s
6′	152.2		152.4		152.5	
6′a	116.6		116.7		116.6	
7′	192.5		192.5		191.2	
8′	64.9	4.83, s	64.6	4.87, d (1.0)	65.5	4.99, s
9′	85.1	4.77, br s	85.6	4.78, d (5.0)	85.2	4.85, s
9′a	49.5		49.7		49.4	
Me-6	24.4	2.98, s	24.6	2.99, s	24.5	3.06, s
Me-6′	23.2	2.48, s	23.2	2.47, s	23.2	2.48, s
1″	78.6	4.12, m	80.2	3.90, m		
2″	68.9	3.72, m	71.3	3.61, m		
3″	15.6	0.75, d (6.5)	18.9	1.14, d (6.5)		
4″	17.2	0.99, d (6.5)	17.9	1.11, d (6.5)		
OH-9′		6.24, d (3.0)		6.31, d (3.0)		

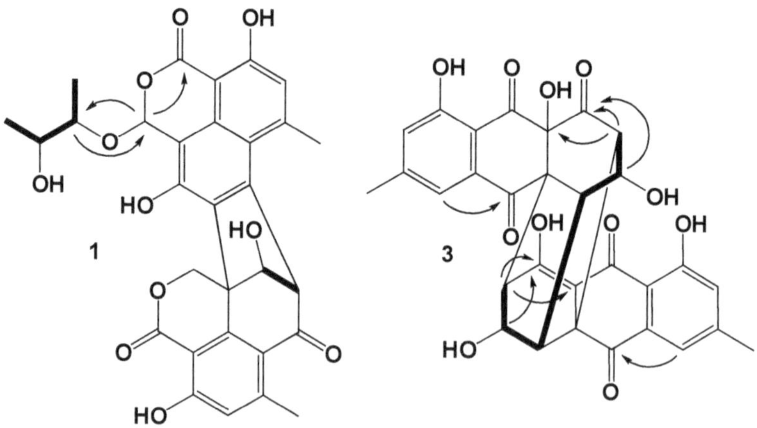

Figure 2. Key ^{1}H-^{1}H COSY and HMBC correlations for **1** and **3**.

Compound **2** was isolated as a light yellow amorphous powder. The molecular formula of **2** was determined to be $C_{30}H_{26}O_{11}$ based on the HRESIMS spectrum (*m*/*z* $[M + H]^+$ 585.1377, calcd for $C_{30}H_{27}O_{11}$, 585.1367), accounting for eighteen degrees of unsaturation (Figure S9). Figure S10 showed the HPLC profile and UV spectrum of **2**. The ^{1}H, ^{13}C and HSQC spectra of **1** (Figures S11–S13, Table 1) showed similar data to those of **2**. The differences are signals for acetal methine [δ_H 6.81/δ_C 99.6 (C-1)] and the substructure attached to C-1 [δ_H 3.90/δ_C 80.2 (C-1″), 3.61/δ_C 71.3 (C-2″), 1.14/δ_C 18.9 (C-3″), 1.11/δ_C 17.9 (C-4″)]. These data revealed the configurations of C-1, C-1″, C-2″ were different from those of **2**, which resulted in the different deshielding effects from the aromatic moiety. Furthermore, the structure was characterized by detailed analysis of 2D NMR spectra (Figure 2 and Figures S13–S15). In the REOSY spectrum (Figure S16), the crossing peaks between H-8′ and H-9′-OH, and between H-9′ and H-1′a revealed the relative configurations of C-8′, 9′ and C-9a′ (Table S2).

Compounds **1** and **2** showed almost the same experimental ECD spectra, which were consistent with the reported bacillisporin I [15] and calculated data (Figure 3). Thus, the configurations of **1** and **2** were determined as 8′*R*, 9′*S*, 9′a*S*, while the configurations of C-1, C-1″ and C-2″ were not determined. Compounds **1** and **2** were named bacillisporins K and L, respectively.

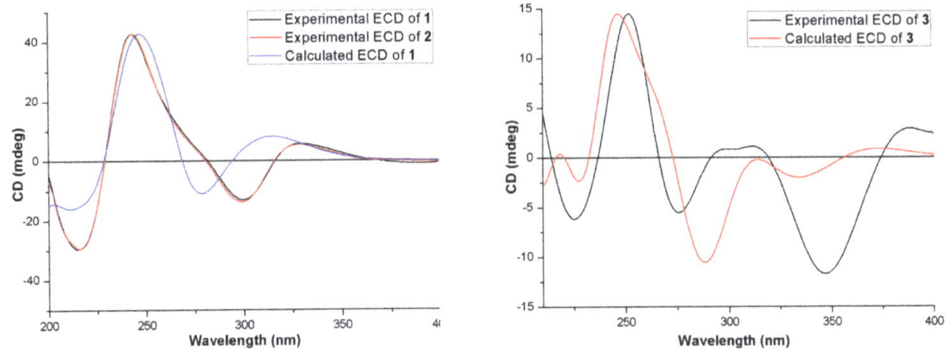

Figure 3. Experimental and calculated ECD spectra of **1**, **2** and **3**.

Compound **3** was isolated as a brown amorphous powder. The molecular formula of **3** was determined to be $C_{30}H_{22}O_{11}$ based on the HRESIMS spectrum (*m/z* [M + H]$^+$ 559.1234, calcd for $C_{30}H_{23}O_{11}$, 559.1235), accounting for twenty degrees of unsaturation (Figure S17). Figure S18 showed the HPLC profile and UV spectrum of **3**. The ^{1}H, ^{13}C, HSQC and ^{1}H- 1H COSY spectra of **2** (Figures S19–S22, Table 2) showed signals for two singlet methyl groups [δ_H 2.44/δ_C 21.6 (C-15), δ_H 2.43/δ_C 21.5 (C-15′)], six sp^3 methine groups [δ_H 2.73/δ_C 55.7 (C-2), δ_H 2.90/δ_C 63.4 (C-2′), δ_H 4.27/δ_C 70.2 (C-3), δ_H 4.56/δ_C 69.0 (C-3′), δ_H 3.46/δ_C 48.1 (C-4), δ_H 3.73/δ_C 44.0 (C-4′)], four aromatic methines [δ_H 7.46/δ_C 120.6 (C-6), δ_H 7.41/δ_C 120.1 (C-6′), 7.24/δ_C 124.0 (C-8), δ_H 7.21/δ_C 123.8 (C-8′)], three sp^3 quaternary carbons including one oxygenated carbons [δ_C 74.6 (C-12′)], five ketone carbonyls [[δ_C 198.8 (C-1′), 184.8 (C-11), [δ_C 192.1 (C-11′), 193.0 (C-13), 192.8 (C-13′)], as well as ten sp^2 quaternary carbons [δ_C 178.1 (C-1), 148.5 (C-7)/148.6 (C-7′), 160.9 (C-9)/161.0 (C-9′), 114.3 (C-10)/113.3 (C-10′), 106.8 (C-12), 132.3 (C-14)/133.5 (C-14′)]. By comparing the NMR data with those of rugulosin A [28], the structure was deduced as an analogue of rugulosin A (**6**). Detailed analysis of the NMR data revealed that the sp^2 quaternary carbons of C-1′ [δ_C 186.7] and C-12′ [δ_C 106.8] in rugulosin A were replaced by one ketone carbonyl [δ_C 198.8] and one oxygenated sp^3 quaternary carbon [δ_C 74.6]. The planar structure of **3** further confirmed by HMBC correlations (Figure 2 and Figure S23, Table S3) from H-2′ to C-1′ and C-12′. The relative configurations of **3** were deduced by comparing the literature data for ^{1}H NMR between rugulosin A and **3**, the chemical shifts of H-3/H-3′ of **3** were δ_H 4.27/4.56 with a coupling constant of 5.0 and 4.5 Hz, which were almost the same as those reported for rugulosin A [28]. In the ROESY spectrum (Figure 4 and Figure S24), correlations between H-3′ and H-2′, H-4/H-4′, and between H-3 and H-2 and H-4′ were observed, which confirmed the relative stereochemistry of **3**. By comparison of experimental and calculated ECD spectra (Figure 4), the absolute configurations of **3** were determined as shown in Figure 1 and named rugulosin D.

Table 2. ^{1}H (500 MHz) and ^{13}C NMR (125 MHz) NMR data of **3** and **6** (in DMSO).

Position	3		6 [16]	
	δ_C	δ_H (*J* in Hz)	δ_C	δ_H (*J* in Hz)
1/1′	178.1/198.8		186.7	
2/2′	55.7/63.4	2.73, d (5.0)/2.90, d (4.5)	59.0	2.77, d (6.0)
3/3′	70.2/69.0	4.27, dd (5.0, 3.0)/4.56, dd (4.5, 4.0)	69.2	4.38, (dd, 6.0, 2.3)
4/4′	48.1/44.0	3.46, brs/3.73, brs	48.4	3.36, brs
5/5′	53.6/63.9		56.3	
6/6′	120.6/120.1	7.46, s/7.41, s	121.2	7.44, d (1.2)
7/7′	148.5/148.6		148.3	
8/8′	124.0/123.8	7.24, s/7.21, s	124.7	7.18, d (1.2)
9/9′	160.9/161.0		160.8	
10/10′	114.3/113.3		114.8	
11/11′	184.8/192.1		181.7	
12/12′	106.8/74.6		106.8	
13/13′	193.0/192.8		194.6	
14/14′	132.3/133.5		132.7	
15/15′	21.6/21.5	2.44, s/2.43, s	22.2	2.41, s
9-OH/	9-OH′	11.71, s/11.04, s		11.4, s

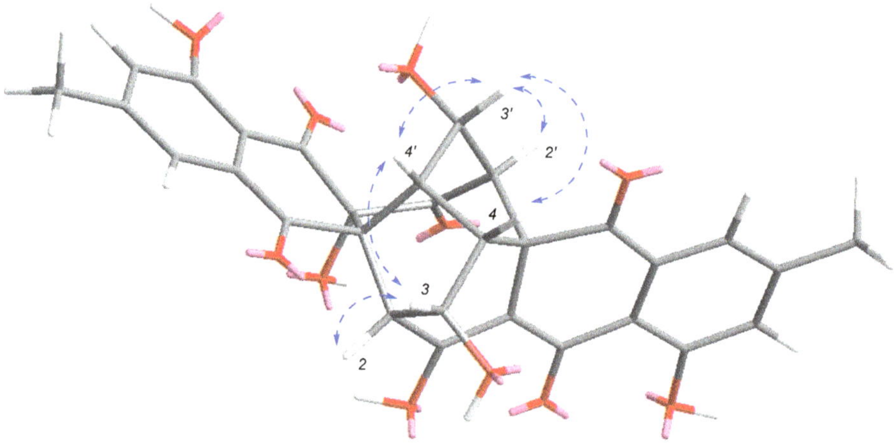

Figure 4. Key ROESY correlations of **3**.

Four known compounds, bacillisporin B [15], macrosporusone D [27], rugulosin A [28] and penicillide [29] were also isolated and characterized by comparing their molecular weight and NMR data with those reported in the literature.

These compounds were tested for antibacterial activities against a panel of pathogens of *S. aureus* ATCC 25923, *Escherichia coli* ATCC 25923 and *Candida albicans* ATCC 10231. Compounds **1**, **2** and **4**–**6** strongly inhibited the growth of *S. aureus* with MIC values of 12.5, 25, 12.5, 6.25, and 0.195 μg/mL (Table 3). None of the tested compounds showed inhibitory effects against *C. albicans* and *E. coli* at concentration of 100 μg/mL.

Table 3. Antibacterial activity of compounds **1–7** (MIC, μg/mL).

Number	1	2	3	4	5	6	7	Control
C. albicans	>100	>100	>100	>100	>100	>100	>100	1 [a]
S. aureus	12.5	25	>100	12.5	6.25	0.195	100	1 [b]
E. coli	>100	>100	>100	>100	>100	>100	>100	1 [c]

[a] Rapamycin, [b] Vancomycin, [c] Ciprofloxacin.

3. Materials and Methods

3.1. General Experimental Procedures

Optical rotations ($[\alpha]_D$) were measured by using an Anton Paar MCP 200 Modular Circular Polarimeter (Austria) in a 100 × 2 mm cell at 25 °C. CD spectra were recorded on Applied Photophysics Chirascan spectropolarimeter (Surrey, UK). NMR experiments were carried on a Bruker Avance 500 spectrometer at 25 °C with residual solvent peaks as references (DMSO-d_6: δ_H 2.50, δ_C 39.52). High resolution ESIMS spectra were measured using an Accurate-Mass-Q-TOF LC/MS 6520 instrument (Santa Clara, CA, USA) in positive ion mode. HPLC was run on an Agilent 1200 Series instrument.

3.2. Microbial Material

Strain BTBU20213036 was isolated from a mud sample collected from the intertidal zones of the Yellow Sea in Qingdao, China, and grown on a potato dextrose agar plate at 28 °C for 10 days. Colonies were about 25 mm diam, texture floccose and funiculose, sporulation abundant, dark greyish green, mycelium yellow, no exudate and soluble pigment, colony reverse brown (Figure S25). The genomic DNA of BTBU20213036 was extracted using DNAquick Plant System (Tiangen, Beijing, China). The ITS sequence of BTBU20213036 was amplified by using a conventional primer pair of ITS4 (5′-TCCTCCGCTTATTGATATGC-3′) and ITS5 (5′-GGAAGTAAAAGTCGTAACAAGG-3′). PCR products were sequenced by Beijing Qingke Biotechnology Co., Ltd. (Beijing, China) and the sequence was deposited in GenBank (accession number, OM049426). Strain BTBU20213036 was identified as *Talaromyces* sp. based on gene sequence analysis of ITS by comparing with sequences from GenBank database using BLAST program (Figure S26). Alignments and calculations of sequence similarity were carried out using CLUSTAL W [30]. The strain was deposited in Beijing Technology and Business University, Beijing, China.

3.3. Fermentation, Extraction and Purification

Talaromyces sp. BTBU20213036 was inoculated on a potato dextrose agar plate and incubated at 28 °C for 7 days. A slice of fungal colony of 1 cm^2 was put into twenty of 1 L conical flasks, each containing 200 g of raw rice, which was soaked in distilled water for 60 min. The inoculated flasks were incubated stationary at 28 °C for 30 days. The fermented materials of *Talaromyces* sp. BTBU20213036 were extracted three times by EtOAc:MeOH (80:20), and the organic solvent was evaporated in vacuo at 45 °C to yield brown crude extract (18.4 g). The crude extract was resuspended into 500 mL distilled water and extracted by 500 mL EtOAc (three times). Then EtOAc was evaporated *in vacuo* at 45 °C to give a dark residue (5.91 g). The EtOAc extract was separated by a reduced pressure silica gel chromatography (50 × 80 mm column, TLC H silica) with a stepwise gradient of 80–100% hexane/CH_2Cl_2 and then 0–90% MeOH/CH_2Cl_2 to afford 15 fractions. The eighth fraction was purified on a Sephadex LH-20 column using an elution of CH_2Cl_2:MeOH (2:1) to give four subfractions. The third subfraction was further separated by HPLC (Agilent ZORBAX SB-C18, 250 × 9.4 mm, 5 μm column, 3.0 mL/min) eluting with 40–50% MeCN/H_2O in 15 min, then to 82% MeCN/H_2O in 20 min to yield compounds **7** (10.2 mg), **5** (3.4 mg), **1** (4.9 mg), **4** (1.2 mg) and **2** (3.4 mg). The ninth fraction was subjected to a Sephadex LH-20 chromatography eluting by CH_2Cl_2:MeOH (2:1) to give four subfractions. The third subfraction was further purified by HPLC (Agilent ZORBAX Eclipse SB-C18, 250 × 9.4 mm, 5 μm column, 3.0 mL/min) eluting by 70% MeOH/H_2O to give compounds

3 (11.2 mg) and **6** (8.4 mg). The procedure for extraction and compounds isolation was shown in Figure S27.

3.3.1. Bacillisporin K (**1**)

Bacillisporin K (**1**): Light yellow amorphous powder; $[\alpha]_D^{25}$ +206.0 (*c* 0.1, MeOH); ^{1}H and ^{13}C NMR data, Table 1; HRESIMS *m/z* 585.1374 $[M + H]^+$ (calcd for $C_{30}H_{27}O_{11}$, 585.1367).

3.3.2. Bacillisporin M (**2**)

Bacillisporin L (**2**): Light yellow amorphous powder; +231.5 (*c* 0.2, MeOH); ^{1}H and ^{13}C NMR data, Table 1; HRESIMS *m/z* 585.1377 $[M + H]^+$ (calcd for $C_{30}H_{27}O_{11}$, 585.1367).

3.3.3. Rugulosin D (**3**)

Rugulosin D (**3**): Brown amorphous powder; +18.0 (*c* 0.2, MeOH); ^{1}H and ^{13}C NMR data, Table 2; HRESIMS *m/z* 559.1234 $[M + H]^+$ (calcd for $C_{30}H_{23}O_{11}$, 559.1235).

3.4. Antibacterial Activity Assays

The antimicrobial activities were performed based on Antimicrobial Susceptibility Testing Standards outlined by the Clinical and Laboratory Standards Institute document M07-A7 (CLSI) [31] and our previous report [26] by using a penal of pathogens of *C. albicans* ATCC 10231, *S. aureus* ATCC 25923 and *E. coli* ATCC 25922. All the tested compounds were dissolved in dimethyl sulfoxide and diluted in two fold. The minimum inhibitory concentrations (MICs) were determined to be the lowest concentration with no visible bacterial in wells.

4. Conclusions

The chemical investigation of a marine-derived fungus *Talaromyces* sp. BTBU20213036 resulted in the isolation of three new compounds (**1–3**), and four previously reported metabolites (**4–7**). Among them, bacillisporins K and L shared dimeric oligophenaleneone scaffold. Rugulosin D (**4**) is a dimer of the emodin-type anthraquinone. The absolute configurations of isolated compounds were determined by quantum chemical calculations of ECD. Compounds **1**, **2** and **4**−**7** displayed antibacterial activities against *S. aureus* with MIC values of 12.5, 25, 12.5, 6.25, 0.195 and 100 μg/mL, respectively. The difference between **3** and **6** is that the hydroxymethine of C-12′ in **3** was replaced by a sp^2 quaternary carbon to form an α,β-unsaturated ketene. The α,β-unsaturated ketene moiety enhanced the antibacterial activity of **6** with 64 folds compared to that of **3**. The antibacterial activity of **6** (MIC = 0.195 μg/mL) is much stronger than the positive control of vancomycin (MIC = 1 μg/mL), which indicates it could be considered as a lead compound for further investigations into the mechanism and development of antibacterial agents.

Supplementary Materials: The following are available online at https://www.mdpi.com/article/10.3390/antibiotics11020222/s1, Figures S1–S24: HRESIMS, HPLC profiles, 1D and 2D NMR spectra for compounds **1–3**, Figures S25 and S26: Colony morphology and Neighbor-joining phylogenetic tree of strain BTBU20213036, Figure S27: Flow chart of the fermentation, extraction and isolation, Tables S1–S3: 1D and 2D NMR data for compounds **1–3**.

Author Contributions: Data curation, F.S., K.Z. and X.X.; Funding acquisition, F.S.; Investigation, F.S., Y.D., S.W., X.Z. and K.Z.; Supervision, X.X.; Writing—original draft, F.S. and X.X.; Writing—review and editing, F.S. and X.X. All authors have read and agreed to the published version of the manuscript.

Funding: This work was supported by grants from the National Natural Science Foundation of China (81973204), the Key Lab of Marine Bioactive Substance and Modern Analytical Technique, SOA (MBSMAT-2019-06), and Research Foundation for Advanced Talents of Beijing Technology and Business University (19008021176).

Data Availability Statement: Data are contained within the text.

Conflicts of Interest: The authors declare no conflict of interest.

References

1. Shen, B. A new golden age of natural products drug discovery. *Cell* **2015**, *163*, 1297–1300. [CrossRef]
2. Jernigan, J.A.; Hatfield, K.M.; Wolford, H.; Nelson, R.E.; Olubajo, B.; Reddy, S.C.; McCarthy, N.; Paul, P.; McDonald, L.C.; Kallen, A.; et al. Multidrug-resistant bacterial infections in U.S. hospitalized patients, 2012–2017. *N. Engl. J. Med.* **2020**, *382*, 1309–1319. [CrossRef] [PubMed]
3. Arendrup, M.C.; Patterson, T.F. Multidrug-resistant *Candida*: Epidemiology, molecular mechanisms, and treatment. *J. Infect. Dis.* **2017**, *216*, S445–S451. [CrossRef] [PubMed]
4. Martínez-Vázquez, A.V.; Vázquez-Villanueva, J.; Leyva-Zapata, L.M.; Barrios-García, H.; Rivera, G.; Bocanegra-García, V. Multidrug resistance of *Escherichia coli* strains isolated from *Bovine Feces* and Carcasses in Northeast Mexico. *Front. Vet. Sci.* **2021**, *8*, 643802. [CrossRef] [PubMed]
5. Hiramatsu, K.; Katayama, Y.; Matsuo, M.; Sasaki, T.; Morimoto, Y.; Sekiguchi, A.; Baba, T. Multi-drug-resistant *Staphylococcus aureus* and future chemotherapy. *J. Infect. Chemother.* **2014**, *20*, 593–601. [CrossRef]
6. Ciloglu, F.U.; Caliskan, A.; Saridag, A.M.; Kilic, I.H.; Tokmakci, M.; Kahraman, M.; Aydin, O. Drug-resistant *Staphylococcus aureus* bacteria detection by combining surface-enhanced Raman spectroscopy (SERS) and deep learning techniques. *Sci. Rep.* **2021**, *11*, 18444. [CrossRef] [PubMed]
7. Carroll, A.R.; Copp, B.R.; Davis, R.A.; Keyzers, R.A.; Prinsep, M.R. Marine natural products. *Nat. Prod. Rep.* **2021**, *38*, 362–413. [CrossRef]
8. Venkatachalam, M.; Gérard, L.; Milhau, C.; Vinale, F.; Dufossé, L.; Fouillaud, M. Salinity and temperature influence growth and pigment production in the marine-derived fungal strain *Talaromyces albobiverticillius* 30548. *Microorganisms* **2019**, *7*, 10. [CrossRef]
9. Mishra, R.C.; Kalra, R.; Dilawari, R.; Deshmukh, S.K.; Barrow, J.; Goel, M. Characterization of an endophytic strain *Talaromyces assiutensis*, CPEF04 with evaluation of production medium for extracellular red pigments having antimicrobial and anticancer properties. *Front. Microbiol.* **2021**, *12*, 665702. [CrossRef]
10. Kumari, M.; Taritla, S.; Sharma, A.; Jayabaskaran, C. Antiproliferative and antioxidative bioactive compounds in extracts of marine-derived endophytic fungus *Talaromyces purpureogenus*. *Front Microbiol.* **2018**, *9*, 1777. [CrossRef]
11. Guo, Z.; Shao, C.; She, Z.; Cai, X.; Liu, F.; Vrijimoed, L.L.P.; Lin, Y. ^{1}H and ^{13}C NMR assignments for two oxaphenalenones bacillosporin C and D from the mangrove endophytic fungus SBE-14. *Magn. Reson. Chem.* **2007**, *45*, 439–441. [CrossRef] [PubMed]
12. Liang, X.; Huang, Z.H.; Shen, W.B.; Lu, X.H.; Zhang, X.X.; Ma, X.; Qi, S.H. Talaromyoxaones A and B: Unusual oxaphenalenone spirolactones as phosphatase inhibitors from the marine-derived fungus *Talaromyces purpureogenus* SCSIO 41517. *J. Org. Chem.* **2021**, *86*, 12831–12839. [CrossRef] [PubMed]
13. Wu, B.; Ohlendorf, B.; Oesker, V.; Wiese, J.; Malien, S.; Schmaljohann, R.; Imhoff, J.F. Acetylcholinesterase inhibitors from a marine fungus *Talaromyces* sp strain LF458. *Mar. Biotechnol.* **2015**, *17*, 110–119. [CrossRef] [PubMed]
14. Wang, M.; Yang, L.; Feng, L.; Hu, F.; Zhang, F.; Ren, J.; Qiu, Y.; Wang, Z. Verruculosins A–B, new oligophenalenone dimers from the soft coral-derived fungus *Talaromyces verruculosus*. *Mar. Drugs* **2019**, *17*, 516. [CrossRef]
15. Dramae, A.; Intaraudom, C.; Bunbamrung, N.; Saortep, W.; Srichomthong, K.; Pittayakhajonwut, P. Heptacyclic oligophenalenones from the soil fungus *Talaromyces bacillisporus* BCC17645. *Tetrahedron* **2020**, *76*, 130980. [CrossRef]
16. Huang, Z.H.; Liang, X.; Li, C.J.; Gu, Q.; Ma, X.; Qi, S.H. Talaromynoids A-I, highly oxygenated meroterpenoids from the marine-derived fungus *Talaromyces purpureogenus* SCSIO 41517 and their lipid accumulation inhibitory activities. *J. Nat. Prod.* **2021**, *84*, 2727–2737. [CrossRef]
17. Wang, W.; Wan, X.; Liu, J.; Wang, J.; Zhu, H.; Chen, C.; Zhang, Y. Two new terpenoids from *Talaromyces purpurogenus*. *Mar. Drugs* **2018**, *16*, 150. [CrossRef]
18. Liu, H.; Yan, C.; Li, C.; You, T.; She, Z. Naphthoquinone derivatives with anti-Inflammatory activity from mangrove-derived endophytic fungus *Talaromyces* sp. SK-S009. *Molecules* **2020**, *25*, 576. [CrossRef]
19. Ma, M.Z.; Yi, W.W.; Qin, L.; Lian, X.Y.; Zhang, Z.Z. Talaromydien a and talaroisocoumarin A, new metabolites from the marine-sourced fungus *Talaromyces* sp. ZZ1616. *Nat. Prod. Res.* **2022**, *36*, 460–465. [CrossRef]
20. Nicoletti, R.; Trincone, A. Bioactive compounds produced by strains of *Penicillium* and *Talaromyces* of marine origin. *Mar. Drugs* **2016**, *14*, 37. [CrossRef]
21. Lan, D.; Wu, B. Chemistry and bioactivities of secondary metabolites from the genus *Talaromyces*. *Chem. Biodivers.* **2020**, *17*, e200022. [CrossRef]
22. Küppers, L.; Ebrahim, W.; El-Neketi, M.; Özkaya, F.C.; Mándi, A.; Kurtán, T.; Orfali, R.S.; Müller, W.E.G.; Hartmann, R.; Lin, W.; et al. Lactones from the sponge-derived fungus *Talaromyces rugulosus*. *Mar. Drugs.* **2017**, *15*, 359. [CrossRef] [PubMed]
23. Noinart, J.; Buttachon, S.; Dethoup, T.; Gales, L.; Pereira, J.A.; Urbatzka, R.; Freitas, S.; Lee, M.; Silva, A.M.S.; Pinto, M.M.M.; et al. A new ergosterol analog, a new bis-anthraquinone and anti-obesity activity of anthraquinones from the marine sponge-associated fungus *Talaromyces stipitatus* KUFA 0207. *Mar. Drugs.* **2017**, *15*, 139. [CrossRef] [PubMed]
24. Song, F.; Lin, R.; Yang, N.; Jia, J.; Wei, S.; Han, J.; Li, J.; Bi, H.; Xu, X. Antibacterial secondary metabolites from marine-derived fungus *Aspergillus* sp. IMCASMF180035. *Antibiotics* **2021**, *10*, 377. [CrossRef] [PubMed]
25. Xu, X.; Li, J.; Zhang, K.; Wei, S.; Lin, R.; Polyak, S.W.; Yang, N.; Song, F. New isocoumarin analogues from the marine-derived fungus *Paraphoma* sp. CUGBMF180003. *Mar. Drugs* **2021**, *19*, 313. [CrossRef]
26. Han, J.; Yang, N.; Wei, S.; Jia, J.; Lin, R.; Li, J.; Bi, H.; Song, F.; Xu, X. Dimeric hexylitaconic acids from the marine-derived fungus *Aspergillus welwitschiae* CUGBMF180262. *Nat. Prod. Res.* **2022**, *36*, 578–585. [CrossRef]

27. Chaiyosang, B.; Kanokmedhakul, K.; Sanmanoch, W.; Boonlue, S.; Hadsadee, S.; Jungsuttiwong, S.; Kanokmedhakul, S. Bioactive oxaphenalenone dimers from the fungus *Talaromyces macrosporus* KKU-1NK8. *Fitoterapia* **2019**, *134*, 429–434. [CrossRef]
28. Yamazaki, H.; Koyama, N.; Omura, S.; Tomoda, H. New rugulosins, anti-MRSA antibiotics, produced by *Penicillium radicum* FKI-3765-2. *Org. Lett.* **2010**, *12*, 1572–1575. [CrossRef]
29. Jeon, H.; Shim, S.H. Chemical constituents of the endophyte *Penicillium* sp. isolated from artemisia princeps. *Chem. Nat. Compd.* **2020**, *56*, 122–124. [CrossRef]
30. Thompson, J.D.; Higgins, D.G.; Gibson, T.J. Clustal-W—Improving the sensitivity of progressive multiple sequence alignment through sequence weighting, position-specific gap penalties and weight matrix choice. *Nucleic. Acids. Res.* **1994**, *22*, 4673–4680. [CrossRef]
31. Clinical and Laboratory Standards Institute. *Methods for Dilution Antimicrobial Susceptibility Tests for Bacteria that Grow Aerobically*, 7th ed.; Approved Standard; Clinical and Laboratory Standards Institute: Wayne, PA, USA, 2008.

antibiotics

Article

Cyclic Tetrapeptides with Synergistic Antifungal Activity from the Fungus *Aspergillus westerdijkiae* Using LC-MS/MS-Based Molecular Networking

Junjie Han [1,†], Hanying Wang [1,2,†], Rui Zhang [1], Huanqin Dai [1,3], Baosong Chen [1], Tao Wang [1], Jingzu Sun [1], Wenzhao Wang [1], Fuhang Song [4], Erwei Li [5], Zhitang Lyu [2,*] and Hongwei Liu [1,3,*]

1 State Key Laboratory of Mycology, Institute of Microbiology, Chinese Academy of Sciences, Beijing 100101, China; hanjj@im.ac.cn (J.H.); whanying2021@163.com (H.W.); raynaymond@outlook.com (R.Z.); daihq@im.ac.cn (H.D.); chenbs@im.ac.cn (B.C.); wangtao@im.ac.cn (T.W.); sunjz@im.ac.cn (J.S.); wangwz@im.ac.cn (W.W.)
2 School of Life Sciences, Institute of Life Science and Green Development, Hebei University, Baoding 071002, China
3 Savaid Medical School, University of Chinese Academy of Sciences, Beijing 100049, China
4 School of Light Industry, Beijing Technology and Business University, Beijing 100048, China; songfuhang@btbu.edu.cn
5 Institutional Center for Shared Technologies and Facilities, Institute of Microbiology, Chinese Academy of Sciences, Beijing 100101, China; liew@im.ac.cn
* Correspondence: lzt325@sina.com (Z.L.); liuhw@im.ac.cn (H.L.); Tel.: +86-10-6480-6074 (H.L.)
† These authors contributed equally to this work.

Abstract: Fungal natural products play a prominent role in the development of pharmaceuticalagents. Two new cyclic tetrapeptides (CTPs), westertide A (**1**) and B (**2**), with eight known compounds (**3–10**) were isolated from the fungus *Aspergillus westerdijkiae* guided by OSMAC (one strain-many compounds) and molecular networking strategies. The structures of new compounds were unambiguously determined by a combination of NMR and mass data analysis, and chemical methods. All of the isolates were evaluated for antimicrobial effects, synergistic antifungal activity, cytotoxic activity, and HDAC inhibitory activity. Compounds **1–2** showed synergistic antifungal activity against *Candida albicans* SC5314 with the presence of rapamycin and weak HDAC (histone deacetylase) inhibitory activity. These results indicate that molecular networking is an efficient approach for dereplication and identification of new CTPs. CTPs might be a good starting point for the development of synergistic antifungal agents.

Keywords: cyclic tetrapeptides; synergistic antifungal activity; molecular networking; *Aspergillus westerdijkiae*

Citation: Han, J.; Wang, H.; Zhang, R.; Dai, H.; Chen, B.; Wang, T.; Sun, J.; Wang, W.; Song, F.; Li, E.; et al. Cyclic Tetrapeptides with Synergistic Antifungal Activity from the Fungus *Aspergillus westerdijkiae* Using LC-MS/MS-Based Molecular Networking. *Antibiotics* **2022**, *11*, 166. https://doi.org/10.3390/antibiotics11020166

Academic Editors: Lucia Nencioni and Barbara Skerlavaj

Received: 22 December 2021
Accepted: 25 January 2022
Published: 27 January 2022

1. Introduction

Fungal natural products play a prominent role in the development of pharmaceutical agents [1,2]. Cyclic tetrapeptides (CTPs) are a type of important bioactive natural product that were found to have a broad range of pharmacological properties, including antimicrobial [3,4], cytotoxic [5–7], and HDAC (histone deacetylase) inhibitory properties [8]. Most of the naturally occurring CTPs are obtained from fungi, such as HC toxin with cytotoxic and antimitogenic activities from *Cochliobolus carbonum* [9], apicidin with antiprotozoan activities from *Fusarium* strains [10], and microsporins A-B with antitumor activity from *Microsporum* cf. *gypseum* [11]. In recent years, some naturally occurring CTPs have been found to inhibit HDAC and regulate gene expression, which are very useful as cancer therapeutics. In addition to use as antineoplastic drugs, HDAC inhibitors (HDACis) also have anti-interstitial fibrosis [12], anti-inflammatory [13], immunomodulatory [14], and metabolic regulation activities [15].

Naturally occurring CTPs are usually produced in low yields, which limits the discovery of new CTPs. MS/MS-based molecular networking paves the way to solving this problem. As a promising strategy, molecular networking can provide guidance and improve efficiency for the discovery of new bioactive analogues with a specific skeleton from complex mixtures. In the field of bioactive peptides discovery, neoantimycin L with excellent cytotoxicity from *Streptomyces conglobatus* RJ8 [16] and thermoactinoamide A with moderate antiproliferative activity from *Thermoactinomyces vulgaris* DSM 43016 [17] were obtained based on molecular networking.

Aspergillus westerdijkiae is an important ochratoxin A (OTA)-producing fungus, whose genome harbors 17 non-ribosomal peptide synthetase (NRPS) genes [18]. However, most NRPS genes are unexpressed under standard laboratory conditions. In this study, we used the one strain-many compounds (OSMAC) method to activate silenced genes and MS/MS-based molecular networking to search for novel and bioactive peptides from *A. westerdijkiae* L1295. As a result, two new cyclic tetrapeptides, westertides A (**1**) and B (**2**), and eight known compounds (**3**–**10**) were obtained (Figure 1). This work describes the details of the isolation, structure elucidation, and biological activities of secondary metabolites from *A. westerdijkiae* L1295.

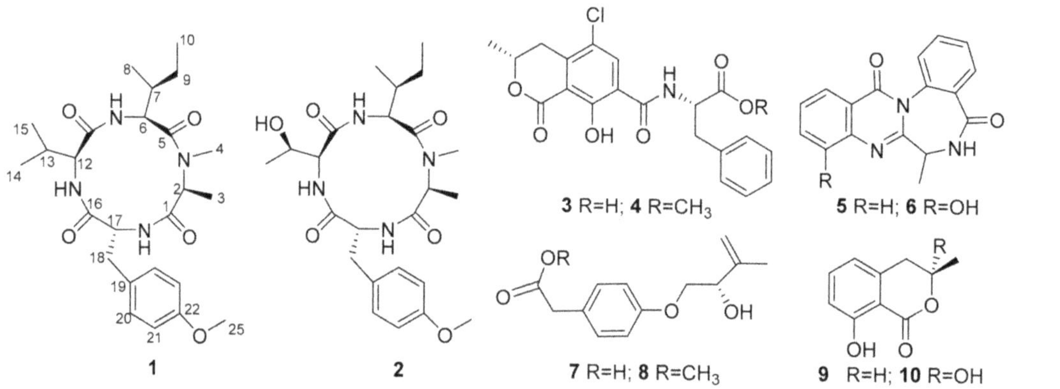

Figure 1. Chemical structures of **1**–**10**.

2. Results

In this study, a molecular networking-OSMAC strategy was applied to accelerate the discovery of cyclic tetrapeptides. First, the fungus *A. westerdijkiae* L1295 was fermented in different culture media and conditions using the OSMAC method (Table S1). Then, the ethyl acetate extracts were further investigated by UPLC-HRMS/MS. The LC-MS/MS data were used to generate a visualized molecular networking that was further annotated by Cytoscape 3.8.2. From the full molecular network, several independent families of molecules were obviously visualized in the crude extracts of *A. westerdijkiae* L1295 fermented on rice, which were different from the other crude extracts (Figures 2 and S1). Further analysis of the molecular network found that a cluster with 19 nodes represented a peptide family, showing MS/MS patterns containing the dipeptide [Ala-Phe] fragment (*m*/*z* 219.1), which has been widely found in the peptide family [19–21] (Figures 2 and S2). Guided by MS/MS and molecular networking, two new cyclic tetrapeptides, westertides A (**1**) and B (**2**), with eight known compounds ochratoxin A (**3**) [22], ochratoxin A methyl ester (**4**) [23], circumdatin F (**5**) [24], circumdatin G (**6**) [25], stachyline B (**7**) [26], westerdijkin A (**8**) [27], mellein (**9**) [28], and 3-hydroxymellein (**10**) [25] were obtained from the solid culture on rice medium and their structure identifications are described below.

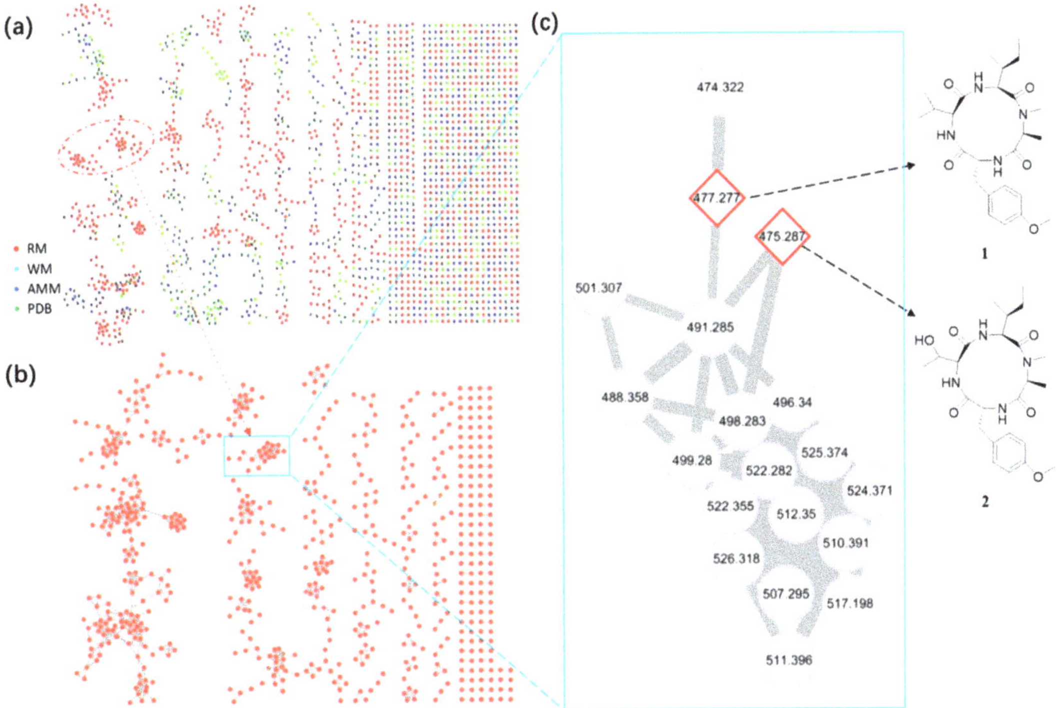

Figure 2. Metabolic analysis of crude extracts from *A. westerdijkiae* L1295. (**a**) Tandem MS/MS-based full molecular networking cluster analysis of different culture extracts of *A. westerdijkiae* L1295. (RM: Rice medium; WM: wheat medium; AMM: *Aspergillus* Minimal Medium; PDB: Potato-Dextrose Broth) of the fungus. (**b**) Molecular networking of *A. westerdijkiae* L1295 fermented on rice. (**c**) The specific subnetwork predicted to contain CTPs in the MS/MS-based molecular networking. The full GNPS network and subnetwork are presented in Figures S1 and S3 in the Supplementary Materials.

Compound **1** was isolated as a pale amorphous solid, which possessed a molecular formula of $C_{25}H_{38}N_4O_5$ (9 degrees of unsaturation) on the basis of HRESIMS and NMR data (Table 1). The ^{1}H, ^{13}C NMR, and HSQC spectra of **1** revealed the presence of 7 methyl groups including 1 *N*-methyl [δ_H/δ_C 3.32 (3H, s)/30.9] and 1 *O*-methyl [δ_H/δ_C 3.78 (3H, s)/55.8], 2 methylene groups [δ_H/δ_C 1.32 (1H, m), 1.82 (1H, m)/25.8; 3.66 (1H, m), 3.91 (1H, m)/35.6], 1 para-disubstituted benzene [δ_H/δ_C 7.05 (2H, d, J = 6.9 Hz)/114.8; 7.29 (2H, d, J = 6.9 Hz)/132.0; δ_C 132.8 and 159.5], 6 methines including 4 characteristic α-methine signals [δ_H/δ_C 4.29 (1H, d, J =5.8 Hz)/65.3; 4.41 (1H, m)/55.8; 4.69 (1H, m)/54.7; 5.17 (1H, dd, J =7.0, 10.0 Hz)/55.3], 3 amide *N*-H protons (δ_H 7.38, 8.83, and 9.88), and 4 amide carbonyls (δ_C 171.2, 173.0, 174.0, and 174.1), suggesting that **1** comprised 4 amino acid residues. HMBC correlations from H_3-3 (δ_H 1.43) to C-2 (δ_C 54.7) and C-1 (δ_C 174.0) and from H_3-4 (δ_H 3.32) to C-2 and C-5 (δ_C 171.2) together with the ^{1}H-^{1}H COSY correlations of H-2-H_3-3 led to the identification of the *N*-Me-Ala residue. The ^{1}H-^{1}H COSY correlations of H_3-10-H_2-9-H-7-H-6 and H_3-8-H-7 together with the HMBC correlations were detected from H-6 (δ_H 5.17) to C-5 (δ_C 171.2), C-7 (δ_C 37.7), C-8 (δ_C 17.6), C-9 (δ_C 25.8), and C-11 (δ_C 173.0); from H-7 (δ_H 2.34) to C-6 (δ_C 55.3); from H_3-8 (δ_H 1.14) to C-6, C-7, and C-9; and from H_3-10 (δ_H 0.98) to C-7 and C-9, which confirmed the presence of the Ile moiety. Similarly, two other amino acid units Val and *O*-Me-Tyr were completely assigned.

Table 1. ^1H (500 MHz) and ^{13}C (125 MHz) NMR data of compound **1** in Pyridine-d_5.

Pos.	1			
	δ_C	δ_H (*J* in Hz)	HMBC	COSY
N-Me-Ala				
1	174.0 C			
2	54.7 CH	4.69 (m)	-	
3	17.1 CH_3	1.43 (d, 4.3)	C1, C2	H-2
4	30.9 CH_3	3.32 (s)	C2, C5	
Ile				
5	171.2 C			
6	55.3 CH	5.17(dd, 10.0, 7.0)	C5, C7, C11	H-7, 6-NH
7	37.7 CH	2.34 (m)	C6, C8, C9	H-6, H-8, H-9
8	17.6 CH_3	1.14 overlapped	C6, C7, C9	H-7
9	25.8 CH_2	1.82 (m), 1.32 (m)	C6, C7, C10	H-7, H-10
10	12.6 CH_3	0.98 (t, 5.5)	C7, C9	H-9
6-NH		8.83 (d, 10.0)		H-6
Val				
11	173.0 C			
12	65.3 CH	4.29 (d, 5.8)	C11, C13, C14, C15, C16	H-13, 12-NH
13	32.6 CH	2.41 (m)	-	H-12, H-14, H-15,
14	20.3 CH_3	1.18 (d, 4.3)	C12, C13	H-13
15	19.2 CH_3	1.14 overlapped	C13	H-13
12-NH		7.38 (brs)		H-12
O-Me-Tyr				
16	174.1 C			
17	55.8 CH	4.41 (m)	-	H-18
18	35.6 CH_2	3.66 (m), 3.91 (m)	C17, C19, C20	H-17
19	132.8 C			
20/24	132.0 CH	7.29 (d, 6.9)	C19, C21/23, C22	H-21/23
21/23	114.8 CH	7.05 (d, 6.9)	C19, C20/24, C22	H-20/24
22	159.5 C			
25	55.8 CH_3	3.78 (s)	C22	
17-NH		9.88 (brs)		

The amino acid sequence of **1** was deduced from the observed key HMBC correlations, NOESY correlations, and MS data. The HMBC correlations from *N*-CH_3 (δ_H 3.32) of Ala to the Ile carbonyl group C-5 (δ_C 171.2), from H-6 (δ_H 5.17) of Ile to the Val carbonyl group C-11 (δ_C 173.0), and from H-12 (δ_H 4.31) of Val to the *O*-Me-Tyr carbonyl group C-16 (δ_C 174.1) suggested a partial sequence of *N*-Me-Ala-Ile-Val-*O*-Me-Tyr (Figure 3). The H_3-4 (δ_H 1.43) of *N*-Me-Ala showed an NOESY correlation with H-20/24 (δ_H 7.29), indicating that **1** was a cyclic peptide, and this conclusion was also confirmed by the 9 degrees of unsaturation and the molecular formula. Additionally, the ESI-MS/MS experimental results (Figures 4 and S4) also confirmed the connections of these residues as cyclo-[*N*-Me-Ala-Ile-Val-*O*-Me-Tyr].

The absolute configuration of the amino acids from compound **1** was established by the advanced Marfey's method [29]. The mixture obtained after hydrolyzing compound **1** and further derivatization with L-FDAA was analyzed by HPLC-DAD. HPLC analyses of the mixture of hydrolysates and appropriate amino acid standards confirmed the D configurations for *O*-Me-Tyr and the L configurations for Tyr, *N*-Me-Ala, and Ile in **1** (Figure 5). Consequently, the structure of **1** was elucidated as cyclo-[L-*N*-Me-Ala-L-Ile-L-Val-D-*O*-Me-Tyr] and named westertide A.

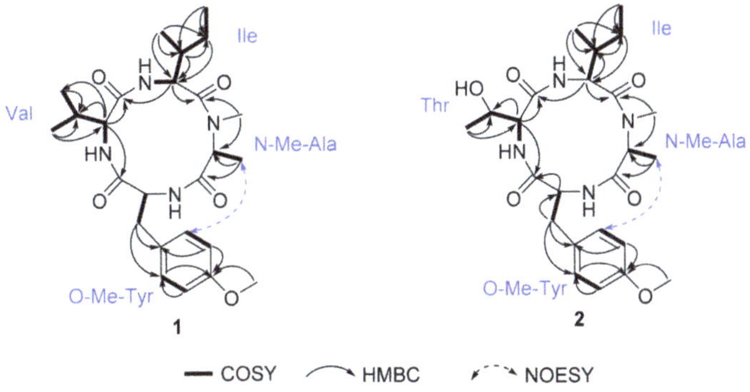

Figure 3. Key ^{1}H-^{1}H COSY, HMBC, and NOESY correlations of **1** and **2**.

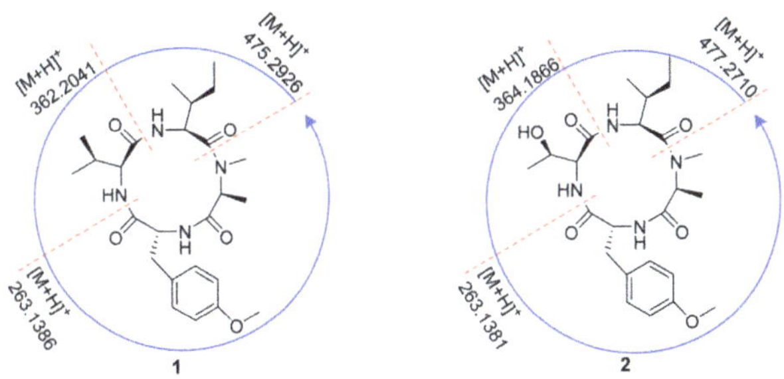

Figure 4. ESI-MS/MS analysis of **1** and **2**.

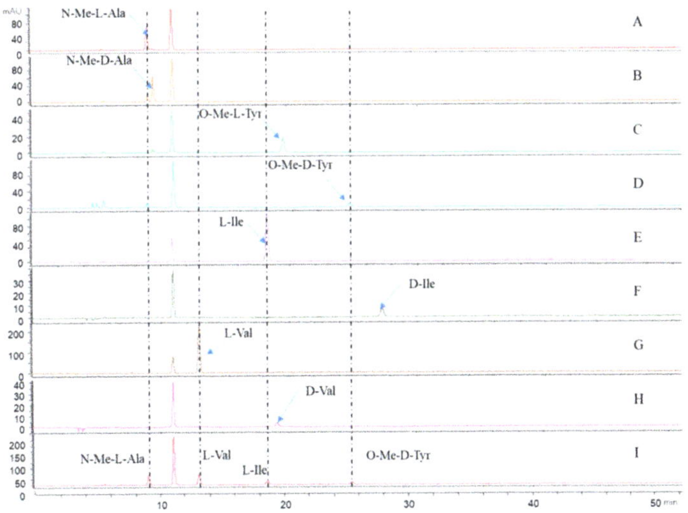

Figure 5. Advanced Marfey's analysis of compound **1**. (**A–H**): The retention times for the FDAA derivatives of *N*-Me-L-Ala, *N*-Me-D-Ala, *O*-Me-L-Tyr, *O*-Me-D-Tyr, L-Ile, D-Ile, L-Val, and D-Val, respectively. (**I**): The FDAA derivatives of the hydrolysate of **1**. The derivatives of the acid hydrolysate and the standard amino acids were subjected to RP HPLC analysis (Kromasil C18 column; 5 μm, 4.6 mm × 250 mm; 1.0 mL/min; UV detection at 340 nm) with a linear gradient of acetonitrile (35–45%) in water (TFA, 0.01%) over 40 min.

Compound **2** was isolated as a white amorphous powder. It was assigned a molecular formula of $C_{24}H_{36}N_4O_6$ (9 degrees of unsaturation) based on its HRESIMS and NMR data (Table 2). The 1D NMR spectroscopic data showed that compound **2** was a cyclic tetrapeptide similar to **1** but bearing a threonine (Thr) residue with signals at δ_H/δ_C 1.38 (3H, d, J = 6.9 Hz)/30.9 (CH_3), δ_H/δ_C 4.42 (1H, overlapped)/65.4 (CH), δ_H/δ_C 4.81 (1H, m)/68.1 (CH), and δ_C 173.5 (C), instead of the valine residue. A comprehensive analysis of its relevant ^1H-^1H COSY, HMQC, HMBC, and NOESY correlations (Figure 3), and the ESI-MS/MS experimental results (Figures 4 and S5) confirmed that **2** has the same planar structure as that of violaceomide A [20]. However, the optical rotation data of **2** ($[\alpha]^{25}_D$ = +249.5, c = 1.0, MeOH) were opposite to that of violaceomide A ($[\alpha]^{25}_D$ = −230.0, c = 0.6, MeOH), implying that they are optical isomers. The HPLC analysis of the acid hydrolysate of **2** after derivatization with L-FDAA revealed that L-*N*-Me-Ala, L-Ile, L-Thr, and D-*O*-Me-Tyr were present in **2** (Figure S6). This result shows that the main difference between compound **2** and violaceomide A is the substitution of L-*O*-Me-Tyr with D-*O*-Me-Tyr. Thus, compound **2** was assigned as cyclo-[L-*N*-Me-Ala-L-Ile-L-Thr-D-*O*-Me-Tyr] and named westertide B.

Table 2. ^1H (500 MHz) and ^{13}C (125 MHz) NMR data of compound **2** in Pyridine-d_5.

Pos.	2			
	δ_C	δ_H (J in Hz)	HMBC	COSY
N-Me-Ala				
1	173.7			
2	54.8	4.75 (m)	C3, C4	
3	17.2	1.46 (d, 7.3)	C1, C2	H-2
4	30.9	3.38 (s)	C2, C5	
Ile				
5	171.3			
6	55.7	5.18 (m)	C5, C7, C11	H-7
7	37.6	2.40 (m)	C6, C8, C9	H-6, H-8, H-9
8	17.8	1.18 (d, 6.5)	C6, C7, C9	H-7
9	25.2	2.10 (m), 1.42 (m)	C6, C7, C10	H-7, H-10
10	12.6	0.93 (t, 7.4)	C7, C9	H-9
6-NH		9.42 (br s) [a]		
Thr				
11	173.5			
12	65.4	4.42 overlapped	C11, C13, C14, C16	H-13, 12-NH
13	68.1	4.81 (m)	-	H-12, H-14, H-15,
14	22.3	1.38 (d, 6.4)	C12, C13	H-13
12-NH		7.38 (br s)		H-12
O-Me-Tyr				
16	174.0			
17	55.8	4.42 overlapped	C16	H-18
18	35.6	3.65 (m), 3.95 (m)	C17, C19, C20	H-17
19	132.8			
20/24	132.1	7.29 (d, 8.1)	C19, C21/23, C22	H-21/23
21/23	114.8	7.06 (d, 8.1)	C19, C20/24, C22	H-20/24
22	159.5			
25	55.8	3.79 (s)	C22	
17-NH		9.91 (br s) [a]		

[a] The position attribution of the active hydrogen refers to the data of violaceomide A [20].

Compounds **1–10** showed no significant bioactivity in the antibacterial, antifungal, and cytotoxicity assays at the dose of 100 μM. In our previous work, we found that peptide-like compounds showed a synergistic antifungal effect with rapamycin [30]. So, we tested whether the new cyclic tetrapeptide compounds could also cause synergistic antifungal activity with rapamycin against *Candida albicans* SC5314. When checkerboard assays were used to obtain the MICs (minimum inhibitory concentrations) with rapamycin for achieving 90% growth inhibition, only 0.008μM of rapamycin was required together with a very low amount (6.25 μM) of compounds **1** and **2**. Based on the fractional inhibitory concentration index (FICI), westertides **1** and **2** showed effective synergism with rapamycin, and the FICI was 0.078 for both compounds **1** and **2** (Table 3). Our results showed that compounds **1** and **2** had strong synergistic antifungal activity with rapamycin. Furthermore, the effects of compounds **1** and **2** on histone deacetylation (HDAC) at the cell level were also evaluated, and compound **1** showed weak HDAC inhibitory activity, with IC_{50} of about 70 μM.

Table 3. MIC values of compounds **1–2** with rapamycin against *C. abicans* SC5314.

Drugs	Anti-Fungal MICs (μM)	Synergistic Anti-Fungal MICs (μM)	FICI [a]	Definition [b]
Rapamycin	0.5	-	-	-
1	>100	6.25	<0.094	S
2	>100	6.25	<0.094	S
Amphotericin B	0.5	0.125	1.25	NS

[a] The concentration of rapamycin in the synergy antifungal screening experiment was 0.008 μM, at which rapamycin does not show antifungal activity. As MIC alone for compounds **1** and **2** > 100 μM, we used 100 to calculate FICI, and the start concentration was 25 μM in the checkerboard assay, and the possible minimal FICI was shown. [b] S: synergism; NS: no synergism.

3. Discussion

With a low molecular weight, low hydrophobicity, and the presence of a hydrogen-bond acceptor and donor, CTPs have been demonstrated to possess diverse pharmacological activities, including antimicrobial [4], cytotoxic [5–7], and HDAC inhibitory bioactivities. In the last decade or so, more than 40 cyclic peptides have been approved by the FDA and EMA, such as vorinostat and romidepsin [31–33].

However, it is relatively difficult to discover CTPs due to their narrow distribution and low yield. As the main natural sources of CTPs, fungi have an abundance of NRPS biosynthetic gene clusters, whereas some of these genes are not expressed under normal experimental conditions. These silent gene clusters outnumber the constitutively expressed clusters by a factor of 5–10 [34]. Hence, strategies that rationally activate silent gene clusters will dramatically enhance our reservoir of potentially therapeutic small molecules [35]. In order to efficiently discover novel cyclic peptides, the molecular network and OSMAC strategy are used in combination with gene mining techniques [36]. Molecular networking can efficiently dereplicate known natural products, thus aiding the discovery of new analogues with a specific skeleton from complex mixtures [37]. The OSMAC strategy can activate some silent genes of target strains to produce more secondary metabolites and obtain novel secondary metabolites [38]. Genome mining is a powerful approach to direct the production of novel and interesting CTPs, which become relevant in the future to search for unculturable microorganisms as a new source of novel bioactive CTPs [39]. In this work, the discovery of two new cyclopeptides from *A. westerdijkiae* using the OSMAC strategy and the MS/MS molecular networking further expanded the structural diversity of the CTPs and the source of CTPs producers.

An estimated 1.2 billion people worldwide suffer from a fungal disease, of which 1.5 to 2 million people die of a fungal infection each year, surpassing those killed by either malaria or tuberculosis [40–42]. About 30% of serious infections are caused by *Candida albicans*, with a mortality rate of up to 40% [43]. Unfortunately, resistance to existing classes of drugs is on the rise due to the limited class of antifungal drugs available and the decline in new drug development. As the process of de novo antifungal discovery fails to meet clinical needs, the approach of repurposing approved drugs has drawn much attention.

Rapamycin, also called sirolimus, is characterized primarily by its antifungal activity against several human fungal pathogens, such as *Candida albicans* [44], *Cryptococcus neoformans* [45], and *Fusarium oxysporum* [46], and potent immunosuppressive activity [47]. The dual effects of rapamycin on antifungus and immunosuppression seem to effectively solve the threat of *Candida* infection when patients are treated with immunosuppressive drugs. However, rapamycin showed weak antifungal activity at the dose used to suppress the immune response in patients. The identification of synergistic actions on rapamycin against fungi can possibly solve this problem. In an early report, Tong et al. showed that some commercial or natural peptide-like compounds synergistically increased the antifungal effect of rapamycin, by targeting the Rbp1 protein (homologue of the FKBP12 protein in mammals) of *C. albicans* to increase the binding of rapamycin-Rbp1 complex with Tor1C protein [30]. In this work, we found two new natural peptide compounds, westertides A and B, showing strong synergistic antifungal activity with rapamycin from *A. westerdijkiae*. The mechanism of their synergistic antifungal effect with rapamycin may be similar to the known peptide compounds, but this requires deep investigation because they showed no antifungal effects alone.

4. Materials and Methods

4.1. General

UV data and optical rotation were recorded on a Thermo Genesys-10S UV-Vis spectrophotometer and Anton Paar MCP 200 Automatic Polarimeter, respectively. High-resolution electrospray ionization mass spectrometry (HRESIMS) data were obtained on an Agilent Accurate-Mass-Q-TOF LC/MS 6520 instrument. NMR spectral data were obtained with a Bruker AVANCE-500 spectrometer (δ_C/δ_H: Pyridine-d_5, 150.4, 135.9, 123.9/8.74, 7.58, 7.22; DMSO, 39.5/2.50). Silica gel (Qingdao Haiyang Chemical Co., Ltd., Qingdao, China, 200–300 mesh), Sephadex LH-20 (GE Healthcare, Uppsala, Sweden), and ODS (50 μm, YMC CO., LTD, YMC Pack, Kyoto, Japan) were used for column chromatography. Semi-preparative HPLC was performed on an Agilent 1200 HPLC system equipped with an Agilent DAD UV−vis spectrometric detector, using a reversed-phase column (C18, 5 μm 9.4 mm × 250 mm, YMC Pack, Kyoto, Japan) with a flow rate of 2.0 mL/min. Biological reagents, chemicals, and media were purchased from standard commercial sources unless stated.

4.2. Fungal Material

A. westerdijkiae was isolated from the mildewed wheat, China, in September 2017. The fungus was identified mainly based on the morphological observation, molecular multilocus phylogeny analysis, and morphological features [48] (Figure S7). The fungus was deposited in China General Microbiological Culture Collection (CGMCC No. 19033).

4.3. Fermentation and Extraction

A. westerdijkiae was cultured on a slant of PDA at 25 °C for 5 days. To prepare inoculum, the spores of the strain on the plate were collected and adjusted to 1×10^6 CFU/mL. A large-scale fermentation was carried out in 40 × 500 mL Fernbach culture flasks, with each flask containing 80 g of rice and 60 mL of distilled water (each with 0.5 mL of spore suspension), incubated at 25 °C for 3 weeks. The fermented rice substrate was extracted repeatedly with ethyl acetate by exhaustive maceration (3 × 4 L), and the organic solvent was evaporated to dryness under vacuum to afford the crude extract (20.1 g).

4.4. LC-MS/MS and Molecular Networking Analysis

LC-MS-MS was performed on an Agilent series 1290 Infinity HPLC instrument, coupled with a Q-TOF Mass spectrometer (Agilent Technologies Inc., Santa Clara, CA, USA), with a YMC C18 column [(YMC Co., Ltd. Kyoto, Japan) YMC-Park, ODS-A,

250 mm × 2.1 mm, S-5 μm, 12 nm, 0.5 mL/min]. The total extracts (0.5 mg/mL, 10 μL) were analyzed by LC-MS with a gradient program of MeCN−H_2O (0.01% TFA) [0–25 min 5–80%, 25–32 min 80–100%, 32–38 min 100%; 0.5 mL/min; MS scan 150–2000 Da] and then with an automated full-dependent MS-MS scan. Mass spectral networks were assembled as described in the reference. Differentiation of the protonated molecules, adducts, and fragment ions was done by identification of the $[M+H]^+$ ion. The All MS/MS data files were converted to ".mzML" format files using MSConver software and uploaded on the GNPS Web platform (http://gnps.ucsd.edu, accessed on 6 August 2021) for MN analysis using Classic mode. For the network creation, a parent mass tolerance of 0.02 Da and a fragment ion tolerance of 0.05 Da were applied. The generated molecular network was visualized in Cytoscape 3.8.2 (www.cytoscape.org, accessed on 6 August 2021) and guided the isolation of **1–8**. The MS/MS molecular network can be browsed and downloaded from the GNPS Web site with the following links: https://gnps.ucsd.edu/ProteoSAFe/status.jsp?task=6794bab0d59245bf875b14c6ebb84ff4 and https://gnps.ucsd.edu/ProteoSAFe/status.jsp?task=84b42a96c887412db918a18f20491b8b (accessed on 6 August 2021).

4.5. Isolation and Characterization Data

The EtOAc fraction was subjected to silica gel column chromatography (CC) using petroleum/ethyl acetate (P/E) in a gradient elution (*v*/*v*, 100:0, 100:1, 100:2, 100:4, 100:10) and dichloromethane/acetone (*v*/*v*, 100:0, 100:2, 100:4, 100:8, 100:12, 100:20, 0:100) to give 16 fractions (*AW*.1–*AW*.16).

Fraction *AW*.6 (0.85 g) was further separated on a silica gel column by elution with increasing concentrations of ethyl acetate in petroleum to give 15 fractions (*AW*.6-1–*AW*.6-15). Compound **9** (4.5 mg) was obtained from subfractions *AW*.6-8 (45 mg) by sephadex LH-20 chromatography eluted with dichloromethane/methanol (*v*/*v*, 1:1). *AW*.6-10 (75 mg) was purified finally by RP-HPLC with acetonitrile/water (50:50) to give **10** (13.5 mg, t_R 22.3 min).

Fraction *AW*.13 (4.3 g) eluted with CH_2Cl_2-Acetone (*v*/*v* 20:1) was first separated by ODS using a gradient of increasing methanol (30%, 45%, 60%, 75%, and 100%) in water to afford 25 subfractions (*AW*.13-1–*AW*.13-25). Compound **6** (30.5 mg, t_R 15.1 min) was obtained from *AW*.13-9 (152 mg) by RP-HPLC using 21% acetonitrile in acidic water (0.005% TFA). Subfractions *AW*.13-11 (170 mg) were followed by RP-HPLC using 32% acetonitrile in water to afford a mixture of **7** (9.1 mg, t_R 22.1 min), **5** (5.6 mg, t_R 31.1 min), and **8** (6.5 mg, t_R 33.5 min). Compounds **1** (2.0 mg, t_R 40.5 min) and **2** (8.0 mg, t_R 32.5 min) were obtained from *AW*.13-15 (55 mg) by RP-HPLC using 45% acetonitrile in acidic water (0.005% TFA). Compound **3** (325.0 mg) was obtained from *AW*.13-21 by recrystallization in acetonitrile. Compound **4** (20.0 mg) was obtained from subfractions and *AW*.13-22 by sephadex LH-20 chromatography eluted with methanol, respectively.

Westertide A (1): pale amorphous solid; $[\alpha]_D^{25}$ +235.57 (*c* 0.5, MeOH); UV (MeOH) λ_{max} (log ε) 222 (4.03), 275 (1.30); Positive HRESIMS: *m*/*z* 475.2926 $[M+H]^+$ (calcd for $C_{25}H_{38}N_4O_5$, 475. 2920). NMR data, see Table 1 and Figures S8–S13.

Westertide B (2): pale amorphous solid, $[\alpha]_D^{25}$ +249.48 (*c* 1.0, MeOH); UV (MeOH) λ_{max} (log ε) 220 (2.78), 275 (1.43) nm; Positive HRESIMS: *m*/*z m*/*z* 477.2710 $[M+H]^+$ (calcd.for $C_{24}H_{37}N_4O_6$, 477.2713). NMR data, see Table 2 and Figures S14–S19.

4.6. Absolute Configuration of Amino Acids

Compound (1.0 mg) was dissolved in 6 N HCl (2.0 mL) and heated at 110 °C for 24 h. The solutions were then evaporated to dryness and placed in a 4 mL reaction vial and treated with a 1 g/100 mL solution of FDAA (200 μL) in acetone, followed by 1.0 M $NaHCO_3$ (40 μL). The reaction mixtures were heated at 45 °C for 90 min, and the reactions were quenched by the addition of HCl (1 M, 40 μL). In a similar fashion, standard *N*-Me-L-Ala, *N*-Me-D-Ala, *O*-Me-L-Tyr, *O*-Me-D-Tyr, L-Ile, D-Ile, L-Val, D-Val, L-Thr, and D-Thr, were derivatized separately. The derivatives of the acid hydrolysate and the standard amino acids were subjected to RP HPLC analysis (Kromasil C18 column; 5 μm, 4.6 mm × 250 mm;

1.0 mL/min; UV detection at 340 nm) with a linear gradient of acetonitrile (35–45%) in water (TFA, 0.01%) over 40 min. The retention times for the FDAA derivatives of *N*-Me-L-Ala, *N*-Me-D-Ala, *O*-Me-L-Tyr, *O*-Me-D-Tyr, L-Ile, D-Ile, L-Val, D-Val, L-Thr, and D-Thr were 9.1, 9.4, 19.0, 25.0, 18.1, 25.8, 12.9, 18.4, 5.2, and 6.1 min, respectively, whereas those for the FDAA derivatives of *N*-Me-Ala, *O*-Me-Tyr, Ile, and Val in the hydrolysate of **1** were 9.1 (*N*-Me-L-Ala), 25.0 (*O*-Me-D-Tyr), 18.1 (L-Ile), and 12.9 (L-Val) min, and *N*-Me-Ala, *O*-Me-Tyr, Ile, and Thr in the hydrolysate of **2** were 9.1 (*N*-Me-L-Ala), 25.0 (*O*-Me-D-Tyr), 18.1 (L-Ile), and 5.2 (L-Thr) min, respectively.

4.7. Evaluation of Biological Activities

4.7.1. Antifungal and Synergistic Antifungal Assay

Candida albicans SC5314 was used as a test strain for the antifungal and synergistic antifungal bioassay. Checkerboard assays were carried out as described previously [29,30]. Overnight cultures were chosen to prepare the strain suspension with RPMI 1640 medium. RPMI 1640 was purchased from Invitrogen, and was used according to the manufacturer's protocol, by supplementing 2% glucose, 3.45% MOS, then adjusting the pH to 7.0. Compounds were dissolved in DMSO. Final concentrations ranged from 0.002 to 2 μg/mL for rapamycin and 0.39 to 25 μg/mL for peptide-like compounds, respectively. Rapamycin was 2-fold diluted from 1 to 11 (column), while selected compounds were 2-fold diluted from A to G (row) of the 96-well microtiter plate. The fractional inhibitory concentration index (FICI) is defined as the sum of the MIC of each drug when used in combination divided by the MIC of the drug used alone. Synergism and antagonism were defined by FICI of $\leq$0.5 and >4, respectively.

4.7.2. Cytotoxicity Assay

Cytotoxicity tests against A549, HepG2, and K562 cell lines were carried out as previously described [49]. Taxol, 5-Flourouracil, and Cisplatin were used as the positive control.

4.7.3. HDAC Activity Assay

The HDAC activity of the compounds was measured using an HDAC8 Deacetylase Fluorometric (Human) Assay Kit (Cat KA4444, Abnova, Taipei, Taiwan) according to the manufacturer's instructions. Fluorescence signal was detected with excitation at 360 nm and emission at 460 nm using a fluorescence microplate reader (Perkin-Elmer, Waltham, MA, USA). Experiments were performed in triplicate and data were analyzed using GraphPad Prism (version 6.0), Kd values were calculated by nonlinear curve fitting using a 1-site binding (hyperbola) model (Y = Bmax*X/(Kd + X).

5. Conclusions

Uncovered by OSMAC and molecular networking strategies, 2 new cyclic tetrapeptides (**1–2**), together with 7 known compounds (**3–10**) were isolated from *A. westerdijkiae*. All of the isolates were evaluated for an antifungal effect, synergistic antifungal activity, cytotoxic activity, and HDAC inhibitory activity. As a result, **1–10** showed no significant bioactivity in the antifungal assays and cytotoxicity assays at the dose of 100 μM. However, compounds **1–2** showed strong synergistic antifungal activity against *C. albicans* with rapamycin. In addtion, compound **1** showed weak HDAC inhibitory activity.

Supplementary Materials: The following are available online at https://www.mdpi.com/article/10.3390/antibiotics11020166/s1, Table S1: Culture media with different compositions and conditions for *A. westerdijkiae*; Figure S1: The molecular network obtained by combining the LC-MS/MS analyses of rice fermentation extract extracts from *A. westerdijkiae* L1295; Figure S2: Cyclictetrapeptides-cluster and the MS/MS spectrum of each node; Figure S3: The cluster corresponding to compounds observed in the molecular networking; Figure S4: The ESI-MS/MS spectrum of **1**; Figure S5: The ESI-MS/MS spectrum of **2**; Figure S6: Advanced Marfey's analysis of compound **2**. Figure S7: Phylogenetic analysis and morphological characters of *A. westerdijkiae* L1295; Figure S8: ^{1}H NMR spectrum of westertide A (**1**) in pyridine-d_5 (500 MHz); Figure S9: ^{13}C NMR spectrum of westertide

A (**1**) in pyridine-d_5 (125 MHz; Figure S10: ^{1}H-^{1}H COSY spectrum of westertide A (**1**) in pyridine-d_5; Figure S11: HSQC spectrum of westertide A (**1**) in pyridine-d_5; Figure S12: HMBC spectrum of westertide A (**1**) in pyridine-d_5; Figure S13: ROESY spectrum of westertide A (**1**) in pyridine-d_5; Figure S14: ^{1}H NMR spectrum of westertide B (**2**) in pyridine-d_5 (500 MHz); Figure S15: ^{13}C NMR spectrum of westertide B (**2**) in pyridine-d_5 (125 MHz); Figure S16: ^{1}H-^{1}H COSY spectrum of westertide B (**2**) in pyridine-d_5; Figure S17: HSQC spectrum of westertide B (**2**) in pyridine-d_5; Figure S18: HMBC spectrum of westertide B (**2**) in pyridine-d_5; Figure S19: ROESY spectrum of westertide B (**2**) in pyridine-d_5.

Author Contributions: Conceptualization, J.H. and H.W.; methodology, J.H. and B.C.; validation and data curation, R.Z., Z.L. and H.D.; formal analysis, R.Z., W.W. and J.H.; investigation, J.H. and H.L.; resources, T.W. and J.S.; writing—original draft preparation, H.W., R.Z. and J.H.; writing—review and editing, Z.L. and H.L.; supervision and project administration, J.H. and H.L.; funding acquisition, E.L., F.S., H.L. and J.H. All authors have read and agreed to the published version of the manuscript.

Funding: This research was funded by the National Special Project for Key Science and Technology of Food Safety (grant No. 2017YFC1601302), and the National Natural Science Foundation (Grant Nos. 22177131, 82073723 and 81872771).

Institutional Review Board Statement: Not applicable.

Informed Consent Statement: Not applicable.

Data Availability Statement: Not applicable.

Acknowledgments: The authors thank the National Special Project and the National Natural Science Foundation for funding.

Conflicts of Interest: The authors declare no conflict of interest.

References

1. Al-Fakih, A.A.; Almaqtri, W.Q.A. Overview on antibacterial metabolites from terrestrial *Aspergillus* spp. *Mycology* **2019**, *10*, 191–209. [CrossRef]
2. Chin, J.M.W.; Puchooa, D.; Bahorun, T.; Jeewon, R. Antimicrobial properties of marine fungi from sponges and brown algae of Mauritius. *Mycology* **2021**, *12*, 231–244. [CrossRef]
3. Fazal, A.; Webb, M.E.; Seipke, R.F. The Desotamide Family of Antibiotics. *Antibiotics* **2020**, *9*, 452. [CrossRef]
4. Chakraborty, S.; Tai, D.F.; Lin, Y.C.; Chiou, T.W. Antitumor and antimicrobial activity of some cyclic tetrapeptides and tripeptides derived from marine bacteria. *Mar. Drugs* **2015**, *13*, 3029–3045. [CrossRef]
5. He, F.; Bao, J.; Zhang, X.Y.; Tu, Z.C.; Shi, Y.M.; Qi, S.H. Asperterrestide A, a cytotoxic cyclic tetrapeptide from the marine-derived fungus *Aspergillus terreus* SCSGAF0162. *J. Nat. Prod.* **2013**, *76*, 1182–1186. [CrossRef]
6. Sun, Y.; Tian, L.; Huang, Y.F.; Sha, Y.; Pei, Y.H. A new cyclotetrapeptide from marine fungus *Trichoderma reesei*. *Pharmazie* **2006**, *61*, 809–810.
7. Gao, C.H.; Chen, Y.N.; Pan, L.X.; Lei, F.; Long, B.; Hu, L.Q.; Zhang, R.C.; Ke, K.; Huang, R.M. Two new cyclic tetrapeptides from deep-sea bacterium Bacillus amyloliquefaciens GAS 00152. *J. Antibiot.* **2014**, *67*, 541–543. [CrossRef]
8. Abdalla, M.A. Medicinal significance of naturally occurring cyclotetrapeptides. *J. Nat. Med.* **2016**, *70*, 708–720. [CrossRef]
9. Walton, J.D. HC-toxin. *Phytochemistry* **2006**, *67*, 1406–1413. [CrossRef]
10. von Bargen, K.W.; Niehaus, E.M.; Bergander, K.; Brun, R.; Tudzynski, B.; Humpf, H.U. Structure Elucidation and Antimalarial Activity of Apicidin F: An Apicidin-like Compound Produced by Fusarium fujikuroi. *J. Nat. Prod.* **2013**, *76*, 2136–2140. [CrossRef]
11. Gu, W.; Cueto, M.; Jensen, P.R.; Fenical, W.; Silverman, R.B. Microsporins A and B: New histone deacetylase inhibitors from the marine-derived fungus Microsporum cf. gypseum and the solid-phase synthesis of microsporin A. *Tetrahedron* **2007**, *63*, 6535–6541. [CrossRef]
12. Davies, E.R.; Haitchi, H.M.; Thatcher, T.H.; Sime, P.J.; Kottmann, R.M.; Ganesan, A.; Packham, G.; O'Reilly, K.M.; Davies, D.E. Spiruchostatin A inhibits proliferation and differentiation of fibroblasts from patients with pulmonary fibrosis. *Am. J. Respir. Cell Mol. Biol.* **2012**, *46*, 687–694. [CrossRef]
13. Leoni, F.; Zaliani, A.; Bertolini, G.; Porro, G.; Pagani, P.; Pozzi, P.; Dona, G.; Fossati, G.; Sozzani, S.; Azam, T.; et al. The antitumor histone deacetylase inhibitor suberoylanilide hydroxamic acid exhibits anti-inflammatory properties via suppression of cytokines. *Proc. Natl. Acad. Sci. USA* **2002**, *99*, 2995–3000. [CrossRef]
14. Roger, T.; Lugrin, J.; Le Roy, D.; Goy, G.; Mombelli, M.; Koessler, T.; Ding, X.C.; Chanson, A.-L.; Reymond, M.K.; Miconnet, I.; et al. Histone deacetylase inhibitors impair innate immune responses to Toll-like receptor agonists and to infection. *Blood* **2011**, *117*, 1205–1217. [CrossRef]

15. Lenoir, O.; Flosseau, K.; Ma, F.X.; Blondeau, B.; Mai, A.; Bassel-Duby, R.; Ravassard, P.; Olson, E.N.; Haumaitre, C.; Scharfmann, R. Specific control of pancreatic endocrine β- and δ-cell mass by class IIa histone deacetylases HDAC4, HDAC5, and HDAC9. *Diabetes* **2011**, *60*, 2861–2871. [CrossRef]
16. Lin, X.; Chai, L.; Zhu, H.R.; Zhou, Y.; Lin., H.W. Applying molecular networking for targeted isolation of depsipeptides. *RSC Adv.* **2021**, *11*, 2774. [CrossRef]
17. Sala, G.D.; Mangoni, A.; Costantino, V.; Teta, R. Identification of the biosynthetic gene cluster of thermoactinoamides and discovery of new congeners by integrated genome mining and MS-Based molecular networking. *Front. Chem.* **2020**, *8*, 397. [CrossRef]
18. Han, X.L.; Chakrabortti, A.; Zhu, J.D.; Liang, Z.X.; Li, J.M. Sequencing and functional annotation of the whole genome of the filamentous fungus *Aspergillus westerdijkiae*. *BMC Genom.* **2016**, *17*, 633. [CrossRef]
19. Hou, X.M.; Li, Y.Y.; Shi, Y.W.; Fang, Y.W.; Chao, R.; Gu, Y.C.; Wang, C.Y.; Shao, C.L. Integrating molecular networking and H NMR to target the isolation of chrysogeamides from a library of marine-derived *Penicillium* fungi. *J. Org. Chem.* **2019**, *84*, 1228–1237. [CrossRef]
20. Liu, J.T.; Gu, B.B.; Yang, L.J.; Yang, F.; Lin, H.W. New Anti-inflammatory cyclopeptides from a sponge-derived fungus *Aspergillus violaceofuscus*. *Front. Chem.* **2018**, *6*, 226. [CrossRef]
21. Yang, X.; Yang, Y.; Peng, T.; Yang, F.; Zhou, H.; Zhao, L.; Xu, L.; Ding, Z. A New cyclopeptide from endophytic *Streptomyces* Sp. Yim 64018. *Nat. Prod. Commun.* **2013**, *8*, 1753–1754. [CrossRef]
22. Jesus, A.D.; Steyn, P.S.; Vleggaar, R.; Wessels, P.L. Carbon-13 nuclear magnetic resonance assignments and biosynthesis of the mycotoxin ochratoxin A. *J. Chem. Soc. Perk. Trans.* **1980**, 52–54. [CrossRef]
23. Xu, X.y.; He, F.; Zhang, X.Y.; Bao, J.; Qi, S.H. New mycotoxins from marine-derived fungus *Aspergillus* sp. SCSGAF0093. *Food Chem. Toxicol.* **2013**, *53*, 46–51. [CrossRef]
24. Joshi, B.K.; Gloer, J.B.; Wicklow, D.T.; Dowd, P.F. Sclerotigenin: A new antiinsectan benzodiazepine from the sclerotia of *Penicillium sclerotigenum*. *J. Nat. Prod.* **1999**, *62*, 650–652. [CrossRef]
25. Dai, J.R.; Carté, B.K.; Sidebottom, P.J.; Sek Yew, A.L.; Ng, S.B.; Huang, Y.C.; Butler, M.S. Circumdatin G, a New alkaloid from the fungus *Aspergillus ochraceus*. *J. Nat. Prod.* **2001**, *64*, 125–126. [CrossRef]
26. Almeida, C.; Part, N.; Bouhired, S.; Kehraus, S.; König, G.M. Stachylines AD from the sponge-derived fungus *Stachylidium* sp. *J. Nat. Prod.* **2011**, *74*, 21–25. [CrossRef]
27. Fredimoses, M.; Zhou, X.F.; Ai, W.; Tian, X.P.; Yang, B.; Lin, X.P.; Xian, J.Y.; Liu, Y.H. Westerdijkin A, a new hydroxyphenylacetic acid derivative from deep sea fungus *Aspergillus westerdijkiae* SCSIO 05233. *Nat. Prod. Res.* **2015**, *29*, 158–162. [CrossRef]
28. Djoukeng, J.D.; Polli, S.; Larignon, P.; Abou-Mansour, E. Identification of phytotoxins from *Botryosphaeria obtusa*, a pathogen of black dead arm disease of grapevine. *Eur. J. Plant Pathol.* **2009**, *124*, 303–308. [CrossRef]
29. Wu, W.; Dai, H.Q.; Bao, L.; Ren, B.; Lu, J.C.; Luo, Y.M.; Guo, L.D.; Zhang, L.X.; Liu, H.W. Isolation and structural elucidation of proline-containing cyclopentapeptides from an endolichenic *Xylaria* sp. *J. Nat. Prod.* **2011**, *74*, 1303–1308. [CrossRef]
30. Tong, Y.; Zhang, J.; Wang, L.; Wang, Q.; Huang, H.; Chen, X.; Zhang, Q.; Li, H.; Sun, N.; Liu, G.; et al. Hyper-synergistic antifungal activity of rapamycin and peptide-like compounds against *Candida albicans* orthogonally via tor1 kinase. *ACS Infect Dis.* **2021**, *7*, 2826–2835. [CrossRef]
31. Sarojini, V.; Cameron, A.J.; Varnava, K.G.; Denny, W.A.; Sanjayan, G. Cyclic Tetrapeptides from nature and design: A review of synthetic methodologies, structure, and function. *Chem. Rev.* **2019**, *119*, 10318–10359. [CrossRef]
32. Campas-Moya, C. Romidepsin for the treatment of cutaneous T-cell lymphoma. *Drugs Today* **2009**, *45*, 787–795. [CrossRef]
33. Marks, P.A.; Breslow, R. Dimethyl sulfoxide to vorinostat: Development of this histone deacetylase inhibitor as an anticancer drug. *Nat. Biotech.* **2007**, *25*, 84–90. [CrossRef]
34. Baltz, R.H. Gifted microbes for genome mining and natural product discovery. *J. Ind. Microbiol. Biotechnol.* **2017**, *44*, 573–588. [CrossRef]
35. Rutledge, P.J.; Challis, G.L. Discovery of microbial natural products by activation of silent biosynthetic gene clusters. *Nat. Rev. Microbiol.* **2015**, *13*, 509–523. [CrossRef]
36. Paulo, B.S.; Sigrist, R.; Angolini, C.F.F.; De Oliveira, L.G. New cyclodepsipeptide derivatives revealed by genome mining and molecular networking. *ChemistrySelect* **2019**, *4*, 7785–7790. [CrossRef]
37. Yang, J.Y.; Sanchez, L.M.; Rath, M.; Liu, X.T.; Boudreau, P.D.; Bruns, N.; Glukhov, E.; Wodtke, A.; de Felicio, R.; Fenner, A.; et al. Molecular Networking as a Dereplication Strategy. *J. Nat. Prod.* **2013**, *76*, 1686–1699. [CrossRef]
38. Romano, S.; Jackson, S.; Patry, S.; Dobson, A. Extending the "One Strain Many Compounds" (OSMAC) Principle to Marine Microorganisms. *Mar. Drugs* **2018**, *16*, 244. [CrossRef]
39. Malek, Z.; Gregory, L.C. Strategies for the discovery of new natural products by genome mining. *ChemBioChem* **2009**, *10*, 625–633.
40. Denning, D.W.; Bromley, M.J. How to bolster the antifungal pipeline. *Science* **2015**, *347*, 1414–1416. [CrossRef]
41. Brown, G.D.; Denning, D.W.; Gow, N.A.R.; Levitz, S.M.; Netea, M.G.; White, T.C. Hidden killers: Human fungal infections. *Sci. Transl. Med.* **2012**, *4*, 165rv13. [CrossRef]
42. Kumaria, A.; Tripathia, A.H.; Gautamb, P.; Gahtoria, R.; Pandec, A.; Singhd, Y.; Madane, T.; Upadhyay, S.K. Adhesins in the virulence of opportunistic fungal pathogens of human. *Mycology* **2021**, *12*, 296–324. [CrossRef]
43. Pfaller, M.A.; Diekema, D.J. Epidemiology of invasive candidiasis: A persistent public health problem. *Clin. Microbiol. Rev.* **2007**, *20*, 133–163. [CrossRef]

44. Singh, K.; Sun, S.; Vezina, C. Rapamycin (AY-22,989), a new antifungal antibiotic. IV. Mechanism of action. *J. Antibiot.* **1979**, *32*, 630–645. [CrossRef]
45. Cruz, M.C.; Cavallo, L.M.; Görlach, J.M.; Cox, G.; Perfect, J.R.; Cardenas, M.E.; Heitman, J. Rapamycin antifungal action is mediated via conserved complexes with FKBP12 and TOR kinase homologs in *Cryptococcus neoformans*. *Mol. Cell. Biol.* **1999**, *19*, 4101–4112. [CrossRef]
46. Wong, G.K.; Griffith, S.; Kojima, I.; Demain, A.L. Antifungal activities of rapamycin and its derivatives, prolylrapamycin, 32-desmethylrapamycin, and 32-desmethoxyrapamycin. *J. Antibiot.* **1998**, *51*, 487–491. [CrossRef]
47. Eng, C.P.; Sehgal, S.N.; Vezina, C. Activity of rapamycin (AY-22,989) against transplanted tumors. *J. Antibiot.* **1984**, *37*, 1231–1237. [CrossRef]
48. Frisvad, J.C.; Frank, J.M.; Houbraken, J.A.M.P.; Kuijpers, A.F.A.; Samson, R.A. New ochratoxin A producing species of *Aspergillus* section Circumdati. *Study Mycol.* **2004**, *50*, 23–43.
49. Han, J.J.; Bao, L.; Tao, Q.Q.; Yao, Y.J.; Liu, X.Z.; Yin, W.B.; Liu, H.W. Gloeophyllins A−J, cytotoxic ergosteroids with various skeletons from a chinese tibet fungus *Gloeophyllum abietinum*. *Org. Lett.* **2015**, *17*, 2538–2541. [CrossRef]

antibiotics

Review

Curcumin: Biological Activities and Modern Pharmaceutical Forms

Maja Urošević *, Ljubiša Nikolić, Ivana Gajić, Vesna Nikolić, Ana Dinić * and Vojkan Miljković

Faculty of Technology, University of Niš, Bulevar Oslobođenja 124, 16000 Leskovac, Serbia; nljubisa@tf.ni.ac.rs (L.N.); ivana@tf.ni.ac.rs (I.G.); nikolicvesna@tf.ni.ac.rs (V.N.); vojkan@tf.ni.ac.rs (V.M.)
* Correspondence: maja@tf.ni.ac.rs (M.U.); anatacic@tf.ni.ac.rs (A.D.)

Abstract: Curcumin (1,7-bis-(4-hydroxy-3-methoxyphenyl)-hepta-1,6-diene-3,5-dione) is a natural lipophilic polyphenol that exhibits significant pharmacological effects in vitro and in vivo through various mechanisms of action. Numerous studies have identified and characterised the pharmacokinetic, pharmacodynamic, and clinical properties of curcumin. Curcumin has an anti-inflammatory, antioxidative, antinociceptive, antiparasitic, antimalarial effect, and it is used as a wound-healing agent. However, poor curcumin absorption in the small intestine, fast metabolism, and fast systemic elimination cause poor bioavailability of curcumin in human beings. In order to overcome these problems, a number of curcumin formulations have been developed. The aim of this paper is to provide an overview of recent research in biological and pharmaceutical aspects of curcumin, methods of sample preparation for its isolation (Soxhlet extraction, ultrasound extraction, pressurised fluid extraction, microwave extraction, enzyme-assisted aided extraction), analytical methods (FTIR, NIR, FT-Raman, UV-VIS, NMR, XRD, DSC, TLC, HPLC, HPTLC, LC-MS, UPLC/Q-TOF-MS) for identification and quantification of curcumin in different matrices, and different techniques for developing formulations. The optimal sample preparation and use of an appropriate analytical method will significantly improve the evaluation of formulations and the biological activity of curcumin.

Keywords: curcumin; metabolism; bioavailability; formulations; pharmacological activities

Citation: Urošević, M.; Nikolić, L.; Gajić, I.; Nikolić, V.; Dinić, A.; Miljković, V. Curcumin: Biological Activities and Modern Pharmaceutical Forms. *Antibiotics* **2022**, *11*, 135. https://doi.org/10.3390/antibiotics11020135

Academic Editors: Fuhang Song and Yunjiang Feng

Received: 10 December 2021
Accepted: 10 January 2022
Published: 20 January 2022

1. Introduction

The main ingredient of the *Curcuma longa* is the rhizome [1], a low-molecular-weight lipophilic molecule that can pass through the cellular membrane easily [2]. By its chemical structure, it belongs to the group of polyphenols [3]. Because of its intensive yellow colour, it is used as a natural food colouring agent [1]. The simple molecular structure and arrangement of functional groups are suitable for examining the relationship between the structure and its activity [4]. The ability of curcumin to interact with different proteins facilitates selective modulation of multiple cellular signalling pathways associated with various chronic diseases [5]. The transcription factors, mediators of inflammation and enzymes such as protein kinase, reductase, and histone acetyltransferase are important molecules for curcumin binding. Curcumin is a powerful epigenetic regulator in many diseases, such as neurological disorders, inflammatory diseases, diabetes, and different types of cancer [6]. Furthermore, it modulates various proteasomal pathways and reduces glycogen metabolism through selective inhibition of phosphorylase kinase enzyme [7]. The studies have shown that curcumin exhibits anti-inflammatory, hypoglycemic, antioxidant, antimicrobial, antiviral, anticancer, neuroprotective, and many other effects [8]. However, the main obstacle to the effective manifestation of the pharmacological activity of curcumin is its poor aqueous solubility and low bioavailability [9–11]. The main factors contributing to the low bioavailability of curcumin in the blood plasma and tissues are its poor absorption, fast metabolism, and rapid systemic elimination [12]. The enhancement of the solubility and bioavailability of this promising molecule is crucial for potential clinical application. Different approaches in developing curcumin formulations can improve its physicochemical characteristics and enable safe and efficient use. For that purpose, formulations including nanoparticles, liposomes, micelles, phospholipid complexes, hydrogels, etc., have been described in the reference sources [11,13]. The aim of this paper is to analyse the factors influencing the bioavailability of curcumin, as well as to review pharmacological activities of curcumin and strategies in order to enhance its bioavailability.

2. Curcumin: Background

Turmeric (*Curcuma longa*) is an aromatic plant from the ginger family (Zingiberaceae). It is grown in the southern and southwestern regions of Asia. It occupies an important place in the cuisine of Iran, Malaysia, India, China, Polynesia, and Thailand. It is used as a spice, and it affects the nature, colour, and taste of food. Curry is the best-known spice that contains turmeric rhizome powder. Curcumin is also used as an ecological dye; it is known as Natural yellow 3 and has been assigned an E number-E100, when used as a food colouring agent [14,15]. Figure 1 shows curcuminoids (curcumin, demethoxycurcumin, and bis-demethoxycurcumin), the main components of turmeric rhizome.

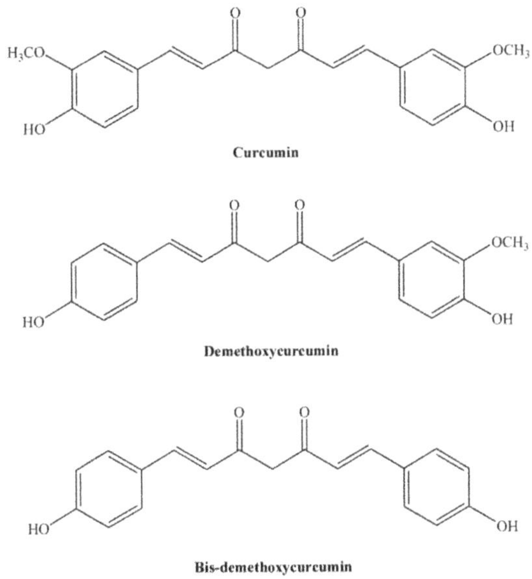

Figure 1. Structural formulae of curcuminoids.

Curcumin, a yellow-orange pigment isolated from turmeric two centuries ago, is a widely studied natural compound that has shown enormous in vitro therapeutic potential. For centuries, it has been used in Ayurvedic medicine and traditional Chinese medicine [11,16]. An overview of the discovery and application of curcumin is shown in Table 1.

Curcumin has been found to possess pleiotropic activities owing to the potential of this polyphenol to modulate multiple signalling molecules. Curcumin exhibits anti-inflammatory, antioxidant, proapoptotic, chemopreventive, chemotherapeutic, antinociceptive, antiproliferative, antiparasitic, and antimalarial effects, and it is used as a wound-healing agent. The research on curcumin and its pharmacological activities has become increasingly important in recent years [9].

Table 1. The History of curcumin.

Year	Discovery	Reference
1815	Vogel and Pelletier were the first to report the "Orange-yellow Substance" isolated from the rhizome of *Curcuma longa* and named it curcumin.	[16]
1842	Vogel Extracted pure preparation of curcumin but did not report its formula.	[17]
1910	Milobedzka and Lampe identified chemical structure of curcumin as diferuloylmethane, or 1,6-heptadiene-3,5-dione-1,7-bis-(4-hidroxy-3-methoxyphenyl)-(1E, 6E).	[18]
1913	The synthesis of curcumin was published.	[19]
1949	Schraufstatter et al. Reported that curcumin is a biologically active compound with antibacterial properties.	[20]
1953	Srinivasan separated and quantified the components of curcumin using chromatography.	[21]
1971	It was discovered that curcumin lowers cholesterol	[22]
1972	It was discovered that curcumin lowers the level of sugar in the blood	[23]
1973	It was discovered that curcumin has an anti-inflammatory effect	[24]
1976	It was discovered that curcumin has an antioxidant effect	[25]
1980	Kuttan et al. demonstrated anticancer activity of curcumin both in vitro and in vivo.	[26]
1995	Curcumin exhibits anti-inflammatory activity by suppressing the proinflammatory transcription factor, nuclear factor-kappa B (NF-κB)	[27]

3. Isolation of Curcumin from Turmeric Rhizome and Methods of Identification

Turmeric rhizome contains two main classes of pharmacologically active secondary metabolites: curcuminoids and essential oil [28]. Curcuminoids (curcumin, demethoxycurcumin and bis-demethoxycurcumin) are most responsible for the biological activity of turmeric [29]. The isolation of curcuminoids from turmeric rhizomes is achieved by applying conventional and modern extraction methods [30]. Soxhlet extraction [31] and maceration [32] are classic extraction methods. Of the modern methods for extraction of curcuminoids, ultrasound extraction [33], enzyme-assisted extraction [34], microwave extraction [35], supercritical fluid extraction [36], and pressurized fluid extraction [37] are used. The most commonly used solvents for curcuminoid extraction are ethanol, dichloromethane, ethyl acetate, isopropanol, methanol, n-butanol, and acetone [29,38,39]. Sahne et al. used acetone as a solvent in conventional and unconventional extraction processes due to its high solubilization capacity [31]. In the paper by Muthukumar et al., various organic solvents for curcumin extraction were examined. The research findings show that acetone is the most efficient extraction solvent [39]. Thin-layer chromatography (TLC) is a classical analytical technique for separating curcumin from the extraction mixture [29,38]. Curcumin is quantified in the extract using high-performance liquid chromatography (HPLC). After extraction, the organic solvents are removed from the extract by evaporation on a vacuum evaporator. The residue (oleoresin) is then dissolved in methanol and subjected to HPLC analysis [40]. The yield and stability of curcumin depend on the extraction method used. Sahne et al. analysed curcumin extraction from the turmeric rhizome using several advanced methods, and the results were compared with the results obtained by Soxhlet extraction, the most commonly used reference method. The result showed that the yield of curcumin extraction obtained using the Soxhlet method (6.9%) was significantly higher than the one obtained by extraction using microwaves (3.72%), ultrasound (3.92%), and enzymes (4.1%). Although modern extraction methods do not show high extraction yields similar to the Soxhlet method, their advantages (low temperature and short extraction time, use of a very small amount of solvent) make them more favourable methods for curcumin extraction [31].

The kinetic degradation of curcumin from a natural mixture of curcuminoids in different conditions (pH, temperature, and dielectric constant of the solvent), as well as the degradation of pure curcumin in defined conditions, were examined in the paper by

Naksuriya et al. An aqueous buffer/methanol 50:50 (*v*/*v*) mixture was used as a standard medium to examine the kinetics of curcumin degradation. The results showed that the degradation of pure curcumin present in the curcuminoid mixture underwent a first-order reaction. An increase in pH, temperature, and dielectric constant of the medium lead to an increase in the rate of curcumin degradation. Curcumin showed rapid degradation due to autoxidation in aqueous buffer pH = 8.0 with a constant rate of 0.28 h^{-1}, which corresponds to a half-life ($t_{1/2}$) of 2.5 h. Curcumin incorporated as a mixture of curcuminoids into ω-methoxypoly(ethylene glycol)-*b*-(*N*-(2-benzoyloxypropyl) methacrylamide) polymer micelles was about 300–500 times more stable than pure curcumin in a mixture of phosphate buffer and methanol.

Incorporating curcumin into polymer micelles is a promising approach for stabilising this compound and developing formulations suitable for further pharmaceutical and clinical trials [41]. Liu et al. examined natural deep eutectic solvents formed from organic acids and sugars for the efficiency of curcuminoid extraction. In optimal conditions (the temperature of 50 °C, 0.1/10 g/mL ratio of solid and liquid components, and extraction time of 30 min), higher extraction yields were achieved when a solvent with a ratio of citric acid and glucose 1:1 and 15% water was used, compared to the conventional extraction solvents. The proposed method is an excellent alternative for extracting natural pigments since it is environmentally friendly and sustainable [42]. During the isolation and purification of curcuminoids from oleoresin, the volatile oil of turmeric dissolves curcumin, thus creating a problem in the recrystallization process. Different organic solvents and their combinations for selective recrystallization of curcuminoids were examined. A mixture of isopropyl alcohol and hexane (1:1.5, *v*/*v*) proved to be the best solvent for recrystallization in the purification of curcuminoids. The total curcumin content in the raw curcuminoid powder was 76.82% *w*/*w*, whereas, in the recrystallized powder, the purity was increased up to 99.45% *w*/*w* [29].

4. Physico-Chemical Properties of Curcumin

Curcumin (1,7-bis-(4-hydroxy-3-methoxyphenyl)-1,6-heptadiene-3,5-dione) [43] or diferuloylmethane [44] is an integral component of turmeric (up to ~5%), a well-known traditional spice [45]. It is a lipophilic compound, insoluble in water, acidic, and neutral solutions, and soluble in ethanol, dimethylsulfoxide, and acetone. Curcumin can be extracted from turmeric rhizomes by using organic solvents. The molecular formula of curcumin is $C_{21}H_{20}O_6$, and the molecular weight is 368.38 g/mol. The melting point of curcumin is 183 °C. Curcumin is a tautomeric compound due to the presence of β-diketone in the molecular structure and shows diketo/keto–enol tautomerism (Figure 2) [46,47].

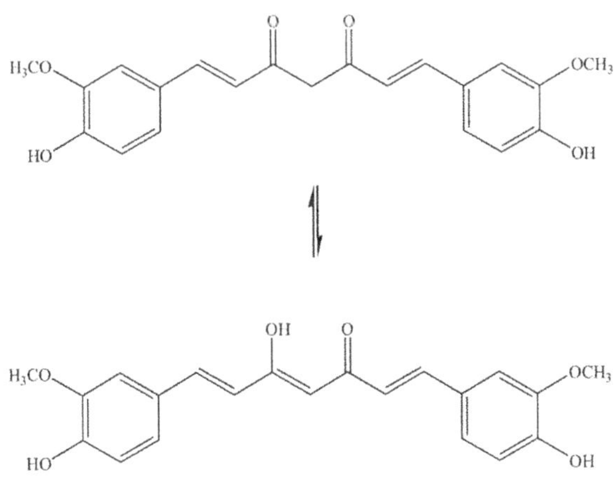

Figure 2. Keto-enol tautomerism of curcumin.

The diketo tautomer can exist in *cis* and *trans* forms. Solvent polarity, pH, and temperature significantly affect curcumin's keto–enol balance [48]. The ratio of keto and enol tautomers of curcumin, on the other hand, strongly influences pharmacological activities [49]. Manolova et al. examined the tautomerism of curcumin in ethanol/water binary mixtures using ultraviolet–visible (UV–VIS) spectroscopy and advanced quantum chemical calculations. The results show that only enol–keto tautomer is present in ethanol. The addition of water leads to the emergence of a new spectral range, which is assigned to the diketo tautomeric form. The diketo form is dominant in the mixture of water and ethanol 90:10 (*v*/*v*). The observed equilibrium shift is explained by quantum chemical calculations, which show that water molecules stabilise the diketo tautomer by forming stable complexes [50]. Kawano et al. analysed keto–enol tautomers of curcumin by using liquid chromatography/mass spectrometry. The research findings show that the enol form is the main form in solution (water/acetonitrile) [51]. In nonpolar solvents (carbon tetrachloride) in the solid state and solution, curcumin exists in enol form [50,51].

Curcumin is unstable in the solution form. It has an intense yellow colour, which changes to dark red in the basic solution [52].

5. Structure, Bioavailability, and Safety of the Application of Curcumin

Curcumin, a polyphenol from the diarylheptanoid group, has two aromatic rings symmetrically substituted by methoxy and a phenolic OH group in the *ortho* position, which are connected to a conjugated seven-membered hydrocarbon chain with an enone part and a 1,3-diketone group (Figure 3). The active functional groups of curcumin are two *o*-methoxy and two phenolic groups, two double bonds in the hydrocarbon chain and the 1,3-keto–enol part of the structure [53].

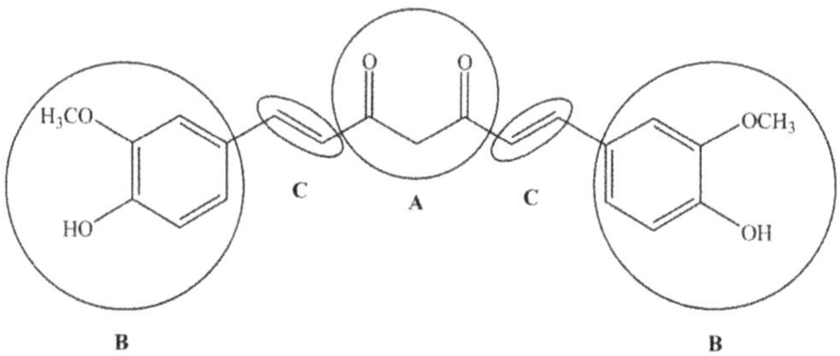

Figure 3. Important functional parts of curcumin: 1,3-keto-enol part (**A**), *o*-methoxy and phenolic groups (**B**) and a double bond (**C**).

Aromatic groups provide hydrophobicity, whereas the α,β-unsaturated β-diketo part of the structure allows flexibility to the molecule. These unique properties of curcumin make it capable of binding to various biomacromolecules. Biologically critical chemical reactions of curcumin are realized through the H-bond of the β-dicarbonyl group and phenolic hydroxyl residues, as well as the ether residue of the methoxy group, and by binding with metals and nonmetals. It has been demonstrated that curcumin binds directly to numerous signalling molecules, such as inflammatory molecules, protein kinase, protein reductase, histone acetyltransferase, histone deacetylase, glyoxalase I, xanthine oxidase, human immunodeficiency virus (HIV1) integrase, HIV1 protease, sarco/endoplasmic reticulum calcium ATPase, deoxyribonucleic acid (DNA) methyltransferase 1, carrier proteins, and metal ions. The diketo group forms chelates with transition metals, reducing metal-induced toxicity, while some of the metal complexes exhibit enhanced antioxidant activity because they mimic enzymes [54]. Curcumin can also bind directly to DNA and ribonucleic acid (RNA). The ability of curcumin to bind to carrier proteins improves its solubility and bioavailability. Curcumin is unstable at physiological pH and degrades rapidly in an autoxidation reaction to the major bicyclopentadione product in which a 7-carbon chain has undergone oxygenation and double cyclization [55]. The alkaline hydrolysis products (ferulic acid, vanillin, ferulaldehyde, and feruloylmethane), as well as oxidation products (such as bicyclopentadione), show biological activity but are significantly less active than curcumin [56].

The clinical trials with curcumin have clearly demonstrated its safety, tolerability, and efficacy against different chronic diseases in humans [8]. The human studies did not indicate any toxic effects when curcumin was administered orally in the dosage of 6 g/day during 4–7 weeks [57]. The study on safety, tolerability, and activity of liposomal curcumin (Lipocurc™) on patients with locally advanced or metastatic cancer was conducted by Greil et al. It demonstrated that 300 mg/m^2 of liposomal curcumin was the maximum safe dosage for patients with cancer treatment [58]. Saghatelyan et al. assessed the efficacy and safety of intravenous infusion of curcumin in combination with paclitaxel in patients with metastatic and advanced breast cancer. After a 12-week treatment, curcumin administered intravenously did not cause any significant health issues, nor did it deteriorate the quality of life [59].

6. The Metabolism of Curcumin

Poor bioavailability of curcumin in humans at a dose of 12 g/day is a consequence of poor absorption in the small intestine, fast metabolism in the liver, and rapid systemic elimination [60]. Most of the orally administered curcumin is excreted through faeces, without metabolism, while a smaller, absorbed part undergoes metabolic modification. The metabolism of curcumin takes place in two stages. The first phase involves the reduction in the presence of reductases, which takes place in enterocytes and hepatocytes. The reduction products are dihydrocurcumin, tetrahydrocurcumin, hexahydrocurcumin, and octahydrocurcumin (hexahydrocurcuminol) [61]. The curcumin reduction reaction is catalyzed by enzymes such as nicotinamide adenine dinucleotide phosphate (NADPH)-dependent reductase, alcohol dehydrogenase, and an unidentified microsomal enzyme [62]. In the paper by Hassaninasab et al., the enzyme for curcumin reduction was purified from *Escherichia coli* and characterized. It was found that the microbial metabolism of curcumin by a purified enzyme involved reduction in two steps, in which, depending on NADPH, curcumin was converted into an intermediate product, dihydrocurcumin, and then into the final product, tetrahydrocurcumin [63]. Curcumin and its reduced metabolites are readily conjugated to glucuronic acid and sulfate in vivo and in vitro. Glucuronidation and sulfation reactions take place in the presence of glucuronyl transferase and sulfotransferase, respectively. Glucuronidation and sulfation of curcumin occur in the liver and intestines of rats and humans [61]. After oral administration in humans, a portion of curcumin is absorbed and can be identified as a water-soluble glucuronide and sulfate conjugate in the plasma. Human phenol sulfotransferase 1A1 (SULT1A1) and human phenol sulfotransferase 1A3 (SULT1A3) are responsible for the sulfation of curcumin in humans and in the intestines of rats, while uridine diphosphate-glucuronosyltransferase (UGT) catalyses glucuronidation of curcumin [54]. The reduction or conjugation of curcumin generates species with a reduced ability to inhibit the expression of cyclooxygenase-2 (COX-2) compared with curcumin. Tetrahydrocurcumin, hexahydrocurcumin, and curcumin sulfate exhibit weaker inhibition of prostaglandin E2, while hexahydrocurcuminol is inactive [64]. The biological activity of curcumin metabolites other than tetrahydrocurcumin is significantly reduced compared with curcumin [65,66]. To enhance the bioavailability of curcumin, piperine which interferes with glucuronidation, curcumin in liposomes, curcumin nanoparticles, curcumin phospholipid complexes, and structural curcumin analogues are used. Figure 4 shows the metabolic and nonmetabolic transformations of curcumin.

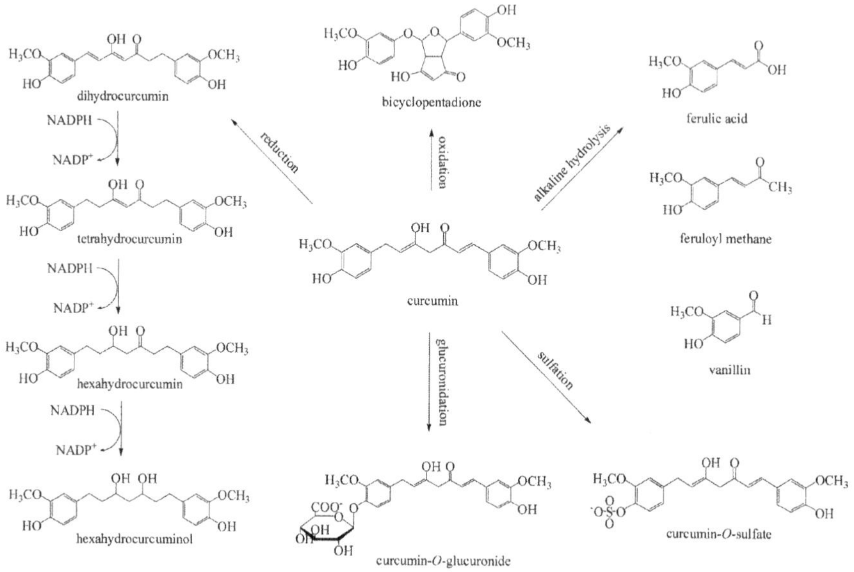

Figure 4. Important metabolic and nonmetabolic transformations of curcumin.

7. Characterization of Curcumin

Curcuminoids are widely used in the food processing and pharmaceutical industries due to their properties. The detection and characterization of curcuminoids in different matrices are of great importance. The choice of the analytical method for curcuminoid analysis depends on the sample type, the purpose of the analysis, and the detection and quantification limits [67,68]. The techniques based on chromatography and electrophoresis are among the selected methods for determining curcuminoids. TLC is one of the methods used for fractioning turmeric extracts [38]. The use of the TLC method for turmeric analysis has declined due to prolonged separation time and poor resolution, although it is selective, easy to perform, and inexpensive. New high-performance thin-layer chromatography (HPTLC) methods that overcome these limitations have been developed [69]. The principle of operation is the same as with TLC. Higher resolution, lower detection limit and improved image scanning are advantages of HPTLC methods [68]. HPLC, in combination with UV–VIS detector, is the most commonly used chromatographic method for the qualitative and quantitative analysis of curcumin due to its high precision, accuracy, and low detection limit. Various HPLC methods have been developed to analyse curcuminoids (Table 2).

Table 2. HPLC methods for curcuminoid analysis.

Matrix Sample	Column	Mobile Phase	λ, nm	Limit of Detection	Reference
Turmeric Powder	RP C_{18}	Acetonitrile and 0.1% Trifluro-Acetic Acid (50:50, *v*/*v*)	420	27.99 ng/mL	[70]
Turmeric Extracts	Alltima C_{18} column	Acetonitrile and 2% Acetic Acid (40:60, *v*/*v*)	425	0.90 µg/mL	[71]
Commercial Samples of Turmeric	C_{18}	Methanol, 2% Acetic acid, and Acetonitrile	425	0.05 µg/mL	[72]
Curcuminoids-Loaded Liposome	Zorbax Eclipse XDB C_{18} (4 × 150mm, 5 µm)	Acetonitrile and 0.1% OrthoPhosphoric Acid (50:50, *v*/*v*)	425	0.124 µg/mL	[73]

Table 2. *Cont.*

Matrix Sample	Column	Mobile Phase	λ, nm	Limit of Detection	Reference
Samples of Turmeric	Zorbax SB-C_{18} column (4.6 × 250 mm, 5 μm)	Acetonitrile and 0.4% Aqueous Acetic Acid	430	0.31 μg/mL	[37]
Extract of Turmeric	C_{18} (4.6 × 150mm, 5 μm)	Acetonitrile and 2% Acetic Acid (55:45, *v*/*v*)	425	0.0738 ppm	[74]
Extract of Turmeric	Waters Xterra MS C_{18} column (4.6 × 250 mm, 5 μm)	Distilled Water and Acetonitrile (65:35, *v*/*v*) Containing 1% Acetic Acid	425	1.13 μg/mL	[75]
Turmeric Rhizome	Brownlee SPP C_{18} column (4.6 × 100 mm, 2.7 μm)	Water and Acetonitrile (70:30, *v*/*v*)	420	1.0 μg/mL	[76]

Liquid chromatography–mass spectrometry (LC/MS) [67] is used to identify and quantify traces of curcumin in biological fluids, food, or other complex matrices. Various LC/MS methods have been developed to detect and quantify curcumin in different matrices [77–79]. A rapid and sensitive, selective high-throughput ultrahigh performance liquid chromatography method with tandem mass spectrometry (UPLC/Q-TOF-MS) was developed and validated to quantify curcuminoids to reduce analysis time and improve sensitivity [80]. The UV–VIS spectroscopy can also quantify curcuminoids if the sample matrix or other present components do not absorb within this range. Curcumin shows an absorption maximum at 425 nm [62,68,81]. Fourier transform infrared spectroscopy (FTIR), near-infrared spectroscopy (NIR), Raman's spectroscopy, nuclear magnetic resonance spectroscopy (NMR), and fluorescence spectroscopy are also used to characterize curcumin [82–85].

Curcumin exists in three polymorphic forms: monoclinic form and two orthorhombic forms. Pandey et al. examined polymorphs using X-ray diffraction (XRD) and differential scanning calorimetry (DSC) and found that curcumin polymorphs were monotropically linked to each other, with the monoclinic form being the most stable [86].

Electron paramagnetic resonance (EPR) spectroscopy is an efficient and noninvasive spectroscopic method for analysing samples with unpaired electrons. It is used to quantify the types of radicals and analyse the antioxidative effects of substances [87]. EPR spectroscopy was applied for determining the potential and capacity of curcumin against free radicals (DPPH, nitric oxide radical (NO·), hydroxyl radical (HO·) and superoxide anion radical (O_2) [88,89]. In the study by Nikolić et al., EPR spectroscopy was used for assessing the antioxidant activity of curcumin-loaded low-energy nanoemulsions according to Tempol stable nitroxide free radical. The research findings show that nanoemulsion with curcumin exhibits swift activity, thus neutralising free radicals within the first five minutes from the beginning of the reaction [90].

8. Formulations

A large number of the curcumin formulations with volatile oil (volatile oil formulation) [91,92], piperine [93], and lecithin [94] have been designed. These formulations increase the absorption of curcumin after oral administration compared with pure curcumin. Liposomes, micelles, phospholipid complexes, cyclodextrins, nanoparticles, emulsions, hydrogels, and phytosomes are new promising curcumin formulations. Such formulations provide more prolonged circulation, better absorption and resistance to metabolic processes, increase absorption from the small intestine, and prolong half-life in the plasma, and thus, increase the efficiency of curcumin [95–97] (Figure 5).

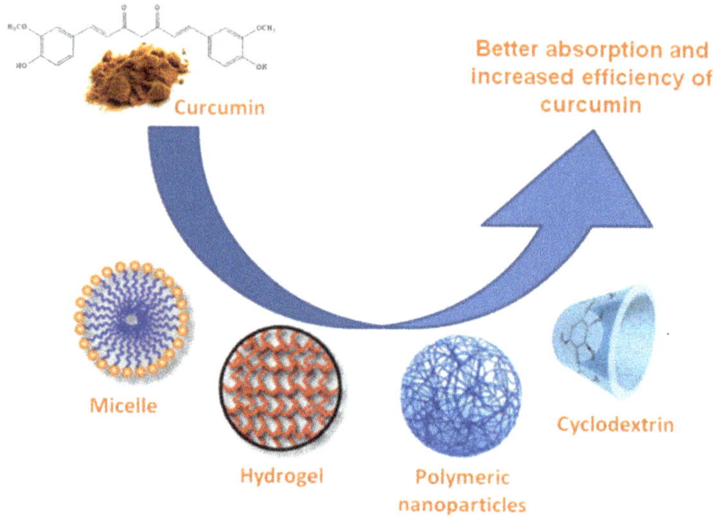

Figure 5. Formulations of curcumin.

Phytosome formulations with curcumin, formulations with volatile oils of turmeric rhizome, and curcumin formulations with a combination of hydrophilic carrier, cellulose derivatives, and natural antioxidants (CHC), compared to a standardized mixture of curcumin (CS), were tested in a study on healthy volunteers. The CHC formulation of curcumin significantly increases the content of curcuminoids in the blood compared with standard curcumin [92]. Cyclodextrins (CDs) can form molecular inclusion complexes with lipophilic compounds, thus enhancing water solubility, dispersion, and absorption of active components. The bioavailability of the curcumin formulation with γ-cyclodextrin was investigated. This formulation was compared with a standardized curcumin extract and two commercially available formulations, the curcumin phytosome formulation (CSL) and the curcumin formulation with rhizome-extracted turmeric essential oils (CEO). The formulation of curcumin with γ-cyclodextrin significantly enhances the absorption of curcuminoids in healthy people [9]. The inclusion complex of curcumin with β-cyclodextrin was prepared using the coprecipitation method. The solubility of curcumin in water increased from 0.00122 to 0.721 mg/mL by forming an inclusion complex. The release of the inclusion complex from nanocomposite and conventional poly (*N*-isopropylacrylamide/sodium alginate) hydrogels cross-linked with nanoclay and *N*,*N*′-methylenebis(acrylamide) (BIS), respectively, was tested under simulated gastrointestinal conditions. At pH = 1.2 and pH = 6.8, hydrogels showed the lowest and the highest release-swelling ratio, respectively. The swelling coefficient and cumulative release decreased with increasing nanoclay content in nanocomposite hydrogels. Conversely, as the BIS ratio in conventional hydrogels increased, the swelling ratio and cumulative release increased [98]. The polyol dilution method was used to formulate liposomes with curcumin in the paper by Kongkaneramit et al. Lipid phase was a mixture of hydrogenated phosphatidylcholine and cholesterol in a molar ratio of 9:1. Propylene glycol, glycerin, and polyethylene glycol 400 were used as polyol solvents. The type and amount of polyol affect both the size of the liposomes and the amount of curcumin incorporated. The preparation temperature is also an important factor in liposome production [99]. Tai et al. studied the stability and release performance of curcumin from liposomes with different contents of hydrogenated phospholipids [100].

Chitosan-coated liposomes may be an alternative carrier for drug delivery in humans. In the work of Cuomo et al., the applicability of chitosan-coated liposomes with curcumin, as well as anionic liposomes with curcumin, was evaluated. The applicability of the formulations was examined in vitro by measuring the bioavailability of ingested curcumin. It has been shown that the presence of a positively charged liposome surface enables better absorption of curcumin in the small intestine, which increases its overall bioavailability [101]. Curcumin nanoemulsion was formulated as a low-energy emulsion and converted to a nanoemulgel using cross-linked polyacrylic acid (Carbopol® 934) as a gelling agent to increase the solubility and absorption of curcumin after topical application to the skin. The nanoemulgel formulation showed faster and earlier wound healing in psoriatic mice compared with curcumin and betamethasone-17-valerate gel. The research findings show that curcumin nanoemugel formulation is a promising candidate for successful long-term treatment of psoriasis [102].

Curcumin nanoemulsions are highly effective in preventing tumour recurrence after surgery and metastasis [103]. A formulation of eye drops (thermosensitive hydrogel containing latanoprost and curcumin nanoparticles) for dual drug delivery has been developed and characterized. The developed dual drug delivery system has shown a prolonged release profile, in vitro and in vivo biocompatibility, reduced levels of inflammation and apoptosis of cells, and protection of trabecular mesh (TM) cells from oxidative damage [104]. PLGA curcumin nanoparticles have shown increased oral and intravenous bioavailability [105]. The oral formulation of nanocurcumin can significantly reduce recovery time in hospitalized patients with COVID-19 [106]. The hybrid curcumin-phospholipid complex was used as a system for oral drug administration to inhibit the metastasis of breast and lung cancer [107]. A high-performance formulation of curcumin phospholipid complex, which can improve the flow, solubility, and oral bioavailability of curcumin, was developed by Wang et al. [108]. Polymer micelles made using block copolymer methoxy-poly(ethylene glycol) (mPEG)-poly(caprolactone) (PCL) enable delayed release of curcumin [109].

In the study by Karavasili et al., the activity of peptide hydrogel with simultaneous delivery of doxorubicin and curcumin in the therapy of head and neck cancer cells was examined. The findings showed the therapeutic utility of a double peptide hydrogel with built-in drugs for the local treatment of head and neck cancer [110]. The amylopectin–chitosan composite hydrogel (LRA–CS) for curcumin delivery was synthesized and tested by Liu et al. The release characteristics of encapsulated curcumin in the simulated gastric and intestinal fluid were observed. The findings showed that LRA–CS hydrogel provided stability of curcumin in the stomach and its release in the small intestine [111]. Chitosan–nanocellulose hydrogel with nonionic surfactant was also used for the delivery of curcumin [112]. Cyclodextrin nanospongoid-based hydrogel (CDNS) was used for transdermal codelivery of curcumin (CUR) and resveratrol (RES). Nanosponges enhanced the in vitro release of curcumin 10 times and the release of resveratrol 2.5 times compared with regular curcumin and resveratrol. The combination of CUR–CDNS and RES–CDNS demonstrated a synergistic cytotoxic effect on MCF-7 cells. A hydrogel base was developed with carbomer and propylene glycol, in which CUR-CDNS and RES-CDNS were incorporated. The photostability of curcumin and resveratrol in the CDNS hydrogel formulation increased almost five and seven times, respectively, compared with the hydrogel formulated without CDNS. Curcumin and resveratrol intake is significantly enhanced when delivered using a CDNS-hydrogel base [113]. In Shef et al., curcumin was incorporated into the oxidized cellulose–polyvinyl alcohol hydrogel system by the freezing process. In vitro studies on rats have shown that this can be an effective method for natural wound healing [114]. In the work of Sahin et al., a new, highly bioavailable formulation of curcumin, advanced ultrasol curcumin (AUC), with improved intestinal stability, was developed. In administered doses, AUC effectively improves the pathophysiology of osteoarthritis in experimentally induced osteoarthritis in rats [115]. An overview of curcumin formulations tested on humans and animals is shown in Tables 3 and 4, respectively.

Table 3. Clinical applications of curcumin.

Disease	Dose	Duration	Patients	Results	Reference
Overweight	80 mg/Day	6 Weeks	48 Overweight Girl Students	Positive antioxidant effect and prevention of lipid peroxidation in overweight individuals.	[116]
Metabolic syndrome(MetS)	80 mg/Day	12 Weeks	50 Patients	Supplementation with Nanomicelle curcumin Significantly improved serum triglyceride in MetS patients.	[117]
Diabetic sensorimotor polyneuropathy	80 mg/Day	8 Weeks	80 Diabetic patients	Nanocurcumin supplementation reduced the severity of diabetic sensorimotor polyneuropathy in patients with type 2 diabetes mellitus.	[118]
Migraine	80 mg/Day	2 Months	80 Patients	Combination of omega-3 fatty acids and nanocurcumin modulates interleukin-6 gene Expression and high-sensitivity C-reactive protein serum levels in patients with migraine.	[119]
Nonalcoholic fatty liver disease (NAFLD)	80 mg/Day	3 Months	84 Patients	Nanocurcumin improves glucose indices, lipids, inflammation, and nesfatin in overweight and obese patients with nonalcoholic fatty liver disease (NAFLD).	[120]
Hemodialysis (HD)	120 mg/Day	12 Weeks	60 Patients	Nanocurcumin shows beneficial effects in lowering inflammation and Hs-CRP levels, as well as adhesion molecules (ICAM-1, VCAM-1), in hemodialysis patients.	[121]
Overweight and obesity	500 mg/Day	10 Weeks	60 Adolescent	Ten weeks of curcumin supplementation had beneficial effects on inflammation and oxidative stress markers among postpubescent overweight and obese girl adolescents.	[122]
Coronavirus disease-2019	1050 mg/Day	14 Days	158 Patients	Curcumin is a safe and natural therapeutic option to prevent post-COVID-19 thromboembolic events.	[123]
Ulcerative colitis (UC)	450 mg/Day	8 Weeks	41 Patients	Low-dose oral curcumin is not effective in inducing remission in mild-to-moderate ulcerative colitis.	[124]
Ulcerative colitis (UC)	1500 mg/Day	8 Weeks	70 Patients	Consumption of the curcumin supplement, along with drug therapy, significant improvement of the clinical outcomes, quality of life, Hs-CRP, and ESR in patients with mild-to-moderate UC.	[125]

Table 4. Application of curcumin in vivo-animal models.

Curcumin Form	Activity	Animal Model	Reference
Curcumin	Antidepressive effect	Sprague–Dawley rats	[126]
Nanocurcumin	Wound-healing Agent	Male Wistar rats	[127]
Curcumin, nanoparticles	Antibacterial and anti-inflammatory agent	Male C57BL/6 mice	[128]
Curcumin	Inhibitors of NF-κB	Mus musculus, C57BL/6J	[129]
Curcumin	Decontaminate and accelerate the Wound contraction	Wistar Rats	[130]
Curcumin, Nanoparticles	Adjuvant agent for the treatment of Hodgkin's Lymphoma	Mice	[131]
Curcumin, Nanoparticles	Contrasting agent	Sprague–Dawley rats	[132]
Curcumin C3 Complex	Cancer prevention	Male C57BL/6 wild-type mice	[133]
Curcumin, Hydrogel	Wound-healing agent	Mus musculus var. albino mice	[134]
PVA/Chitosan/Curcumin Patches	Wound-healing agent	Wistar Rats	[135]

A study was conducted to compare the oral bioavailability of the newly developed formulation of curcumin Curene® with a formulation of curcumin containing volatile turmeric oil (CP-01) and standard curcuminoids 95%, on healthy volunteers. It was found that the oral bioavailability of Curene® is significantly higher compared with CP-01 and standard curcuminoids (95%) and that it is safe to be administered to healthy people in trial conditions [136]. The anti-inflammatory activity of Longvida® Optimized Curcumin (LC) was examined on two-month-old wild-type mice and GFAP-IL6. LC can alleviate inflammation and minimize neurodegeneration and motoric defects in GFAP-IL6 mice [137]. There is a range of commercial formulations of curcumin with defined bioavailability and pharmacokinetics such as Meriva®, LongVida®, CurQfenTM, MicroActive curcumin, Micronized curcumin, NovaSol® (micellar curcumin) CurcuWin®, BiocurcumaxTM Curcumin C3 Complex®+Bioperine, Cavacurmin® Theracurmin™. Of the commercial formulations, NovaSol® (185), Curcuwin® (136) and LongVida® (100) stand out as they show bioavailability over 100 times higher than the reference curcumin [11]. The formulations CurcuminRich, Biomor, Liposomal curcumin mango, Liposomal curcumin, and Dr. Mercola Curcumin Advanced are also available on the market for oral administration of curcumin [95].

9. Biological Activities of Curcumin

This section describes in detail the biological activities of curcumin, with a special emphasis on its antimicrobial activity (Figure 6).

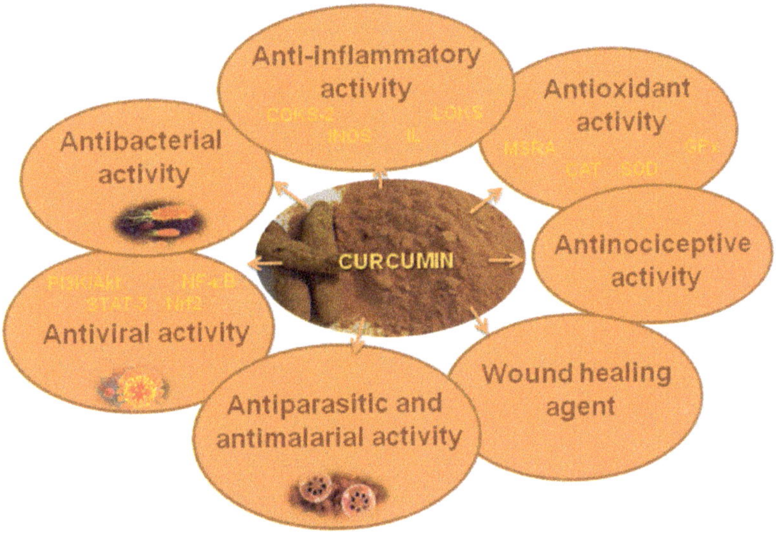

Figure 6. Biological activities of curcumin.

9.1. *Antimicrobial Activity*

9.1.1. Antibacterial Activity

Curcumin shows a wide range of antibacterial effects. It causes membrane damage in the cells of Gram-positive (*Staphylococcus aureus* and *Enterococcus faecalis*) and Gram-negative (*Escherichia coli* and *Pseudomonas aeruginosa*) bacteria [138]. It blocks the growth of bacteria owing to its structural characteristics and the formation of antioxidant products, inhibits bacterial virulence factors and the formation of bacterial biofilm, and prevents bacterial adhesion to host receptors. As a photosensitizer, curcumin induces phototoxicity and inhibits bacterial growth under blue light [139]. In the study by Adamczak et al., the effectiveness of curcumin was tested in vitro on over 100 strains of pathogens within 19 species. The antimicrobial activity was determined by the broth microdilution method and by calculating the minimum inhibitory concentration (MIC). The results confirmed a much higher susceptibility to Gram-positive than to Gram-negative bacteria. The MIC was also high in *Staphylococcus aureus*, *Staphylococcus haemolyticus*, *Escherichia coli*, and *Proteus mirabilis* resistant to a large number of drugs ($\geq$2000 µg/mL). However, curcumin was effective against some species and strains: *Streptococcus pyogenes* (mean MIC = 31.25 µg/mL), *Staphylococcus aureus* sensitive to methicillin (250 µg/mL), *Acinetobacter lwoffii* (250 µg/mL) and single strains of *Enterococcus faecalis* and *Pseudomonas aeruginosa* (62.5 µg/mL). Curcumin shows poor activity against clinical isolates of *Candida species*. Curcumin can be considered a promising antibacterial agent but with very selective activity [140]. The antimicrobial activity of curcumin against pathogens in burn wounds is shown in Table 5.

Table 5. Minimum inhibitory concentrations (MIC) of curcumin and fractional inhibitory concentration indices (FICIs) for potentially important pathogens of burn wounds [141].

Isolate	Genes	Curcumin MIC μg/mL	FICI
Klebsiella pneumonie	DHA	128	0.5
Pseudomonas aeruginosa	VEB	128	0.5
Acinetobacter baumanni	OXA-23, OXA-24	128	0.37
Acinetobacter baumanni	OXA-23, OXA-24	128	1
Pseudomonas aeruginosa	IMP-1	128	1
Enterococcus faecalis ATCC 29212	Type strain	128	0.26
Pseudomonas aeruginosa	GES	128	0.75
Acinetobacter baumanni	OXA-23, OXA-24	512	0.25
Acinetobacter baumanni ATCC19606	Type Strain	512	0.5
Acinetobacter baumanni	OXA-23, OXA-24	512	0.25
Pseudomonas aeruginosa	IMP-1	512	0.064
Pseudomonas aeruginosa	VIM-1	512	0.064
Escherichia coli ATCC 25922	Type Strain	256	0.4
Klebsiella pneumonie ATCC700603	Type Strain	256	0.5
Klebsiella pneumonie	NDM-6	256	0.28
Klebsiella pneumonie	NDM-1	256	0.56
Klebsiella pneumonie	NDM-6	256	0.56
Pseudomonas aeruginosa	IMP-2	256	0.56

9.1.2. Antiviral Activity

The antiviral effect of curcumin is manifested through interference with virus replication or through suppression of cellular signalling pathways that are essential for virus replication, such as the phosphatidylinositol 3-kinase/protein kinase B (PI3K/Akt) and NF-κB [142]. Curcumin exhibits antiviral activity against DNA and RNA viruses [143]. Jeong et al. established a mechanism by which curcumin pretreatment controlled the early stage of viral haemorrhagic septicaemia virus (VHSV) infection in fathead minnow cells. By rearranging the F-actin/G-actin ratio through reduced regulation of heat shock cognate 71 (HSC71), virus entry into cells is suppressed [144]. Ferreira et al. found that curcumin significantly reduced the replication of HIV-1 and herpes simplex virus type 2 (HSV-2) in chronically infected T cells and human primary genital epithelial cells [145]. Curcumin is an inhibitor of the redox function apurinic/apyrimidinic endonuclease 1 (APE1), affecting many genes, thus accounting for the wide range of curcumin effects on various human diseases. Curcumin effectively blocks the replication of the herpes virus associated with sarcoma and inhibits the pathogenic processes of angiogenesis and cell invasion [146]. Curcumin also exhibits antiviral activity against Zika and chikungunya viruses, dengue virus, hepatitis C virus [147], coxsackievirus, human papilloma virus, severe acute respiratory syndrome coronavirus 2 (SARS-CoV-2), etc., [147–149].

SARS-CoV-2 is an infectious virus that causes coronavirus disease-2019 (COVID-19) [150]. The disease with significant mortality worldwide poses a global threat due to the difficulties in treatment because there is currently no approved antiviral drug with proven efficacy and minor adverse effects [151,152]. The severity of the pandemic has prompted scientists to examine existing drugs with the potential for treating COVID-19 [150]. Studies show that curcumin is a good candidate for treating the COVID-19 virus and preventing fatal complications of this disease due to its thoroughly tested and confirmed anti-inflammatory, antiviral, antinociceptive immunomodulatory, antipyretic, antifibrotic, pulmoprotective, and antifatigue effects. Curcumin can interact with spike proteins or angiotensin 2 (ACE2) proteins in the signalling pathway induced by COVID-19. Curcumin also inhibits several important signalling pathways in viral infection, such as transcription factors (NF-κB, signal transducer and activator of transcription 3 (STAT-3), Vnt/b-catenin, nuclear factor E2-related factor (Nrf2), p38/MAPK, and virus-induced inflammation by modulating the manifestation of various factors (IL-10, Interleukin-18 (IL-18), IL-6, tumour necrosis factor (TNF) α/β and COX-2) in COVID-19 [153,154]. Valizadeha et al. investigated the effect of nanocurcumin on the modulation of inflammatory cytokines in patients with COVID-19. Messenger ribonucleic acid (mRNA) expression and cytokine secretion levels of Interleukin-1 β (IL-1β), IL-6, TNF-α, and IL-18 were assessed by polymerase chain reaction (PCR) in real-time and enzyme-linked immunosorbent assay (ELISA), respectively. The results showed that the expression of IL-1β and IL-6 mRNA was dramatically reduced after nanocurcumin administration. This study suggests that by regulating the inflammatory response, nanocurcumin can be used as an innovative therapeutic agent for patients with COVID-19 [155].

9.1.3. Antiparasitic and Antimalarial Activity

Curcumin shows activity against different types of parasites in vitro and in vivo. The antiprotozoal activity of curcumin *is shown* against *Leishmania major*, *Leishmania donovani* [156,157], *Trichomonas vaginalis* [158], *Entamoeba histolytica* [159], *Giardia lamblia* [160], *Toxoplasma gondii* [161], *Neospora caninum* [162], etc. The combinations of netilmicin and curcumin and metronidazole and curcumin show a synergistic effect and can be used to treat leishmaniasis and amoebiasis, respectively [156,159]. Curcumin also exhibits anthelmintic activity on the nematode *Ascaridia galli* and the cestode *Raillietina cesticillus* [163,164]. The studies show that curcumin is used to treat malaria and that it can increase the effectiveness of antimalarial drugs [165]. Busari et al. examined the in vivo antiplasmodial activity and the assessment of the toxicity of curcumin incorporated into poly(lactic-*co*-glycolic) nanoparticles. The formulated drug with nanoparticles demonstrated better activity against the malaria parasite than free curcumin. The antimalarial activity of the drug is better at lower concentrations. In vivo toxicity studies have confirmed the safety of the formulated drug at the tested doses [166]. Novaes et al. evaluated the efficacy of curcumin as a complementary strategy in benznidazole-based chemotherapy in mice acutely infected with *Trypanosoma cruzi*. The results of the research show that the combination of benznidazole with curcumin may be a relevant therapy in the treatment of Chagas' disease caused by *T. cruzi* because it reduces the toxic effects of benznidazole and increases its antiparasitic activity [165]. An overview of the recent research of curcumin's antiviral, antiparasitic, and antimalarial activities is shown in Table 6.

Table 6. Antiviral, antiparasitic and antimalarial activity of curcumin.

Activity	Substance	Type of Microorganism	Therapeutic Effect	Reference
Antiviral				
	Curcumin, nanomicelles	Hepatitis C virus	The antiviral effects of curcumin nanomicelles on hepatitis C virus.	[167]
	Curcumin	Vesicular stomatitis virus	Determination of curcumin effects on vesicular stomatitis virus Dicer-1 expression.	[168]
	Curcumin	Chikungunya virus, zika virus	Antiviral activity of curcumin against zika and chikungunya virus.	[147]
	Curcumin, nanoparticles	Human immunodeficiency virus 1 (HIV-1)	Immunomodulatory activities of curcumin-stabilized silver nanoparticles on HIV-1.	[169]
	Curcumin	Enterovirus 71 (EV71)	Antiviral effects of curcumin on EV71.	[170]
	Curcumin	Human T lymphotropic virus 1 (HTLV-1)	Determination of curcumin on the expression of c-FLIP in HTLV-1-associated myelopathy/tropical spastic paraparesis (HAM/TSP) patients.	[171]
	Curcumin	Kaposi's sarcoma-associated herpesvirus (KSHV or HHV8)	Antiviral activity of curcumin against KSHV replication and pathogenesis.	[146]
	Curcumin	Human immunodeficiency virus 1 (HIV-1)	Multifunctional mesoporous curcumin encapsulated iron phenanthroline Nanocluster on HIV-1.	[172]

Table 6. *Cont.*

Activity	Substance	Type of Microorganism	Therapeutic Effect	Reference
	Curcumin	Zika virus	Inhibitory effects of novel natural products against zika virus.	[173]
	Curcumin, nanocurcumin	Dengue virus	Antiviral activity of curcumin encapsulated in nanoemulsion against dengue virus serotypes.	[174]
	Curcumin	Transmissible gastroenteritis virus	Antiviral effects of curcumin on transmissible gastroenteritis virus.	[175]
	Curcumin	Human parainfluenza virus type 3	Evaluation of curcumin on replication of human parainfluenza virus type 3.	[176]
	Curcumin	Hepatitis B virus	Evaluation of curcumin on viral entry of hepatitis B.	[156]
Antiparasitic and Antimalarial				
	Curcumin and netilmicin	*Leishmania major, Leishmania donovani*	Antileishmanial activity of netilmicin combined with curcumin significantly enhanced compared with when used alone.	[177]
	Nanoformulation of curcumin and miltefosine	*Leishmania donovani*	Combination therapy of curcumin with miltefosine exhibited a synergistic effect on both promastigotes and amastigotes under in vitro conditions.	[166]
	Curcumin Encapsulated to PLGA	*Plasmodium berghei*	Encapsulation of curcumin in PLGA led to increased parasite suppression about 56.8% at 5 mg/kg of nanoformulation, which was higher than in free curcumin (40.5%) at 10 mg/kg.	[178]
	Curcumin alone	*Giardia lamblia*	Curcumin inhibited giardia proliferation disrupted the cytoskeletal structures of trophozoites in the dose-dependent mode.	[160]
	Curcumin alone	*Fasciola gigantica*	A significant decrease was observed in the expression of glutathione-S-transferase and superoxide dismutase.	[179]
	Curcumin alone	*Cryptosporidium parvum*	The anticryptosporidial and antioxidant activity of curcumin against C. parvum were confirmed.	[180]
	Nanotized curcumin-benzothiophene conjugate	*Plasmodium falciparum*	The improved oral bioavailability of the nanotized formulation lowered the dosage at which the pharmacological effect was achieved while avoiding any observable adverse side effects.	[181]
	Curcumin, nanocomposite	*Plasmodium falciparum*	The antiparasitic effect of the nanocomposite on the metabolites of plasmodium falciparum	[182]

9.2. Anti-Inflammatory Activity

Reactive oxygen species (ROS) play a key role in enhancing inflammation by activating transcription factors NF-κB, the activator protein 1 (AP-1), in acetylation and deacetylation of nuclear histone in a range of inflammatory diseases [183]. The anti-inflammatory effect of curcumin is based on its ability to inhibit COX-2, lipoxygenase (LOX), inducible nitric oxide synthase (iNOS) [184], arachidonic acid metabolism [185], cytokines (interleukins) [186,187], and tumour necrosis factor [188], NF-κB [184] and the release of steroid hormones [189]. COX-2, LOX, and iNOS are important enzymes that mediate inflammatory processes. Improper regulation of COX-2 and iNOS has been associated with the pathophysiology of certain types of cancer in humans, as well as with inflammatory disorders [184]. Curcumin and rutin downregulate COX-2 and reduce tumour-associated inflammation in HPV16-transgenic mice [190] The findings of preclinical studies in animal models with invasive pneumonia showed that curcumin exhibits a protective effect. It regulates the expression of pro and anti-inflammatory factors (interleukin-6 (IL-6), interleukin-8 (IL-8), interleukin-10 (IL-10), and COX-2), induces apoptosis of polymorphonuclear neutrophilic (PMN) cells, and removes ROS, thus improving the inflammatory response. These studies indicate that curcumin can be used as a therapeutic agent against pneumonia and acute lung injury (ALI)/fatal acute respiratory distress syndrome (ARDS) in humans, resulting from coronavirus infection [191].

9.3. Antioxidant Activity

ROS and reactive nitrogen species (RNS) are generated in the human body in various endogenous systems, in pathophysiological conditions or exposure to various physical and chemical factors. Free radicals can change lipids (lipid peroxidation), proteins (loss of enzyme activity), and DNA (mutagenesis and carcinogenesis); they contribute to ageing and many human diseases. Natural protective antioxidant mechanisms include superoxide dismutase (SOD), catalase (CAT), glutathione (GSH), glutathione peroxidase (GPx), and reductase, vitamin E (tocopherols and tocotrienols), vitamin C, etc. [4]. Curcumin also shows strong antioxidant activity. The antioxidant property is attributed to the presence of various functional groups, including methoxy, phenoxy, and carbon–carbon double bonds in its structure. Curcumin is a classic phenolic antioxidant that donates H atoms from phenolic groups [192]. In the work of Samarghandian et al., it was found that curcumin can inhibit oxidative damage caused by stress in the brain, liver, and kidneys of rats [193]. Lipid peroxidation is significantly reduced in rats treated with curcumin before applying γ-radiation [194]. Curcumin increases enzymatic antioxidant activity by increasing the expression of methionine sulfoxide reductase (MSRA) and increasing the levels of the enzymes MSRA, SOD, CAT and GPx [195]. Curcumin may act as an antioxidant against oxidative stress in rats with diabetes mellitus by increased SOD expression in cochlear fibroblasts [196]. The antioxidant activity of curcumin was assessed using the 1,1-diphenyl-2-picryl hydrazyl (DPPH) radical assay compared to ascorbic acid, a known antioxidant. The percentage of free radical removal using curcumin and ascorbic acid was 69 and 62%, respectively, at a concentration of 0.1 mM. No significant difference was observed between curcumin and ascorbic acid in antioxidant potential [197]. Curcumin has shown a large capacity to remove smaller oxidative molecules such as H_2O_2, HO•, ROO•. Curcumin can be used as an effective antioxidant to protect against ROS in the cytoplasm of cells [198]. Curcumin formulations with different carriers that are stable and protected from various influences are used as antioxidants [199,200].

9.4. Antinociceptive Activity

Preclinical studies have shown that curcumin has an antinociceptive effect on inflammatory and neuropathic pain. The effects of curcumin on postoperative pain in rats were investigated in the work of Zhu et al. The results of the study show that curcumin can alleviate postoperative pain and accelerate recovery from surgery. However, treatment with curcumin before surgery did not affect the threshold of postoperative pain and recovery rate [201]. The antinociceptive effect of poly(d,l-lactide-*co*-glycolide) nanovesicles with curcumin (PLGA-CUR) administered intravenously or intrathecally in mice in small and high doses was tested using formalin test, zymosan-induced hyperalgesia and sciatic nerve ligation that causes neuropathic allodynia and hyperalgesia. PLGA-CUR administered intravenously managed to reduce the response to nociceptive stimuli in the formalin test and zymosan-induced hyperalgesia, while pure curcumin was inactive. The low doses of intrathecally administered PLGA-CUR significantly reduced allodynia produced by sciatic nerve ligation. Long-lasting antinociceptive effects were observed when high doses of PLGA-CUR were administered intrathecally. At high doses, intrathecally applied pure curcumin had only rapid and transient antinociceptive effects. Measuring cytokine levels and brain-derived neurotrophic factor (BDNF) in the spinal cord of neuropathic mice shows that the antinociceptive effects of PLGA-CUR depend on the decline in the release of cytokines and BDNF in the spinal cord. The study results show the efficacy of PLGA-CUR and suggest that the nanoformulation of PLGA-CUR could be a new potential drug in the treatment of pain [202].

9.5. Wound Healing Agent

Curcumin has strong modulating effects on the wound healing process. The wound healing process consists of four phases: coagulation, inflammation, proliferation, and tissue remodelling. Curcumin induces apoptosis of inflammatory cells during the early phase of wound healing, inhibits the activity of the transcription factor NF-κB, reduces the production of cytokines (TNF-α and IL-1), removes ROS, affects the production of antioxidant enzymes and thus reduces inflammation and shortens the inflammatory phase in the wound healing process. The studies have shown that during the proliferation phase, curcumin enhances fibroblast migration, enhances granulation tissue formation, collagen deposition, and re-epithelialization. In the final phase of wound healing, by increasing the production of the transforming growth factor β, curcumin enhances wound contractions and therefore increases fibroblast proliferation [203]. Various topical curcumin formulations such as films, fibres, emulsions, hydrogels, and various nanoformulations have been developed for targeted delivery of curcumin at the wounds [203–205]. Sodium alginate-g-poly(*N*-isopropylacrylamide), (Alg-pNIPAM), a thermosensitive hydrogel with incorporated curcumin as an in vivo wound dressing was synthesized by Zakerikhoob et al. In vivo studies have shown accelerated collagenesis, re-epithelialization, and wound contraction using the Alg-pNIPAM formulation with curcumin. The formulation showed a more significant anti-inflammatory effect than the free curcumin solution. Given the antioxidant and anti-inflammatory properties of curcumin and the concomitant effect of alginate in keeping wound areas moist, the developed thermosensitive formulation of curcumin could help accelerate wound healing [204]. An overview of the recent research about the biological activities of curcumin is shown in Table 7.

Table 7. Biological activities of curcumin.

Activity	Substance	Target	Therapeutic Effect	Reference
Anti-inflammatory				
	Curcumin	COX-2 NF-κB p-IκB ROS	Attenuates colistin-induced neurotoxicity in N2a cells via anti-inflammatory activity, suppression of oxidative stress, and a apoptosis.	[206]
	Curcumin	NF-κB COX-2	Attenuates airway inflammation and airway remoulding in cigarette smoke-induced COPD mice.	[207]
	Curcumin and rutin	COX-2	Reduce tumour-associated inflammation in HPV16-transgenic mice.	[189]
	Curcumin, curcumin and capsaicin	COX-2 IL-6 TGF-β	Combined curcumin and capsaicin are efficient against the lipopolysaccharide Induced expression of proinflammatory cytokines in peripheral blood mononuclear cells.	[208]
Antioxidant				
	Curcumin-loaded sodium alginate/ZnO hydrogel beads	DPPH Assay	Composite hydrogel beads have protected curcumin from light degradation can therefore prolong its antioxidant activity.	[209]
	Curcumin	MDA SOD GSH-Px	Curcumin protects the liver, kidneys and brain from the oxidative damage caused by irradiation.	[210]
	Curcumin, curcumin and piperine	MDA SOD Catalase	Curcumin may be used as an adjunct therapy in individuals with oxidative stress.	[211]
	Curcumin	MDA total Antioxidant Capacity (TAC)	Pure curcumin reduces MDA concentration and increases total antioxidant capacity.	[212]
Antinociceptive				
	Curcumin	DRG Neurons β-Endorphin and Enkephalin	The curcumin attenuates cancer-induced bone pain	[213]
	Curcumin	γ-Aminobutyric Acid (GABA) and Opioid Receptors	Antinociception of curcumin	[214]

Table 7. *Cont.*

Activity	Substance	Target	Therapeutic Effect	Reference
Wound healing agent	Curcumin-loaded PLGA nanovesicles (PLGA-CUR)	Cytokine and BDNF	Antinociceptive effects of PLGA-CUR	[201]
	Curcumin	The acid-Sensing Ion Channels (ASICs)	Antinociceptive Effects of Curcumin	[215]
	PVA/chitosan/curcumin patches	Cell Line Studies and MTT Assay	Antibacterial activity of PVA/Chi/Cur against four major bacterial strains commonly found in wound sites and water retainability indicates it to be a perfect material for wound treatment.	[135]
	Nanocurcumin	Fibroblast, Collagen, Reepithelization	Curcumin nanoformulation enhanced wound repair by inhibiting the inflammatory response, stimulating angiogenesis, inducing fibroblast proliferation and enhancing reepithelization and synthesis of collagen.	[127]
	Curcumin, Hydrogel	L929 Fibroblast Cells	Curcumin incorporation accelerates full-thickness skin wound healing	[114]

10. Conclusions

Curcumin is a widely studied natural compound that has exhibited enormous in vitro and in vivo therapeutic potential. Curcumin has anti-inflammatory, antioxidant, antiviral, proapoptotic, chemopreventive, chemotherapeutic, antinociceptive, antiproliferative, antiparasitic, and antimalarial effects and is used as a wound-healing agent. Poor absorption of curcumin in the small intestine, rapid metabolism, and rapid systemic elimination cause poor bioavailability of curcumin in humans. Different curcumin formulations are used for more prolonged circulation, better permeability, and resistance to metabolic processes, and thus to increase the efficacy of curcumin. Liposomes, micelles, phospholipid complexes, cyclodextrins, nanoparticles, emulsions, hydrogels, and phytosomes have been described in the reference sources as promising formulations that improve the physicochemical properties of curcumin and enable its safe and efficient use.

Author Contributions: Conceptualization, V.N.; methodology, V.N. and L.N.; investigation, M.U., I.G., A.D. and V.M.; writing—original draft preparation, M.U., I.G., A.D. and V.M.; writing—review and editing, V.N., L.N., M.U., I.G., A.D. and V.M.; visualization, V.N. and L.N.; supervision, V.N. and L.N. All authors have read and agreed to the published version of the manuscript.

Funding: The Republic of Serbia-Ministry of Education, Science, and Technological Development, Program for financing scientific research work, number 451-03-9/2021-14/200133.

Institutional Review Board Statement: Not applicable.

Informed Consent Statement: Not applicable.

Data Availability Statement: Data available in a publicly accessible repository.

Acknowledgments: Republic of Serbia-Ministry of Education, Science and Technological Development, Program for financing scientific research work, number 451-03-9/2021-14/200133.

Conflicts of Interest: The authors declare no conflict of interest.

References

1. Hosseini, A.; Hosseinzadeh, H. Antidotal or protective effects of *Curcuma longa* (turmeric) and its active ingredient, curcumin, against natural and chemical toxicities: A review. *Biomed. Pharmacother.* **2018**, *99*, 411–421. [CrossRef] [PubMed]
2. Toden, S.; Goel, A. The Holy Grail of Curcumin and its Efficacy in Various Diseases: Is Bioavailability Truly a Big Concern? *J. Restor. Med.* **2017**, *6*, 27–36. [CrossRef] [PubMed]
3. Den Hartogh, D.J.; Gabriel, A.; Tsiani, E. Antidiabetic Properties of Curcumin II: Evidence from In Vivo Studies. *Nutrients* **2020**, *12*, 58. [CrossRef]
4. Arshad, L.; Haque, M.A.; Abbas Bukhari, S.N.; Jantan, I. An overview of structure–activity relationship studies of curcumin analogs as antioxidant and anti-inflammatory agents. *Future Med. Chem.* **2017**, *9*, 605–626. [CrossRef] [PubMed]

5. Hassan, F.U.; Rehman, M.S.-u.; Khan, M.S.; Ali, M.A.; Javed, A.; Nawaz, A.; Yang, C. Curcumin as an Alternative Epigenetic Modulator: Mechanism of Action and Potential Effects. *Front. Genet.* **2019**, *10*, 514. [CrossRef] [PubMed]
6. Giordano, A.; Tommonaro, G. Curcumin and Cancer. *Nutrients* **2019**, *11*, 2376. [CrossRef] [PubMed]
7. Heng, M. Phosphorylase Kinase Inhibition Therapy in Burns and Scalds. *BioDiscovery* **2017**, *20*, e11207. [CrossRef]
8. Kunnumakkara, A.B.; Bordoloi, D.; Padmavathi, G.; Monisha, J.; Roy, N.K.; Prasad, S.; Aggarwal, B.B. Curcumin, the golden nutraceutical: Multitargeting for multiple chronic diseases. *Br. J. Pharmacol.* **2017**, *174*, 1325–1348. [CrossRef]
9. Purpura, M.; Lowery, R.P.; Wilson, J.M.; Mannan, H.; Münch, G.; Razmovski-Naumovski, V. Analysis of different innovative formulations of curcumin for improved relative oral bioavailability in human subjects. *Eur. J. Nutr.* **2018**, *57*, 929–938. [CrossRef]
10. Hewlings, S.J.; Kalman, D.S. Curcumin: A Review of Its Effects on Human Health. *Foods* **2017**, *6*, 92. [CrossRef]
11. Jamwal, R. Bioavailable curcumin formulations: A review of pharmacokinetic studies in healthy volunteers. *J. Integr. Med.* **2018**, *16*, 367–374. [CrossRef]
12. Ma, Z.; Wang, N.; He, H.; Tang, X. Pharmaceutical strategies of improving oral systemic bioavailability of curcumin for clinical application. *J. Control. Release* **2019**, *316*, 359–380. [CrossRef]
13. Zheng, B.; McClements, D.J. Formulation of More Efficacious Curcumin Delivery Systems Using Colloid Science: Enhanced Solubility, Stability, and Bioavailability. *Molecules* **2020**, *25*, 2791. [CrossRef]
14. Kocaadam, B.; Şanlier, N. Curcumin, an active component of turmeric (*Curcuma longa*), and its effects on health. *Crit. Rev. Food Sci. Nutr.* **2017**, *57*, 2889–2895. [CrossRef]
15. Almeida, H.H.; Barros, L.; Barreira, J.C.; Calhelha, R.C.; Heleno, S.A.; Sayer, C.; Miranda, C.G.; Leimann, F.V.; Barreiro, M.F.; Ferreira, I.C. Bioactive evaluation and application of different formulations of the natural colorant curcumin (E100) in a hydrophilic matrix (yogurt). *Food Chem.* **2018**, *261*, 224–232. [CrossRef]
16. Vogel, H.; Pelletier, J. Curcumin-biological and medicinal properties. *J. Pharma* **1815**, *2*, 24–29.
17. Vogel, A. Memoire sur la Curcumine. *J. Pharm. Chim.* **1842**, *3*, 20–27.
18. Milobedzka, J.; Kostanecki, S.; Lampe, V. Zur Kenntnis des Curcumins. *Ber. Deut. Chem. Ges.* **1910**, *43*, 2163–2170. [CrossRef]
19. Lampe, V.; Milobedzka, J. Studien über Curcumin. *Ber. Deut. Chem. Ges.* **1913**, *46*, 2235–2240. [CrossRef]
20. Schraufstätter, E.; Bernt, H. Antibacterial Action of Curcumin and Related Compounds. *Nature* **1949**, *164*, 456–457. [CrossRef]
21. Srinivasan, K.R. A Chromatographic Study of the Curcuminoids in *Curcuma Longa*, L. *Pharm. Pharmacol.* **1953**, *5*, 448–457. [CrossRef] [PubMed]
22. Patil, T.N.; Srinivasan, M. Hypocholesteremic effect of curcumin in induced hypercholesteremic rats. *Indian J. Exp. Boil.* **1971**, *9*, 167–169.
23. Srinivasan, M. Effect of curcumin on blood sugar as seen in a diabetic subject. *Indian J. Med. Sci.* **1972**, *26*, 269–270. [PubMed]
24. Srimal, R.C.; Dhawan, B.N. Pharmacology of diferuloyl methane (curcumin), a non-steroidal anti-inflammatory agent. *J. Pharm. Pharmacol.* **1973**, *25*, 447–452. [CrossRef] [PubMed]
25. Sharma, O.P. Antioxidant activity of curcumin and related compounds. *Biochem. Pharmacol.* **1976**, *25*, 1811–1812. [CrossRef]
26. Kuttan, R.; Bhanumathy, P.; Nirmala, K.; George, M.C. Potential anticancer activity of turmeric (*Curcuma longa*). *Cancer Lett.* **1985**, *29*, 197–202. [CrossRef]
27. Singh, S.; Aggarwal, B.B. Activation of Transcription Factor NF-κB Is Suppressed by Curcumin (Diferuloylmethane). *J. Biol. Chem.* **1995**, *270*, 24995–25000. [CrossRef]
28. Guimarães, A.F.; Vinhas, A.C.A.; Gomes, A.F.; Souza, L.H.; Krepsky, P.B. Essential Oil of *Curcuma longa* L. Rhizomes Chemical Composition, Yield Variation and Stability. *Quím. Nova* **2020**, *43*, 909–913. [CrossRef]
29. Pawar, H.A.; Gavasane, A.J.; Choudhary, P.D. A Novel and Simple Approach for Extraction and Isolation of Curcuminoids from Turmeric Rhizomes. *Nat. Prod. Chem. Res.* **2018**, *6*, 1–4. [CrossRef]
30. Tripathy, S.; Verma, D.K.; Thakur, M.; Patel, A.R.; Srivastav, P.P.; Singh, S.; Gupta, A.K.; Chávez-González, M.L.; Aguilar, C.N.; Chakravorty, N.; et al. Curcumin Extraction, Isolation, Quantification and Its Application in Functional Foods: A Review with a Focus on Immune Enhancement Activities and COVID-19. *Front. Nutr.* **2021**, *8*, 675. [CrossRef]
31. Sahne, F.; Mohammadi, M.; Najafpour, G.D.; Moghadamnia, A.A. Extraction of bioactive compound curcumin from turmeric (*Curcuma longa* L.) via different routes: A comparative study. *Pak. J. Biotechnol.* **2016**, *13*, 173–180.
32. Monton, C.; Settharaksa, S.; Luprasong, C.; Songsak, T. An optimization approach of dynamic maceration of *Centella asiatica* to obtain the highest content of four centelloids by response surface methodology. *Rev. Bras. Farmacogn.* **2019**, *29*, 254–261. [CrossRef]
33. Patil, S.S.; Rathod, V.K. Synergistic Effect of Ultrasound and Three Phase Partitioning for the Extraction of Curcuminoids from *Curcuma longa* and its Bioactivity Profile. *Process Biochem.* **2020**, *93*, 85–93. [CrossRef]
34. Sahne, F.; Mohammadi, M.; Najafpour, G.D.; Moghadamnia, A.A. Enzyme-assisted ionic liquid extraction of bioactive compound from turmeric (*Curcuma longa* L.): Isolation, purification and analysis of curcumin. *Ind. Crops Prod.* **2017**, *95*, 686–694. [CrossRef]
35. Liang, H.; Wang, W.; Xu, J.; Zhang, Q.; Shen, Z.; Zeng, Z.; Li, Q. Optimization of ionic liquid-based microwave-assisted extraction technique for curcuminoids from *Curcuma longa* L. *Food Bioprod. Process.* **2017**, *104*, 57–65. [CrossRef]
36. Nagavekar, N.; Singhal, R.S. Supercritical fluid extraction of *Curcuma longa* and *Curcuma amada* oleoresin: Optimization of extraction conditions, extract profiling, and comparison of bioactivities. *Ind. Crops Prod.* **2019**, *134*, 134–145. [CrossRef]

37. Chao, I.-C.; Wang, C.-M.; Marcotullio, M.C.; Lin, L.-G.; Ye, W.-C.; Zhang, Q.-W. Simultaneous Quantification of Three Curcuminoids and Three Volatile Components of Curcuma longa Using Pressurized Liquid Extraction and High-Performance Liquid Chromatography. *Molecules* **2018**, *23*, 1568. [CrossRef]
38. Nurjanah, N.; Saepudin, E. Curcumin isolation, synthesis and characterization of curcumin isoxazole derivative compound. *AIP Conf. Proc.* **2019**, *2168*, 020065. [CrossRef]
39. Muthukumar, V.P.; Vaishnavi, M.; Theepapriys, S.; Saravanaraj, A. Process Development for the Effective Extraction of Curcumin from *Curcuma Longa* L (Turmeric). *Int. J. Eng. Technol.* **2018**, *7*, 151–155. [CrossRef]
40. Yadav, D.K.; Sharma, K.; Dutta, A.; Kundu, A.; Awasthi, A.; Goon, A.; Banerjee, K.; Saha, S. Purity Evaluation of Curcuminoids in the Turmeric Extract Obtained by Accelerated Solvent Extraction. *J. AOAC Int.* **2017**, *100*, 586–591. [CrossRef]
41. Naksuriya, O.; Van Steenbergen, M.J.; Toraño, J.S.; Okonogi, S.; Hennink, W.E. A Kinetic Degradation Study of Curcumin in Its Free Form and Loaded in Polymeric Micelles. *AAPS J.* **2016**, *18*, 777–787. [CrossRef] [PubMed]
42. Liu, Y.; Li, J.; Fu, R.; Zhang, L.; Wang, D.; Wang, S. Enhanced extraction of natural pigments from *Curcuma longa* L. using natural deep eutectic solvents. *Ind. Crop. Prod.* **2019**, *140*, 111620. [CrossRef]
43. Nair, D.S.; Krishnakumar, K.; Krishnan, B. Pharmacological profile of curcumin: A review. *J. Biol. Innov.* **2017**, *6*, 533–541.
44. Rathore, S.; Mukim, M.; Sharma, P.; Devi, S.; Nagar, J.C.; Khalid, M. Curcumin: A Review for Health Benefits. *Int. J. Res. Rev.* **2020**, *7*, 273–290.
45. Nelson, K.M.; Dahlin, J.L.; Bisson, J.; Graham, J.; Pauli, G.F.; Walters, M.A. The Essential Medicinal Chemistry of Curcumin: Miniperspective. *J. Med. Chem.* **2017**, *60*, 1620–1637. [CrossRef]
46. Zielińska, A.; Alves, H.; Marques, V.; Durazzo, A.; Lucarini, M.; Alves, T.; Morsink, M.; Willemen, N.; Eder, P.; Chaud, M.; et al. Properties, Extraction Methods, and Delivery Systems for Curcumin as a Natural Source of Beneficial Health Effects. *Medicina* **2020**, *56*, 336. [CrossRef]
47. Angelini, G.; Pasc, A.; Gasbarri, C. Curcumin in silver nanoparticles aqueous solution: Kinetics of keto-enol tautomerism and effects on AgNPs. *Colloids Surf. A Physicochem. Eng. Asp.* **2020**, *603*, 125235. [CrossRef]
48. Girardon, M.; Parant, S.; Monari, A.; Dehez, F.; Chipot, C.; Rogalska, E.; Canilho, N.; Pasc, A. Triggering Tautomerization of Curcumin by Confinement into Liposomes. *ChemPhotoChem* **2019**, *3*, 1034–1041. [CrossRef]
49. Rege, S.A.; Megha, A.; Momin, S.A. Mini review on Keto-Enol ratio of curcuminoids. *Ukr. J. Food Sci.* **2019**, *7*, 27–32. [CrossRef]
50. Manolova, Y.; Deneva, V.; Antonov, L.; Drakalska, E.; Momekova, D.; Lambov, N. The effect of the water on the curcumin tautomerism: A quantitative approach. *Spectrochim. Acta Part A Mol. Biomol. Spectrosc.* **2014**, *132*, 815–820. [CrossRef]
51. Kawano, S.-I.; Inohana, Y.; Hashi, Y.; Lin, J.-M. Analysis of keto-enol tautomers of curcumin by liquid chromatography/mass spectrometry. *Chin. Chem. Lett.* **2013**, *24*, 685–687. [CrossRef]
52. Liu, J.; Wang, H.; Wang, P.; Guo, M.; Jiang, S.; Li, X.; Jiang, S. Films based on κ-carrageenan incorporated with curcumin for freshness monitoring. *Food Hydrocoll.* **2018**, *83*, 134–142. [CrossRef]
53. Yang, H.; Du, Z.; Wang, W.; Song, M.; Sanidad, K.; Sukamtoh, E.; Zheng, J.; Tian, L.; Xiao, H.; Liu, Z.; et al. Structure–Activity Relationship of Curcumin: Role of the Methoxy Group in Anti-inflammatory and Anticolitis Effects of Curcumin. *J. Agric. Food Chem.* **2017**, *65*, 4509–4515. [CrossRef] [PubMed]
54. Heger, M.; Van Golen, R.F.; Broekgaarden, M.; Michel, M.C. The Molecular Basis for the Pharmacokinetics and Pharmacodynamics of Curcumin and Its Metabolites in Relation to Cancer. *Pharmacol. Rev.* **2014**, *66*, 222–307. [CrossRef] [PubMed]
55. Gordon, O.N.; Luis, P.B.; Sintim, H.O.; Schneider, C. Unraveling Curcumin Degradation. *J. Biol. Chem.* **2015**, *290*, 4817–4828. [CrossRef] [PubMed]
56. Zhu, J.; Sanidad, K.Z.; Sukamtoh, E.; Zhang, G. Potential roles of chemical degradation in the biological activities of curcumin. *Food Funct.* **2017**, *8*, 907–914. [CrossRef]
57. Soleimani, V.; Sahebkar, A.; Hosseinzadeh, H. Turmeric (*Curcuma longa*) and its major constituent (curcumin) as nontoxic and safe substances: Review. *Phytother. Res.* **2018**, *32*, 985–995. [CrossRef]
58. Greil, R.; Greil-Ressler, S.; Weiss, L.; Schönlieb, C.; Magnes, T.; Radl, B.; Bolger, G.T.; Vcelar, B.; Sordillo, P.P. A phase 1 dose-escalation study on the safety, tolerability and activity of liposomal curcumin (Lipocurc™) in patients with locally advanced or metastatic cancer. *Cancer Chemother. Pharmacol.* **2018**, *82*, 695–706. [CrossRef]
59. Saghatelyan, T.; Tananyan, A.; Janoyan, N.; Tadevosyan, A.; Petrosyan, H.; Hovhannisyan, A.; Hayrapetyan, L.; Arustamyan, M.; Arnhold, J.; Rotmann, A.-R.; et al. Efficacy and safety of curcumin in combination with paclitaxel in patients with advanced, metastatic breast cancer: A comparative, randomized, double-blind, placebo-controlled clinical trial. *Phytomedicine* **2020**, *70*, 153218. [CrossRef]
60. Prasad, S.; Tyagi, A.K.; Aggarwal, B.B. Recent Developments in Delivery, Bioavailability, Absorption and Metabolism of Curcumin: The Golden Pigment from Golden Spice. *Cancer Res. Treat.* **2014**, *46*, 2–18. [CrossRef]
61. Dei Cas, M.; Ghidoni, R. Dietary Curcumin: Correlation between Bioavailability and Health Potential. *Nutrients* **2019**, *11*, 2147. [CrossRef]
62. Kotha, R.R.; Luthria, D.L. Curcumin: Biological, Pharmaceutical, Nutraceutical, and Analytical Aspects. *Molecules* **2019**, *24*, 2930. [CrossRef]
63. Hassaninasab, A.; Hashimoto, Y.; Tomita-Yokotani, K.; Kobayashi, M. Discovery of the curcumin metabolic pathway involving a unique enzyme in an intestinal microorganism. *Proc. Natl. Acad. Sci. USA* **2011**, *108*, 6615–6620. [CrossRef]

64. Ireson, C.; Orr, S.; Jones, D.J.; Verschoyle, R.; Lim, C.K.; Luo, J.L.; Howells, L.; Plummer, S.; Jukes, R.; Williams, M.; et al. Characterization of metabolites of the chemopreventive agent curcumin in human and rat hepatocytes and in the rat in vivo, and evaluation of their ability to inhibit phorbol ester-induced prostaglandin E2 production. *Cancer Res.* **2001**, *61*, 1059–1064.
65. Schneider, C.; Gordon, O.N.; Edwards, R.L.; Luis, P.B. Degradation of Curcumin: From Mechanism to Biological Implications. *J. Agric. Food Chem.* **2015**, *63*, 7606–7614. [CrossRef]
66. Aggarwal, B.B.; Deb, L.; Prasad, S. Curcumin Differs from Tetrahydrocurcumin for Molecular Targets, Signaling Pathways and Cellular Responses. *Molecules* **2015**, *20*, 185–205. [CrossRef]
67. Zaghary, W.; Hanna, E.; Zanoun, M.; Abdallah, N.; Sakr, T. Curcumin: Analysis and Stability. *J. Adv. Pharm. Res.* **2019**, *3*, 47–58. [CrossRef]
68. Kotra, V.S.R.; Satyabanta, L.; Goswami, T.K. A critical review of analytical methods for determination of curcuminoids in turmeric. *J. Food Sci. Technol.* **2019**, *56*, 5153–5166. [CrossRef]
69. Kushwaha, P.; Shukla, B.; Dwivedi, J.; Saxena, S. Validated high-performance thin-layer chromatographic analysis of curcumin in the methanolic fraction of *Curcuma longa* L. rhizomes. *Futur. J. Pharm. Sci.* **2021**, *7*, 178. [CrossRef]
70. Jadhav, B.-K.; Mahadik, K.-R.; Paradkar, A.-R. Development and Validation of Improved Reversed Phase-HPLC Method for Simultaneous Determination of Curcumin, Demethoxycurcumin and Bis-Demethoxycurcumin. *Chromatographia* **2007**, *65*, 483–488. [CrossRef]
71. Wichitnithad, W.; Jongaroonngamsang, N.; Pummangura, S.; Rojsitthisak, P. A simple isocratic HPLC method for the simultaneous determination of curcuminoids in commercial turmeric extracts. *Phytochem. Anal.* **2009**, *20*, 314–319. [CrossRef]
72. Jayaprakasha, G.K.; Rao, L.J.M.; Sakariah, K.K. Improved HPLC Method for the Determination of Curcumin, Demethoxycurcumin, and Bisdemethoxycurcumin. *J. Agric. Food Chem.* **2002**, *50*, 3668–3672. [CrossRef]
73. Thorat, B.; Jangle, R. Reversed-phase High-performance Liquid Chromatography Method for Analysis of Curcuminoids and Curcuminoid-loaded Liposome Formulation. *Indian J. Pharm. Sci.* **2013**, *75*, 60–66. [CrossRef]
74. Radha, A.; Ragavendran, P.; Thomas, A.; Kumar, D.S. A cost effective hplc method for the analysis of curcuminoids. *Hygeia J. Drugs Med.* **2016**, *8*, 1–15. [CrossRef]
75. Nugroho, A.; Lukitaning, E.; Rakhmawati, N.; Rohman, A. Analysis of Curcumin in Ethanolic Extract of *Curcuma longa* Linn. and Curcuma xanthorriza Roxb. Using High Performance Liquid Chromatography with UV-Detection. *Res. J. Phytochem.* **2015**, *9*, 188–194. [CrossRef]
76. Hwang, K.-W.; Son, D.; Jo, H.-W.; Kim, C.H.; Seong, K.C.; Moon, J.-K. Levels of curcuminoid and essential oil compositions in turmerics (*Curcuma longa* L.) grown in Korea. *J. Appl. Biol. Chem.* **2016**, *59*, 209–215. [CrossRef]
77. Na Bhuket, P.R.; Niwattisaiwong, N.; Limpikirati, P.; Khemawoot, P.; Towiwat, P.; Ongpipattanakul, B.; Rojsitthisak, P. Simultaneous determination of curcumin diethyl disuccinate and its active metabolite curcumin in rat plasma by LC–MS/MS: Application of esterase inhibitors in the stabilization of an ester-containing prodrug. *J. Chromatogr. B Anal. Technol. Biomed. Life Sci.* **2016**, *1033–1034*, 301–310. [CrossRef]
78. Ma, W.; Wang, J.; Guo, Q.; Tu, P. Simultaneous determination of doxorubicin and curcumin in rat plasma by LC–MS/MS and its application to pharmacokinetic study. *J. Pharm. Biomed. Anal.* **2015**, *111*, 215–221. [CrossRef]
79. Ramalingam, P.; Ko, Y.T. A validated LC-MS/MS method for quantitative analysis of curcumin in mouse plasma and brain tissue and its application in pharmacokinetic and brain distribution studies. *J. Chromatogr. B Anal. Technol. Biomed. Life Sci.* **2014**, *969*, 101–108. [CrossRef]
80. Ashraf, K.; Mujeeb, M.; Ahmad, A.; Ahmad, N.; Amir, M. Determination of Curcuminoids in *Curcuma longa* Linn. by UPLC/Q-TOF–MS: An Application in Turmeric Cultivation. *J. Chromatogr. Sci.* **2015**, *53*, 1346–1352. [CrossRef]
81. Van Nong, H.; Hung, L.X.; Thang, P.N.; Chinh, V.D.; Van Vu, L.; Dung, P.T.; Van Trung, T.; Nga, P.T. Fabrication and vibration characterization of curcumin extracted from turmeric (*Curcuma longa*) rhizomes of the northern Vietnam. *SpringerPlus* **2016**, *5*, 1147. [CrossRef]
82. Sathisaran, I.; Dalvi, S.V. Crystal Engineering of Curcumin with Salicylic Acid and Hydroxyquinol as Coformers. *Cryst. Growth Des.* **2017**, *17*, 3974–3988. [CrossRef]
83. Thangavel, K.; Dhivya, K. Determination of curcumin, starch and moisture content in turmeric by Fourier transform near infrared spectroscopy (FT-NIR). *Eng. Agric. Environ. Food* **2019**, *12*, 264–269. [CrossRef]
84. Pöppler, A.-C.; Lübtow, M.M.; Schlauersbach, J.; Wiest, J.; Meinel, L.; Luxenhofer, R. Loading-Dependent Structural Model of Polymeric Micelles Encapsulating Curcumin by Solid-State NMR Spectroscopy. *Angew. Chem. Int. Ed.* **2019**, *58*, 18540–18546. [CrossRef]
85. Ali, Z.; Saleem, M.; Atta, B.M.; Khan, S.S.; Hammad, G. Determination of curcuminoid content in turmeric using fluorescence spectroscopy. *Spectrochim. Acta Part A Mol. Biomol. Spectrosc.* **2019**, *213*, 192–198. [CrossRef]
86. Pandey, K.U.; Dalvi, S.V. Understanding stability relationships among three curcumin polymorphs. *Adv. Powder Technol.* **2019**, *30*, 266–276. [CrossRef]
87. Iravani, S.; Soufi, G.J. Electron paramagnetic resonance (EPR) spectroscopy: Food, biomedical and pharmaceutical analysis. *Biomed. Spectrosc. Imaging* **2020**, *9*, 165–182. [CrossRef]
88. Dudylina, A.L.; Ivanova, M.V.; Shumaev, K.B.; Ruuge, E.K. Superoxide Formation in Cardiac Mitochondria and Effect of Phenolic Antioxidants. *Cell Biochem. Biophys.* **2018**, *77*, 99–107. [CrossRef]

89. Morales, N.P.; Sirijaroonwong, S.; Yamanont, P.; Phisalaphong, C. Electron Paramagnetic Resonance Study of the Free Radical Scavenging Capacity of Curcumin and Its Demethoxy and Hydrogenated Derivatives. *Biol. Pharm. Bull.* **2015**, *38*, 1478–1483. [CrossRef]
90. Nikolić, I.; Mitsou, E.; Damjanović, A.; Papadimitriou, V.; Antić-Stanković, J.; Stanojevic, B.; Xenakis, A.; Savić, S. Curcumin-loaded low-energy nanoemulsions: Linking EPR spectroscopy-analysed microstructure and antioxidant potential with in vitro evaluated biological activity. *J. Mol. Liq.* **2020**, *301*, 112479. [CrossRef]
91. Gopi, S.; Ac, K.V.; Varma, K.; Jude, S.; Amalraj, A.; Arundhathy, C.; George, R.; Sreeraj, T.; Divya, C.; Kunnumakkara, A.B.; et al. Comparative Oral Absorption of Curcumin in a Natural Turmeric Matrix with Two Other Curcumin Formulations: An Open-label Parallel-arm Study. *Phytother. Res.* **2017**, *31*, 1883–1891. [CrossRef]
92. Jäger, R.; Lowery, R.P.; Calvanese, A.V.; Joy, J.M.; Purpura, M.; Wilson, J.M. Comparative absorption of curcumin formulations. *Nutr. J.* **2014**, *13*, 11. [CrossRef]
93. Baspinar, Y.; Üstündas, M.; Bayraktar, O.; Sezgin, C. Curcumin and piperine loaded zein-chitosan nanoparticles: Development and in-vitro characterisation. *Saudi Pharm. J.* **2018**, *26*, 323–334. [CrossRef]
94. Kim, L.; Kim, J.Y. Chondroprotective effect of curcumin and lecithin complex in human chondrocytes stimulated by IL-1β via an anti-inflammatory mechanism. *Food Sci. Biotechnol.* **2019**, *28*, 547–553. [CrossRef]
95. Henriques, M.C.; Faustino, M.A.F.; Braga, S.S. Curcumin Innovative Delivery Forms: Paving the "Yellow Brick Road" of Antitumoral Phytotherapy. *Appl. Sci.* **2020**, *10*, 8990. [CrossRef]
96. Liu, W.; Zhai, Y.; Heng, X.; Che, F.Y.; Chen, W.; Sun, D.; Zhai, G. Oral bioavailability of curcumin: Problems and advancements. *J. Drug Target.* **2016**, *24*, 694–702. [CrossRef]
97. Stohs, S.J.; Chen, O.; Ray, S.D.; Ji, J.; Bucci, L.R.; Preuss, H.G. Highly Bioavailable Forms of Curcumin and Promising Avenues for Curcumin-Based Research and Application: A Review. *Molecules* **2020**, *25*, 1397. [CrossRef]
98. Kasapoglu-Calik, M.; Ozdemir, M. Synthesis and controlled release of curcumin-β-cyclodextrin inclusion complex from nanocomposite poly(N-isopropylacrylamide/sodium alginate) hydrogels. *J. Appl. Polym. Sci.* **2019**, *136*, 47554. [CrossRef]
99. Kongkaneramit, L.; Aiemsumang, P.; Kewsuwan, P. Development of curcumin liposome formulations using polyol dilution method. *Songklanakarin J. Sci. Technol.* **2016**, *38*, 605–610.
100. Tai, K.; Rappolt, M.; Mao, L.; Gao, Y.; Yuan, F. Stability and release performance of curcumin-loaded liposomes with varying content of hydrogenated phospholipids. *Food Chem.* **2020**, *326*, 126973. [CrossRef]
101. Cuomo, F.; Cofelice, M.; Venditti, F.; Ceglie, A.; Miguel, M.; Lindman, B.; Lopez, F. In-vitro digestion of curcumin loaded chitosan-coated liposomes. *Colloids Surf. B Biointerfaces* **2018**, *168*, 29–34. [CrossRef]
102. Algahtani, M.S.; Ahmad, M.Z.; Ahmad, J. Nanoemulsion loaded polymeric hydrogel for topical delivery of curcumin in psoriasis. *J. Drug Deliv. Sci. Technol.* **2020**, *59*, 101847. [CrossRef]
103. Guerrero, S.; Inostroza-Riquelme, M.; Contreras-Orellana, P.; Diaz-Garcia, V.; Lara, P.; Vivanco-Palma, A.; Cárdenas, A.; Miranda, V.; Robert, P.; Leyton, L.; et al. Curcumin-loaded nanoemulsion: A new safe and effective formulation to prevent tumor reincidence and metastasis. *Nanoscale* **2018**, *10*, 22612–22622. [CrossRef]
104. Cheng, Y.-H.; Ko, Y.-C.; Chang, Y.-F.; Huang, S.-H.; Liu, C.J.-L. Thermosensitive chitosan-gelatin-based hydrogel containing curcumin-loaded nanoparticles and latanoprost as a dual-drug delivery system for glaucoma treatment. *Exp. Eye Res.* **2019**, *179*, 179–187. [CrossRef]
105. Gera, M.; Sharma, N.; Ghosh, M.; Huynh, D.L.; Lee, S.J.; Min, T.; Kwon, T.; Jeong, D.K. Nanoformulations of curcumin: An emerging paradigm for improved remedial application. *Oncotarget* **2017**, *8*, 66680–66698. [CrossRef]
106. Saber-Moghaddam, N.; Salari, S.; Hejazi, S.; Amini, M.; Taherzadeh, Z.; Eslami, S.; Rezayat, S.M.; Jaafari, M.R.; Elyasi, S. Oral nano-curcumin formulation efficacy in management of mild to moderate 28 hospitalized coronavirus disease -19 patients: An open label nonrandomized clinical trial. *Phytother. Res.* **2021**, *35*, 2616–2623. [CrossRef]
107. Liu, Y.; Huang, P.; Hou, X.; Yan, F.; Jiang, Z.; Shi, J.; Xie, X.; Shen, J.; Fan, Q.; Wang, Z.; et al. Hybrid curcumin–phospholipid complex-near-infrared dye oral drug delivery system to inhibit lung metastasis of breast cancer. *Int. J. Nanomed.* **2019**, *14*, 3311–3330. [CrossRef]
108. Wang, J.; Wang, L.; Zhang, L.; He, D.; Ju, J.; Li, W. Studies on the curcumin phospholipid complex solidified with Soluplus®. *J. Pharm. Pharmacol.* **2018**, *70*, 242–249. [CrossRef]
109. Gupta, A.; Costa, A.P.; Xu, X.; Lee, S.-L.; Cruz, C.N.; Bao, Q.; Burgess, D.J. Formulation and characterization of curcumin loaded polymeric micelles produced via continuous processing. *Int. J. Pharm.* **2020**, *583*, 119340. [CrossRef]
110. Karavasili, C.; Andreadis, D.A.; Katsamenis, O.L.; Panteris, E.; Anastasiadou, P.; Kakazanis, Z.; Zoumpourlis, V.; Markopoulou, C.K.; Koutsopoulos, S.; Vizirianakis, I.S.; et al. Synergistic Antitumor Potency of a Self-Assembling Peptide Hydrogel for the Local Co-delivery of Doxorubicin and Curcumin in the Treatment of Head and Neck Cancer. *Mol. Pharm.* **2019**, *16*, 2326–2341. [CrossRef]
111. Liu, K.; Huang, R.-L.; Zha, X.-Q.; Li, Q.-M.; Pan, L.-H.; Luo, J.-P. Encapsulation and sustained release of curcumin by a composite hydrogel of lotus root amylopectin and chitosan. *Carbohydr. Polym.* **2020**, *232*, 115810. [CrossRef]
112. Gunathilake, T.M.S.U.; Ching, Y.C.; Chuah, C.H.; Illias, H.A.; Ching, K.Y.; Singh, R.; Nai-Shang, L. Influence of a nonionic surfactant on curcumin delivery of nanocellulose reinforced chitosan hydrogel. *Int. J. Biol. Macromol.* **2018**, *118*, 1055–1064. [CrossRef]

113. Pushpalatha, R.; Selvamuthukumar, S.; Kilimozhi, D. Cyclodextrin nanosponge based hydrogel for the transdermal co-delivery of curcumin and resveratrol: Development, optimization, in vitro and ex vivo evaluation. *J. Drug Deliv. Sci. Technol.* **2019**, *52*, 55–64. [CrossRef]
114. Shefa, A.A.; Sultana, T.; Park, M.K.; Lee, S.Y.; Gwon, J.-G.; Lee, B.-T. Curcumin incorporation into an oxidized cellulose nanofiber-polyvinyl alcohol hydrogel system promotes wound healing. *Mater. Des.* **2020**, *186*, 108313. [CrossRef]
115. Sahin, K.; Orhan, C.; Er, B.; Durmus, A.S.; Ozercan, I.H.; Sahin, N.; Padigaru, M.; Morde, A.; Rai, D. Protective Effect of a Novel Highly Bioavailable Formulation of Curcumin in Experimentally Induced Osteoarthritis Rat Model. *Curr. Dev. Nutr.* **2020**, *4* (Suppl. 2), 1765. [CrossRef]
116. Fakhri, S.; Shakeryan, S.; Alizadeh, A.; Shahryari, A. Effect of 6 Weeks of High Intensity Interval Training with Nano curcumin Supplement on Antioxidant Defense and Lipid Peroxidation in Overweight Girls- Clinical Trial. *Iran. J. Diabetes Obes.* **2020**, *11*, 173–180. [CrossRef]
117. Bateni, Z.; Rahimi, H.R.; Hedayati, M.; Afsharian, S.; Goudarzi, R.; Sohrab, G. The effects of nano-curcumin supplementation on glycemic control, blood pressure, lipid profile, and insulin resistance in patients with the metabolic syndrome: A randomized, double-blind clinical trial. *Phytother. Res.* **2021**, *35*, 3945–3953. [CrossRef] [PubMed]
118. Asadi, S.; Gholami, M.S.; Siassi, F.; Qorbani, M.; Khamoshian, K.; Sotoudeh, G. Nano curcumin supplementation reduced the severity of diabetic sensorimotor polyneuropathy in patients with type 2 diabetes mellitus: A randomized double-blind placebo-controlled clinical trial. *Complement. Ther. Med.* **2019**, *43*, 253–260. [CrossRef] [PubMed]
119. Abdolahi, M.; Sarraf, P.; Javanbakht, M.H.; Honarvar, N.M.; Hatami, M.; Soveyd, N.; Tafakhori, A.; Sedighiyan, M.; Djalali, M.; Jafarieh, A.; et al. A Novel Combination of ω-3 Fatty Acids and Nano-Curcumin Modulates Interleukin-6 Gene Expression and High Sensitivity C-reactive Protein Serum Levels in Patients with Migraine: A Randomized Clinical Trial Study. *CNS Neurol. Disord. Drug Targets* **2018**, *17*, 430–438. [CrossRef] [PubMed]
120. Jazayeri-Tehrani, S.A.; Rezayat, S.M.; Mansouri, S.; Qorbani, M.; Alavian, S.M.; Daneshi-Maskooni, M.; Hosseinzadeh-Attar, M.J. Nano-curcumin improves glucose indices, lipids, inflammation, and Nesfatin in overweight and obese patients with non-alcoholic fatty liver disease (NAFLD): A double-blind randomized placebo-controlled clinical trial. *Nutr. Metab.* **2019**, *16*, 8. [CrossRef] [PubMed]
121. Afshar, G.V.; Rasmi, Y.; Yagmayee, P.; Khadem-Ansari, M.-H.; Makhdomii, K.; Rasooli, J. The Effects of Nano-curcumin Supplementation on Serum Level of hs-CRP, Adhesion Molecules, and Lipid Profiles in Hemodialysis Patients, A Randomized Controlled Clinical Trial. *Iran. J. Kidney Dis.* **2020**, *14*, 52–61.
122. Saraf-Bank, S.; Ahmadi, A.; Paknahad, Z.; Maracy, M.; Nourian, M. Effects of curcumin supplementation on markers of inflammation and oxidative stress among healthy overweight and obese girl adolescents: A randomized placebo-controlled clinical trial. *Phytother. Res.* **2019**, *33*, 2015–2022. [CrossRef]
123. Pawar, K.S.; Mastud, R.N.; Pawar, S.K.; Pawar, S.S.; Bhoite, R.R.; Bhoite, R.R.; Kulkarni, M.V.; Deshpande, A.R. Oral Curcumin with Piperine as Adjuvant Therapy for the Treatment of COVID-19: A Randomized Clinical Trial. *Front. Pharmacol.* **2021**, *12*, 669362. [CrossRef]
124. Kedia, S.; Bhatia, V.; Thareja, S.; Garg, S.; Mouli, V.P.; Bopanna, S.; Tiwari, V.; Makharia, G.; Ahuja, V. Low dose oral curcumin is not effective in induction of remission in mild to moderate ulcerative colitis: Results from a randomized double blind placebo controlled trial. *World J. Gastrointest. Pharmacol. Ther.* **2017**, *8*, 147–154. [CrossRef]
125. Sadeghi, N.; Mansoori, A.; Shayesteh, A.; Hashemi, S.J. The effect of curcumin supplementation on clinical outcomes and inflammatory markers in patients with ulcerative colitis. *Phytother. Res.* **2019**, *34*, 1123–1133. [CrossRef]
126. Zhang, W.-Y.; Guo, Y.-J.; Han, W.-X.; Yang, M.-Q.; Wen, L.-P.; Wang, K.-Y.; Jiang, P. Curcumin relieves depressive-like behaviors via inhibition of the NLRP3 inflammasome and kynurenine pathway in rats suffering from chronic unpredictable mild stress. *Int. Immunopharmacol.* **2019**, *67*, 138–144. [CrossRef]
127. Hamam, F.; Nasr, A. Curcumin-loaded mesoporous silica particles as wound-healing agent: An In vivo study. *Saudi J. Med. Med. Sci.* **2020**, *8*, 17–24. [CrossRef]
128. Peng, K.-T.; Chiang, Y.-C.; Huang, T.-Y.; Chen, P.-C.; Chang, P.-J.; Lee, C.-W. Curcumin nanoparticles are a promising anti-bacterial and anti-inflammatory agent for treating periprosthetic joint infections. *Int. J. Nanomed.* **2019**, *14*, 469–481. [CrossRef]
129. Doukas, S.G.; Doukas, P.G.; Sasaki, C.T.; Vageli, D. The in vivo preventive and therapeutic properties of curcumin in bile reflux-related oncogenesis of the hypopharynx. *J. Cell. Mol. Med.* **2020**, *24*, 10311–10321. [CrossRef]
130. Paolillo, F.R.; Rodrigues, P.G.S.; Bagnato, V.S.; Alves, F.; Pires, L.; Corazza, A.V. The effect of combined curcumin-mediated photodynamic therapy and artificial skin on Staphylococcus aureus–infected wounds in rats. *Lasers Med. Sci.* **2020**, *36*, 1219–1226. [CrossRef]
131. Guorgui, J.; Wang, R.; Mattheolabakis, G.; Mackenzie, G.G. Curcumin formulated in solid lipid nanoparticles has enhanced efficacy in Hodgkin's lymphoma in mice. *Arch. Biochem. Biophys.* **2018**, *648*, 12–19. [CrossRef]
132. Sherin, S.; Balachandran, S.; Abraham, A. Curcumin incorporated titanium dioxide nanoparticles as MRI contrasting agent for early diagnosis of atherosclerosis- rat model. *Veter. Anim. Sci.* **2020**, *10*, 100090. [CrossRef]
133. Guo, Y.; Wu, R.; Gaspar, J.M.; Sargsyan, D.; Su, Z.-Y.; Zhang, C.; Gao, L.; Cheng, D.; Li, W.; Wang, C.; et al. DNA methylome and transcriptome alterations and cancer prevention by curcumin in colitis-accelerated colon cancer in mice. *Carcinogenesis* **2018**, *39*, 669–680. [CrossRef]

134. Pham, L.; Dang, L.H.; Truong, M.D.; Nguyen, T.H.; Le, L.; Le, V.T.; Nam, N.D.; Bach, L.G.; Nguyen, V.T.; Tran, N.Q. A dual synergistic of curcumin and gelatin on thermal-responsive hydrogel based on Chitosan-P123 in wound healing application. *Biomed. Pharmacother.* **2019**, *117*, 109183. [CrossRef]
135. Niranjan, R.; Kaushik, M.; Prakash, J.; Venkataprasanna, K.S.; Arpana, C.; Balashanmugam, P.; Venkatasubbu, G.D. Enhanced wound healing by PVA/Chitosan/Curcumin patches: In vitro and in vivo study. *Colloids Surf. B Biointerfaces* **2019**, *182*, 110339. [CrossRef]
136. Panda, S.K.; Parachur, V.A.; Mohanty, N.; Swain, T.; Sahu, S. A Comparative Pharmacokinetic Evaluation of a Bioavailable Curcumin Formulation Curene®with Curcumin Formulation Containing Turmeric Volatile Oil and Standard Curcuminoids 95% in Healthy Human Subjects. *Funct. Foods Health Dis.* **2019**, *9*, 134–144. [CrossRef]
137. Ullah, F.; Asgarov, R.; Venigalla, M.; Liang, H.; Niedermayer, G.; Münch, G.; Gyengesi, E. Effects of a solid lipid curcumin particle formulation on chronic activation of microglia and astroglia in the GFAP-IL6 mouse model. *Sci. Rep.* **2020**, *10*, 100090. [CrossRef] [PubMed]
138. Tyagi, P.; Singh, M.; Kumari, H.; Kumari, A.; Mukhopadhyay, K. Bactericidal Activity of Curcumin I Is Associated with Damaging of Bacterial Membrane. *PLoS ONE* **2015**, *10*, e0121313. [CrossRef] [PubMed]
139. Zheng, D.; Huang, C.; Huang, H.; Zhao, Y.; Khan, M.R.U.; Zhao, H.; Huang, L. Antibacterial Mechanism of Curcumin: A Review. *Chem. Biodivers.* **2020**, *17*, e2000171. [CrossRef] [PubMed]
140. Adamczak, A.; Ożarowski, M.; Karpiński, T.M. Curcumin, a Natural Antimicrobial Agent with Strain-Specific Activity. *Pharmaceuticals* **2020**, *13*, 153. [CrossRef] [PubMed]
141. Sharahi, J.Y.; Ahovan, Z.; Maleki, D.T.; Rad, Z.R.; Rad, Z.R.; Goudarzi, M.; Shariati, A.; Bostanghadiri, N.; Abbasi, E.; Hashemi, A. In vitro antibacterial activity of curcumin-meropenem combination against extensively drug-resistant (XDR) bacteria isolated from burn wound infections. *Avicenna J. Phytomedicine* **2020**, *10*, 3–10. [CrossRef]
142. Mathew, D.; Hsu, W.-L. Antiviral potential of curcumin. *J. Funct. Foods* **2018**, *40*, 692–699. [CrossRef]
143. Balasubramanian, A.; Pilankatta, R.; Teramoto, T.; Sajith, A.M.; Nwulia, E.; Kulkarni, A.; Padmanabhan, R. Inhibition of dengue virus by curcuminoids. *Antivir. Res.* **2019**, *162*, 71–78. [CrossRef]
144. Jeong, E.-H.; Vaidya, B.; Cho, S.-Y.; Park, M.-A.; Kaewintajuk, K.; Kim, S.R.; Oh, M.-J.; Choi, J.-S.; Kwon, J.; Kim, D. Identification of regulators of the early stage of viral hemorrhagic septicemia virus infection during curcumin treatment. *Fish Shellfish. Immunol.* **2015**, *45*, 184–193. [CrossRef]
145. Ferreira, V.H.; Nazli, A.; Dizzell, S.E.; Mueller, K.; Kaushic, C. The Anti-Inflammatory Activity of Curcumin Protects the Genital Mucosal Epithelial Barrier from Disruption and Blocks Replication of HIV-1 and HSV-2. *PLoS ONE* **2015**, *10*, e0124903. [CrossRef]
146. Li, H.; Zhong, C.; Wang, Q.; Chen, W.; Yuan, Y. Curcumin is an APE1 redox inhibitor and exhibits an antiviral activity against KSHV replication and pathogenesis. *Antivir. Res.* **2019**, *167*, 98–103. [CrossRef]
147. Mounce, B.C.; Cesaro, T.; Carrau, L.; Vallet, T.; Vignuzzi, M. Curcumin inhibits Zika and chikungunya virus infection by inhibiting cell binding. *Antivir. Res.* **2017**, *142*, 148–157. [CrossRef]
148. Teymouri, M.; Pirro, M.; Johnston, T.P.; Sahebkar, A. Curcumin as a multifaceted compound against human papilloma virus infection and cervical cancers: A review of chemistry, cellular, molecular, and preclinical features. *BioFactors* **2017**, *43*, 331–346. [CrossRef]
149. Babaei, F.; Nassiri-Asl, M.; Hosseinzadeh, H. Curcumin (a constituent of turmeric): New treatment option against COVID-19. *Food Sci. Nutr.* **2020**, *8*, 5215–5227. [CrossRef]
150. Soni, V.K.; Mehta, A.; Ratre, Y.K.; Tiwari, A.K.; Amit, A.; Singh, R.P.; Sonkar, S.C.; Chaturvedi, N.; Shukla, D.; Vishvakarma, N.K. Curcumin, a traditional spice component, can hold the promise against COVID-19? *Eur. J. Pharmacol.* **2020**, *886*, 173551. [CrossRef]
151. Dourado, D.; Freire, D.T.; Pereira, D.T.; Amaral-Machado, L.; Alencar, N.; de Barros, A.L.B.; Egito, E.S.T. Will curcumin nanosystems be the next promising antiviral alternatives in COVID-19 treatment trials? *Biomed. Pharmacother.* **2021**, *139*, 111578. [CrossRef]
152. Zahedipour, F.; Hosseini, S.A.; Sathyapalan, T.; Majeed, M.; Jamialahmadi, T.; Al-Rasadi, K.; Banach, M.; Sahebkar, A. Potential effects of curcumin in the treatment of COVID -19 infection. *Phytother. Res.* **2020**, *34*, 2911–2920. [CrossRef]
153. Subhan, F.; Khalil, A.A.K.; Zeeshan, M.; Haider, A.; Tauseef, I.; Haleem, S.K.; Ibrahim, A.S. Curcumin: From Ancient Spice to Modern Anti-Viral Drug in COVID-19 Pandemic. *Life Sci.* **2020**, *1*, 69–73. [CrossRef]
154. Thimmulappa, R.K.; Mudnakudu-Nagaraju, K.K.; Shivamallu, C.; Subramaniam, K.; Radhakrishnan, A.; Bhojraj, S.; Kuppusamy, G. Antiviral and immunomodulatory activity of curcumin: A case for prophylactic therapy for COVID-19. *Heliyon* **2021**, *7*, e06350. [CrossRef]
155. Valizadeh, H.; Abdolmohammadi-Vahid, S.; Danshina, S.; Gencer, M.Z.; Ammari, A.; Sadeghi, A.; Roshangar, L.; Aslani, S.; Esmaeilzadeh, A.; Ghaebi, M.; et al. Nano-curcumin therapy, a promising method in modulating inflammatory cytokines in COVID-19 patients. *Int. Immunopharmacol.* **2020**, *89*, 107088. [CrossRef]
156. Khanra, S.; Kumar, Y.P.; Dash, J.; Banerjee, R. In vitro screening of known drugs identified by scaffold hopping techniques shows promising leishmanicidal activity for suramin and netilmicin. *BMC Res. Notes* **2018**, *51*, 990–997. [CrossRef]
157. Bafghi, A.F.; Haghirosadat, B.F.; Yazdian, F.; Mirzaei, F.; Pourmadadi, M.; Pournasir, F.; Hemati, M.; Pournasir, S. A novel delivery of curcumin by the efficient nanoliposomal approach against Leishmania major. *Prep. Biochem. Biotechnol.* **2021**, *51*, 990–997. [CrossRef]

158. Mallo, N.; Lamas, J.; Sueiro, R.A.; Leiro, J.M. Molecular Targets Implicated in the Antiparasitic and Anti-Inflammatory Activity of the Phytochemical Curcumin in Trichomoniasis. *Molecules* **2020**, *25*, 5321. [CrossRef] [PubMed]
159. Rangel-Castañeda, I.A.; Hernández-Hernández, J.M.; Pérez-Rangel, A.; González-Pozos, S.; Carranza-Rosales, P.; Charles-Niño, C.L.; Tapia-Pastrana, G.; Ramírez-Herrera, M.A.; Castillo-Romero, A. Amoebicidal activity of curcumin on Entamoeba histolytica trophozoites. *J. Pharm. Pharmacol.* **2018**, *70*, 426–433. [CrossRef] [PubMed]
160. Gutiérrez-Gutiérrez, F.; Palomo-Ligas, L.; Hernández-Hernández, J.M.; Pérez-Rangel, A.; Aguayo-Ortiz, R.; Hernández-Campos, A.; Castillo, R.; González-Pozos, S.; Cortés-Zárate, R.; Ramírez-Herrera, M.A.; et al. Curcumin alters the cytoskeleton and microtubule organization on trophozoites of Giardia lamblia. *Acta Trop.* **2017**, *172*, 113–121. [CrossRef] [PubMed]
161. El-Shafey, A.A.M.; Hegab, M.H.A.; Seliem, M.M.E.; Barakat, A.M.A.; Mostafa, N.E.; Abdel-Maksoud, H.A.; Abdelhameed, R.M. Curcumin@metal organic frameworks nano-composite for treatment of chronic toxoplasmosis. *J. Mater. Sci. Mater. Med.* **2020**, *31*, 1–13. [CrossRef]
162. Qian, W.; Wang, H.; Shan, D.; Li, B.; Liu, J.; Liu, Q. Activity of several kinds of drugs against Neospora caninum. *Parasitol. Int.* **2015**, *64*, 597–602. [CrossRef]
163. Bazh, E.K.A.; El-Bahy, N.M. In vitro and in vivo screening of anthelmintic activity of ginger and curcumin on Ascaridia galli. *Parasitol. Res.* **2013**, *112*, 3679–3686. [CrossRef]
164. El-Bahy, N.M.; Bazh, E.K.A. Anthelmintic activity of ginger, curcumin, and praziquentel against Raillietina cesticillus (in vitro and in vivo). *Parasitol. Res.* **2015**, *114*, 2427–2434. [CrossRef]
165. Novaes, R.D.; Sartini, M.V.P.; Rodrigues, J.P.F.; Gonçalves, R.V.; Santos, E.C.; Souza, R.L.M.; Caldas, I.S. Curcumin Enhances the Anti-Trypanosoma cruzi Activity of Benznidazole-Based Chemotherapy in Acute Experimental Chagas Disease. *Antimicrob. Agents Chemother.* **2016**, *60*, 3355–3364. [CrossRef]
166. Busari, Z.A.; Dauda, K.A.; Morenikeji, O.A.; Afolayan, F.; Oyeyemi, O.T.; Meena, J.; Sahu, D.; Panda, A.K. Antiplasmodial Activity and Toxicological Assessment of Curcumin PLGA-Encapsulated Nanoparticles. *Front. Pharmacol.* **2017**, *8*, 622. [CrossRef]
167. Naseri, S.; Darroudi, M.; Aryan, E.; Gholoobi, A.; Rahimi, H.R.; Ketabi, K.; Movaqar, A.; Abdoli, M.; Gouklani, H.; Tei-mourpour, R. The antiviral effects of curcumin nanomicelles on the attachment and entry of hepatitis C virus. *Iran. J. Virol.* **2017**, *11*, 29–35.
168. Ahmed, J.; Tan, Y.; Ambegaokar, S. Effects of Curcumin on Vesicular Stomatitis Virus (VSV) Infection and Dicer-1 Expression. *FASEB J.* **2017**, *31*, 622.11. [CrossRef]
169. Sharma, R.K.; Cwiklinski, K.; Aalinkeel, R.; Reynolds, J.L.; Sykes, D.E.; Quaye, E.; Oh, J.; Mahajan, S.D.; Schwartz, S.A. Immunomodulatory activities of curcumin-stabilized silver nanoparticles: Efficacy as an antiretroviral therapeutic. *Immunol. Investig.* **2017**, *46*, 833–846. [CrossRef]
170. Huang, H.-I.; Chio, C.-C.; Lin, J.-Y. Inhibition of EV71 by curcumin in intestinal epithelial cells. *PLoS ONE* **2018**, *13*, e0191617. [CrossRef]
171. Poursina, Z.; Mohammadi, A.; Yazdi, S.Z.; Humpson, I.; Vakili, V.; Boostani, R.; Rafatpanah, H. Curcumin increased the expression of c-FLIP in HTLV-1-associated myelopathy/tropical spastic paraparesis (HAM/TSP) patients. *J. Cell. Biochem.* **2019**, *120*, 15740–15745. [CrossRef] [PubMed]
172. Sharma, A.; Yadav, A.; Gupta, N.; Sharma, S.; Kakkar, R.; Cwiklinski, K.; Quaye, E.; Mahajan, S.D.; Schwartz, S.A.; Sharma, R.K. Multifunctional mesoporous curcumin encapsulated iron-phenanthroline nanocluster: A new Anti-HIV agent. *Colloids Surf. B Biointerfaces* **2019**, *180*, 289–297. [CrossRef] [PubMed]
173. Nabila, N.; Suada, N.K.; Denis, D.; Yohan, B.; Adi, A.C.; Veterini, A.S.; Anindya, A.L.; Sasmono, R.T.; Rachmawati, H. Antiviral Action of Curcumin Encapsulated in Nanoemulsion against Four Serotypes of Dengue Virus. *Pharm. Nanotechnol.* **2020**, *8*, 54–62. [CrossRef] [PubMed]
174. Li, Y.; Wang, J.; Liu, Y.; Luo, X.; Lei, W.; Xie, L. Antiviral and virucidal effects of curcumin on transmissible gastroenteritis virus in vitro. *J. Gen. Virol.* **2020**, *101*, 1079–1084. [CrossRef]
175. Zhang, C.; Zhang, K.; Zang, G.; Chen, T.; Lu, N.; Wang, S.; Zhang, G. Curcumin Inhibits Replication of Human Parainfluenza Virus Type 3 by Affecting Viral Inclusion Body Formation. *BioMed Res. Int.* **2021**, *2021*, 13. [CrossRef]
176. Thongsri, P.; Pewkliang, Y.; Borwornpinyo, S.; Wongkajornsilp, A.; Hongeng, S.; Sa-Ngiamsuntorn, K. Curcumin inhibited hepatitis B viral entry through NTCP binding. *Sci. Rep.* **2021**, *11*, 19125. [CrossRef]
177. Tiwari, B.; Pahuja, R.; Kumar, P.; Rath, S.K.; Gupta, K.C.; Goyal, N. Nanotized Curcumin and Miltefosine, a Potential Combination for Treatment of Experimental Visceral Leishmaniasis. *Antimicrob. Agents Chemother.* **2017**, *61*, e01169–16. [CrossRef]
178. Ullah, R.; Rehman, A.; Zafeer, M.F.; Rehman, L.; Khan, Y.A.; Khan, M.A.H.; Khan, S.N.; Khan, A.U.; Abidi, S.M.A. Anthelmintic Potential of Thymoquinone and Curcumin on Fasciola gigantica. *PLoS ONE* **2017**, *12*, e0171267. [CrossRef]
179. Asadpour, M.; Namazi, F.; Razavi, S.M.; Nazifi, S. Comparative efficacy of curcumin and paromomycin against Cryptosporidium parvum infection in a BALB/c model. *Veter. Parasitol.* **2018**, *250*, 7–14. [CrossRef]
180. Ghosh, A.; Banerjee, T. Nanotized curcumin-benzothiophene conjugate: A potential combination for treatment of cerebral malaria. *IUBMB Life* **2020**, *72*, 2637–2650. [CrossRef]
181. Elmi, T.; Ardestani, M.S.; Hajialiani, F.; Motevalian, M.; Mohamadi, M.; Sadeghi, S.; Zamani, Z.; Tabatabaie, F. Novel chloroquine loaded curcumin based anionic linear globular dendrimer G2: A metabolomics study on Plasmodium falciparum in vitro using 1H NMR spectroscopy. *Parasitology* **2020**, *147*, 747–759. [CrossRef]

182. Wang, J.; Zhao, L.; Wei, Z.; Zhang, X.; Wang, Y.; Li, F.; Fu, Y.; Liu, B. Inhibition of histone deacetylase reduces lipopolysaccharide-induced-inflammation in primary mammary epithelial cells by regulating ROS-NF-κB signaling pathways. *Int. Immunopharmacol.* **2018**, *56*, 230–234. [CrossRef]
183. Shehzad, A.; Qureshi, M.; Anwar, M.N.; Lee, Y.S. Multifunctional Curcumin Mediate Multitherapeutic Effects. *J. Food Sci.* **2017**, *82*, 2006–2015. [CrossRef]
184. Banik, U.; Parasuraman, S.; Adhikary, A.K.; Othman, N.H. Curcumin: The spicy modulator of breast carcinogenesis. *J. Exp. Clin. Cancer Res.* **2017**, *36*, 98. [CrossRef]
185. Chai, Y.-S.; Chen, Y.-Q.; Lin, S.-H.; Xie, K.; Wang, C.-J.; Yang, Y.-Z.; Xu, F. Curcumin regulates the differentiation of naïve CD4+T cells and activates IL-10 immune modulation against acute lung injury in mice. *Biomed. Pharmacother.* **2020**, *125*, 109946. [CrossRef]
186. Hui, S.; Liu, K.; Zhu, X.; Kang, C.; Mi, M.T. Effect of Curcumin on IL-6 and IL-8: A Meta-analysis and Systematic Review. *J. Nutr. Food Sci.* **2018**, *8*, 1–7. [CrossRef]
187. Yu, Y.; Shen, Q.; Lai, Y.; Park, S.Y.; Ou, X.; Lin, D.; Jin, M.; Zhang, W. Anti-inflammatory Effects of Curcumin in Microglial Cells. *Front. Pharmacol.* **2018**, *9*, 386. [CrossRef]
188. Castaño, P.R.; Parween, S.; Pandey, A.V. Bioactivity of Curcumin on the Cytochrome P450 Enzymes of the Steroidogenic Pathway. *Int. J. Mol. Sci.* **2019**, *20*, 4606. [CrossRef]
189. Moutinho, M.S.; Aragão, S.; Carmo, D.; Casaca, F.; Silva, S.; Ribeiro, J.; Sousa, H.; Pires, I.; Queiroga, F.; Colaço, B.; et al. Curcumin and Rutin Down-regulate Cyclooxygenase-2 and Reduce Tumor-associated Inflammation in HPV16-Transgenic Mice. *Anticancer Res.* **2018**, *38*, 1461–1466. [CrossRef]
190. Liu, Z.; Ying, Y. The Inhibitory Effect of Curcumin on Virus-Induced Cytokine Storm and Its Potential Use in the Associated Severe Pneumonia. *Front. Cell Dev. Biol.* **2020**, *8*, 479. [CrossRef]
191. Pizzino, G.; Irrera, N.; Cucinotta, M.; Pallio, G.; Mannino, F.; Arcoraci, V.; Squadrito, F.; Altavilla, D.; Bitto, A. Oxidative Stress: Harms and Benefits for Human Health. *Oxid. Med. Cell. Longev.* **2017**, *2017*, 8416763. [CrossRef]
192. Samarghandian, S.; Azimi-Nezhad, M.; Farkhondeh, T.; Samini, F. Anti-oxidative effects of curcumin on immobilization-induced oxidative stress in rat brain, liver and kidney. *Biomed. Pharmacother.* **2017**, *87*, 223–229. [CrossRef]
193. Jagetia, G.C.; Rajanikant, G.K. Curcumin Stimulates the Antioxidant Mechanisms in Mouse Skin Exposed to Fractionated γ-Irradiation. *Antioxidants* **2015**, *4*, 25–41. [CrossRef] [PubMed]
194. Meshkibaf, M.H.; Maleknia, M.; Noroozi, S. Effect of curcumin on gene expression and protein level of methionine sulfoxide reductase A (MSRA), SOD, CAT and GPx in Freund's adjuvant inflammation-induced male rats. *J. Inflamm. Res.* **2019**, *12*, 241–249. [CrossRef] [PubMed]
195. Haryuna, T.S.H.; Munir, D.; Maria, A.; Bashiruddin, J. The Antioxidant Effect of Curcumin on Cochlear Fibroblasts in Rat Models of Diabetes Mellitus. *Iran. J. Otorhinolaryngol.* **2017**, *29*, 197–202. [CrossRef]
196. Asouri, M.; Ataee, R.; Ahmadi, A.A.; Amini, A.; Moshaei, M.R. Antioxidant and Free Radical Scavenging Activities of Curcumin. *Asian J. Chem.* **2013**, *25*, 7593–7595. [CrossRef]
197. Barzegar, A.; Moosavi-Movahedi, A.A. Intracellular ROS Protection Efficiency and Free Radical-Scavenging Activity of Curcumin. *PLoS ONE* **2011**, *6*, e26012. [CrossRef] [PubMed]
198. Chen, S.; Wu, J.; Tang, Q.; Xu, C.; Huang, Y.; Huang, D.; Luo, F.; Wu, Y.; Yan, F.; Weng, Z.; et al. Nano-micelles based on hydroxyethyl starch-curcumin conjugates for improved stability, antioxidant and anticancer activity of curcumin. *Carbohydr. Polym.* **2020**, *228*, 115398. [CrossRef]
199. Ma, Q.; Ren, Y.; Wang, L. Investigation of antioxidant activity and release kinetics of curcumin from tara gum/polyvinyl alcohol active film. *Food Hydrocoll.* **2017**, *70*, 286–292. [CrossRef]
200. Zhu, Q.; Sun, Y.; Yun, X.; Ou, Y.; Zhang, W.; Li, J.-X. Antinociceptive effects of curcumin in a rat model of postoperative pain. *Sci. Rep.* **2014**, *4*, 4932. [CrossRef]
201. Pieretti, S.; Ranjan, A.P.; Di Giannuario, A.; Mukerjee, A.; Marzoli, F.; Di Giovannandrea, R.; Vishwanatha, J.K. Curcumin-loaded Poly (d,l-lactide-co-glycolide) nanovesicles induce antinociceptive effects and reduce pronociceptive cytokine and BDNF release in spinal cord after acute administration in mice. *Colloids Surf. B Biointerfaces* **2017**, *158*, 379–386. [CrossRef]
202. Barchitta, M.; Maugeri, A.; Favara, G.; Lio, R.M.S.; Evola, G.; Agodi, A.; Basile, G. Nutrition and Wound Healing: An Overview Focusing on the Beneficial Effects of Curcumin. *Int. J. Mol. Sci.* **2019**, *20*, 1119. [CrossRef]
203. Mohanty, C.; Sahoo, S.K. Curcumin and its topical formulations for wound healing applications. *Drug Discov. Today* **2017**, *22*, 1582–1592. [CrossRef]
204. Zakerikhoob, M.; Abbasi, S.; Yousefi, G.; Mokhtari, M.; Noorbakhsh, M.S. Curcumin-incorporated crosslinked sodium alginate-g-poly (N-isopropyl acrylamide) thermo-responsive hydrogel as an in-situ forming injectable dressing for wound healing: In vitro characterization and in vivo evaluation. *Carbohydr. Polym.* **2021**, *271*, 118434. [CrossRef]
205. Krausz, A.E.; Adler, B.L.; Cabral, V.; Navati, M.; Doerner, J.; Charafeddine, R.A.; Chandra, D.; Liang, H.; Gunther, L.; Clendaniel, A.; et al. Curcumin-encapsulated nanoparticles as innovative antimicrobial and wound healing agent. *Nanomed. Nanotechnol. Biol. Med.* **2015**, *11*, 195–206. [CrossRef]
206. Dai, C.; Ciccotosto, G.D.; Cappai, R.; Tang, S.; Li, D.; Xie, S.; Xiao, X.; Velkov, T. Curcumin Attenuates Colistin-Induced Neurotoxicity in N2a Cells via Anti-inflammatory Activity, Suppression of Oxidative Stress, and Apoptosis. *Mol. Neurobiol.* **2018**, *55*, 421–434. [CrossRef]

207. Yuan, J.; Liu, R.; Ma, Y.; Zhang, Z.; Xie, Z. Curcumin Attenuates Airway Inflammation and Airway Remolding by Inhibiting NF-κB Signaling and COX-2 in Cigarette Smoke-Induced COPD Mice. *Inflammation* **2018**, *41*, 1804–1814. [CrossRef]
208. Vasanthkumar, T.; Hanumanthappa, M.; Lakshminarayana, R. Curcumin and capsaicin modulates LPS induced expression of COX-2, IL-6 and TGF-β in human peripheral blood mononuclear cells. *Cytotechnology* **2019**, *71*, 963–976. [CrossRef]
209. Wang, H.; Gong, X.; Guo, X.; Liu, C.; Fan, Y.-Y.; Zhang, J.; Niu, B.; Li, W. Characterization, release, and antioxidant activity of curcumin-loaded sodium alginate/ZnO hydrogel beads. *Int. J. Biol. Macromol.* **2019**, *121*, 1118–1125. [CrossRef]
210. Özçelik, M.; Erisir, M.; Guler, O.; Baykara, M.; Kirman, E. The effect of curcumin on lipid peroxidation and selected antioxidants in irradiated rats. *Acta Veter. Brno* **2019**, *87*, 379–385. [CrossRef]
211. Alizadeh, M.; Kheirouri, S. Curcumin reduces malondialdehyde and improves antioxidants in humans with diseased conditions: A comprehensive meta-analysis of randomized controlled trials. *BioMedicine* **2019**, *9*, 23. [CrossRef] [PubMed]
212. Jakubczyk, K.; Drużga, A.; Katarzyna, J.; Skonieczna-Żydecka, K. Antioxidant Potential of Curcumin—A Meta-Analysis of Randomized Clinical Trials. *Antioxidants* **2020**, *9*, 1092. [CrossRef] [PubMed]
213. Zhao, G.; Shi, Y.; Gong, C.; Liu, T.; Nan, W.; Ma, L.; Wu, Z.; Da, C.; Zhou, K.; Zhang, H. Curcumin Exerts Antinociceptive Effects in Cancer-Induced Bone Pain via an Endogenous Opioid Mechanism. *Front. Neurosci.* **2021**, *15*, 696861. [CrossRef] [PubMed]
214. Ju, J.; Shin, J.Y.; Yoon, J.J.; Yin, M.; Yoon, M.H. Differential expression of spinal γ-aminobutyric acid and opioid receptors modulates the analgesic effects of intrathecal curcumin on postoperative/inflammatory pain in rats. *Anesth. Pain Med.* **2018**, *13*, 82–92. [CrossRef]
215. Wu, Y.; Qin, D.; Yang, H.; Fu, H. Evidence for the Participation of Acid-Sensing Ion Channels (ASICs) in the Antinociceptive Effect of Curcumin in a Formalin-Induced Orofacial Inflammatory Model. *Cell. Mol. Neurobiol.* **2016**, *37*, 635–642. [CrossRef]

antibiotics

Review

Antimicrobial *Bacillus*: Metabolites and Their Mode of Action

Charlie Tran [1], Ian E. Cock [2], Xiaojing Chen [3] and Yunjiang Feng [1,*]

[1] Griffith Institute for Drug Discovery (GRIDD), Griffith University, Nathan, QLD 4111, Australia; charlie.tran@griffithuni.edu.au
[2] School of Environment and Science, Griffith University, Nathan, QLD 4111, Australia; i.cock@griffith.edu.au
[3] Bioproton Pty Ltd., Acacia Ridge, QLD 4110, Australia; wendy@bioproton.com
* Correspondence: author: y.feng@griffith.edu.au

Abstract: The agricultural industry utilizes antibiotic growth promoters to promote livestock growth and health. However, the World Health Organization has raised concerns over the ongoing spread of antibiotic resistance transmission in the populace, leading to its subsequent ban in several countries, especially in the European Union. These restrictions have translated into an increase in pathogenic outbreaks in the agricultural industry, highlighting the need for an economically viable, non-toxic, and renewable alternative to antibiotics in livestock. Probiotics inhibit pathogen growth, promote a beneficial microbiota, regulate the immune response of its host, enhance feed conversion to nutrients, and form biofilms that block further infection. Commonly used lactic acid bacteria probiotics are vulnerable to the harsh conditions of the upper gastrointestinal system, leading to novel research using spore-forming bacteria from the genus *Bacillus*. However, the exact mechanisms behind *Bacillus* probiotics remain unexplored. This review tackles this issue, by reporting antimicrobial compounds produced from *Bacillus* strains, their proposed mechanisms of action, and any gaps in the mechanism studies of these compounds. Lastly, this paper explores omics approaches to clarify the mechanisms behind *Bacillus* probiotics.

Keywords: antimicrobials; *Bacillus*; probiotic; animal feed; omics

check for updates

Citation: Tran, C.; Cock, I.E.; Chen, X.; Feng, Y. Antimicrobial *Bacillus*: Metabolites and Their Mode of Action. *Antibiotics* **2022**, *11*, 88. https://doi.org/10.3390/antibiotics11010088

Academic Editor: William N. Setzer

Received: 14 December 2021
Accepted: 8 January 2022
Published: 12 January 2022

1. Introduction

Probiotics are live microorganisms that can be consumed by its host to confer a range of health benefits. These benefits include the production of antimicrobial metabolites, restoration of the host microbiota, modulation of the immune system, and the release of digestive enzymes to improve nutrient uptake [1]. For example, *Bacillus subtilis* MA139 restored microbiota diversity in finishing pigs, improved their resistance to pathogenic illnesses, and promoted animal health and growth [2]. This increase in animal production makes probiotics a suitable alternative to antibiotic use in animals, due to the WHO advocating for its restricted use and its subsequent ban by the EU in 2006 [3].

Probiotics are commonly used in animal feed production, which do not contribute to antibiotic resistance and may even reduce it [4]. Selective probiotic bacteria have been used to treat antibiotic-associated diarrhea (AAD), a common side-effect of antibiotic use. Antibiotics elevate the risk of AAD by disrupting the diversity of the gut biota, allowing the proliferation of opportunistic pathogens such as *Clostridium difficile* [5]. This issue can be tackled through the use of probiotics, which inhibit pathogen growth and restabilize the intestinal microbiota back to normal levels [6]. Furthermore, probiotics can bind to the intestinal walls of its host and competitively exclude competing pathogens. Additionally, these probiotics produce a plethora of antimicrobial compounds that target pathogenic bacteria, which has driven the search for a potent probiotic strain for industrial use.

The issue lies in the presence of antibiotic resistance genes, with the commonly used *Lactobacillus* showing frequent resistance to vancomycin, ciprofloxacin, and aminoglycosides [7]. This development has driven the research into other probiotic genera not yet explored such as *Bacillus*. *Bacillus* probiotics are pore-forming bacteria that can survive the harsh conditions needed for pelletizing and can tolerate aerobic conditions for industrial use, unlike *Lactobacillus* and *Bifidobacterium* [8].

Several review papers have been published in the literature summarizing *Bacillus* metabolites, structural classes, and their antimicrobial activities [9–11]. However, no literature is available investigating the mechanisms of action of the antimicrobial metabolites from *Bacillus*. In this review, we summarized 47 antimicrobial compounds based on their molecular targets in the cell wall, plasma membrane, intracellular processes, and other emerging targets.

2. A Glance of Bioactive *Bacillus* and Their Antimicrobial Metabolites

To gain a good understanding of antimicrobial *Bacillus* sp., and hence their potential as a probiotic supplement, we conducted a literature review on antimicrobial Bacillus. Google-Scholar, PubMed, Scopus, and Science-Direct electronic databases were used to identify original scientific research papers. The terms 'antimicrobial Bacillus' and 'mechanism of action' were used as filters, with the earliest possible time range. Our literature search revealed that 1389 *Bacillus* strains have been reported for antimicrobial activity, composed of 27 different species (Figure 1). The most commonly reported species included *subtilis* (n = 348), *amyloliquefaciens* (n = 214), *licheniformis* (n = 114), *circulans* (n = 89), *thuringiensis* (n = 73), *pumilus* (n = 61), *velezensis* (n = 60), *megaterium* (n = 17), and *mojavensis* (n = 17) (Figure 1). The literature review also suggested that a substantial number of *Bacillus* species were not identified (n = 293). From the antimicrobial *Bacillus* sp., 47 metabolites have been identified and their mechanisms of actions reported [12]. We herein report the chemical structures of the metabolites, their antimicrobial activity, and mechanism of action. Details regarding these compounds, including source strain, anti-microbial activity, molecular target, and references are provided in Supplementary Table S1.

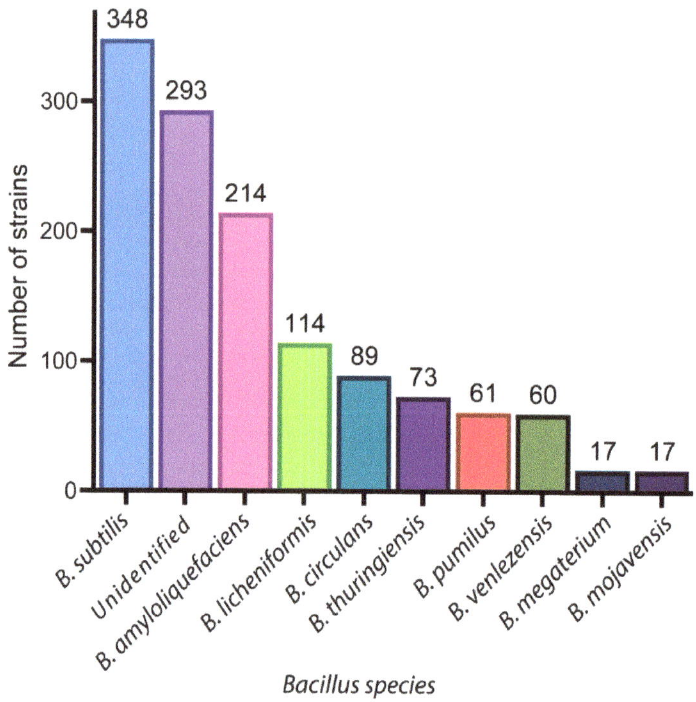

Figure 1. The number of *Bacillus* strains reported for each species.

3. Antimicrobial Metabolites and Their Mechanism of Action

3.1. Metabolites Targeting the Cell Wall

The cell wall is a selectively permeable layer that has a distinct layer of polysaccharides, peptidoglycans, and fungi-specific chitins and glucans [13]. This structure is located outside the plasma membrane and acts as a permeable barrier, which regulates the entry of metabolites into the cell and protects it against external stresses (Figure 2a). The cell wall is a promising target for drug development due to its absence in mammalian cells, and several *Bacillus* strains have been shown to target this structure by releasing enzymes (amylase, cellulase, chitinase, chitosanase, glucanase, and protease) and antimicrobial metabolites. From the reported 47 compounds with clearly defined mechanisms, 9 compounds target the cell wall (Figure 2a).

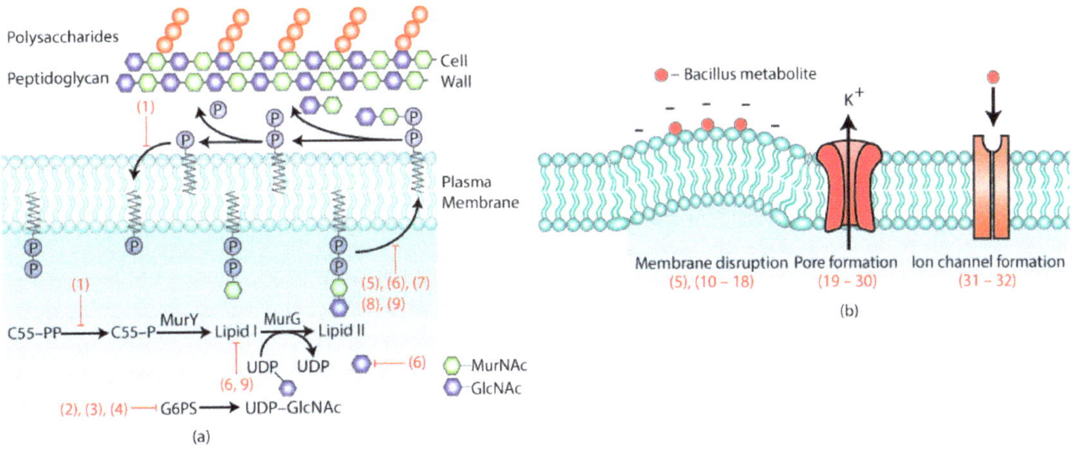

Figure 2. Metabolites targeting (**a**) cell wall and (**b**) plasma membrane.

The peptidoglycan layer provides integrity and protection to the cell. This layer is comprised of linear glycan strands, which alternate between N-acetylglucosamine (GlcNAc) and N-acetylmuramic acid (MurNAc) residues linked by β-1-4 bonds [14]. Bacitracin, an antibiotic first isolated from *B. licheniformis,* primarily acts on gram-positive bacteria such as *Streptococcus mutans* (MIC = 78.12 μg/mL) [15,16]. This antibiotic is comprised of a mixture of compounds, which include bacitracin A (1), B and C. Bacitracin A (Figure 3) prevents the dephosphorylation of undecaprenyl pyrophosphate (C55-PP) to undecaprenyl phosphate (C55-P), which prevents the formation of lipid I/II and the eventual disruption of the peptidoglycan layer [17]. Additionally, recent scanning-electron microscopy (SEM) analysis has shown that bacitracin inhibits the formation of biofilm by *Streptococcus mutans* by downregulating several genes related to cell division and biofilm [16].

Glucosamine-6-phosphate synthetase (G6PS) is an enzyme that catalyzes the production of uridine diphosphate N-acetylglucosamine (UDP-GlcNAc), which is a precursor for peptidoglycan synthesis [18]. Bacilysin (2) and its chlorinated derivative chlorotetaine (3) (Figure 3) were first isolated from *B. subtilis* A14 and *B. amyloliquefaciens* ZJU-2011, respectively [19,20]. Both compounds are active against a broad range of bacteria, with bacilysin inhibiting *E. coli* at MIC = 0.001 μg/mL and chlorotetaine inhibiting *Candidas* spp. and *Aspergillus niger* at an MIC value of 1.8–7.8 μg/mL [20,21]. Bacilysin first enters the cell by binding to a transmembrane transport protein and is subsequently hydrolyzed to anticapsin, a G6PS inhibitor [22]. Kanosamine (4) (Figure 3) produced from *B. cereus* UW85 inhibits a wide array of plant-related pathogens (i.e., *Phytophthora medicaginis* M2913 with an MIC = 25 μg/mL) [23]. Kanosmine inhibits *Candida albicans* by utilizing the glucose transport system to transport itself into the cell, where it is subsequently phosphorylated

ε-Poly-L-lysine electrostatically attaches to the phospholipid layer of the plasma membrane, which disturbs the membrane permeability to eventually lead to cell death [43,44]. Plantazolicin (11) (Figure 5), a product of *B. velezensis* FZB42, has been identified as a bacteriocin of interest, due to its restrictive spectrum against clinically relevant pathogens, such as *B. anthracis*, with an MIC value of 1–16 μg/mL [45]. This is highly relevant due to the very serious nature of anthrax. Further mechanism studies revealed that plantazolicin induces higher membrane fluidity and increases the proportion of cardiolipin, a cholesterol associated with higher osmotic stress [45]. Octapeptins are a class of lipooctapeptide antibiotics that were first isolated from *B. circulans* and that primarily inhibit gram-negative bacteria, with weaker activity on gram-positive bacteria and fungi [46]. Membrane microscopy studies show that octapeptin B (12) (*E. coli* SC 9251 MIC = 0.3 μg/mL) (Figure 5), produced from *B. circulans* ATCC 21656, disrupts the ion permeability of the membrane, which reduces the membrane proton gradient [47]. This translates into extensive membrane damage, the efflux of charged metabolites, and cell lysis. The aurantinins B-D (13–15) (Figure 5), a class of metabolites isolated from *B. subtilis* FMB60, exhibit similar MIC value for certain clinically relevant strains (i.e., *Clostridium sporogenes* CICC 10385 with a MIC $\leq$ 0.78 μg/mL, methicillin-resistant *Staphylococcus aureus* (MRSA) with an MIC = 6.25 μg/mL) [48]. SEM and transmission electron microscopy (TEM) studies show that the aurantinins cause plasma membrane lysis, leading to the efflux of metabolites from the cytoplasm [48]. However, these compounds require further structural elucidation to determine their precise stereochemistry. Myriocin (16) (Figure 5), produced from *B. amyloliquefaciens* LZN01, exerts antifungal activities against *Candidas albicans* (MIC = 1.0 μg/mL) [49]. SEM and TEM microscopy studies have indicated that myriocin binds to serine palmitoyl transferase and disrupts the plasma membrane, causing leakage and eventual pore formation [50]. Further omics analysis has revealed that myriocin alters the expression changes related to sphingolipid metabolism, glycerophospholipid metabolism, steroid biosynthesis, ABC transporters, and protein processing [51]. These genes are all relevant to the plasma membrane, suggesting that myriocin may target the expression of DNA. Gramcidins are a class of antibiotic decapeptides synthesized by *Aneurinibacillus migulanus* (formerly *B. brevis*) and consist of linear gramicidin A, B, C, and the circular gramicidin S. Gramicidin A (17) (Figure 5), a 15 amino-acid peptide, destroys gram-positive bacteria (*Streptococcus pyogenes* with a MIC = 33 nM) [52]. Unlike other antimicrobial metabolites, gramicidin A forms a single ion channel, which distorts the membrane and allows the passage of cations across the membrane [53]. Once inside, gramicidin A can also induce the formation of reactive oxygen species (ROS), which damages the intracellular DNA, mitochondria and triggers necrosis [54]. The gram-positive bacteria *Aneurinibacillus migulanus* (formerly *B. brevis natto*) inhibits several gram-positive, gram-negative, and fungi microbials (e.g., *Staphylococcus aureus* with a MIC value of 3.9 μg/mL) by producing gramicidin S (18) (Figure 5) [55]. Gramicidin S interacts with the plasma membrane by forming oligomeric β-barrel pores, which destroys the barrier properties of the membrane [56,57]. Further in vivo studies have shown that gramicidin S binds to the DNA and inhibits transcription and cell growth [58].

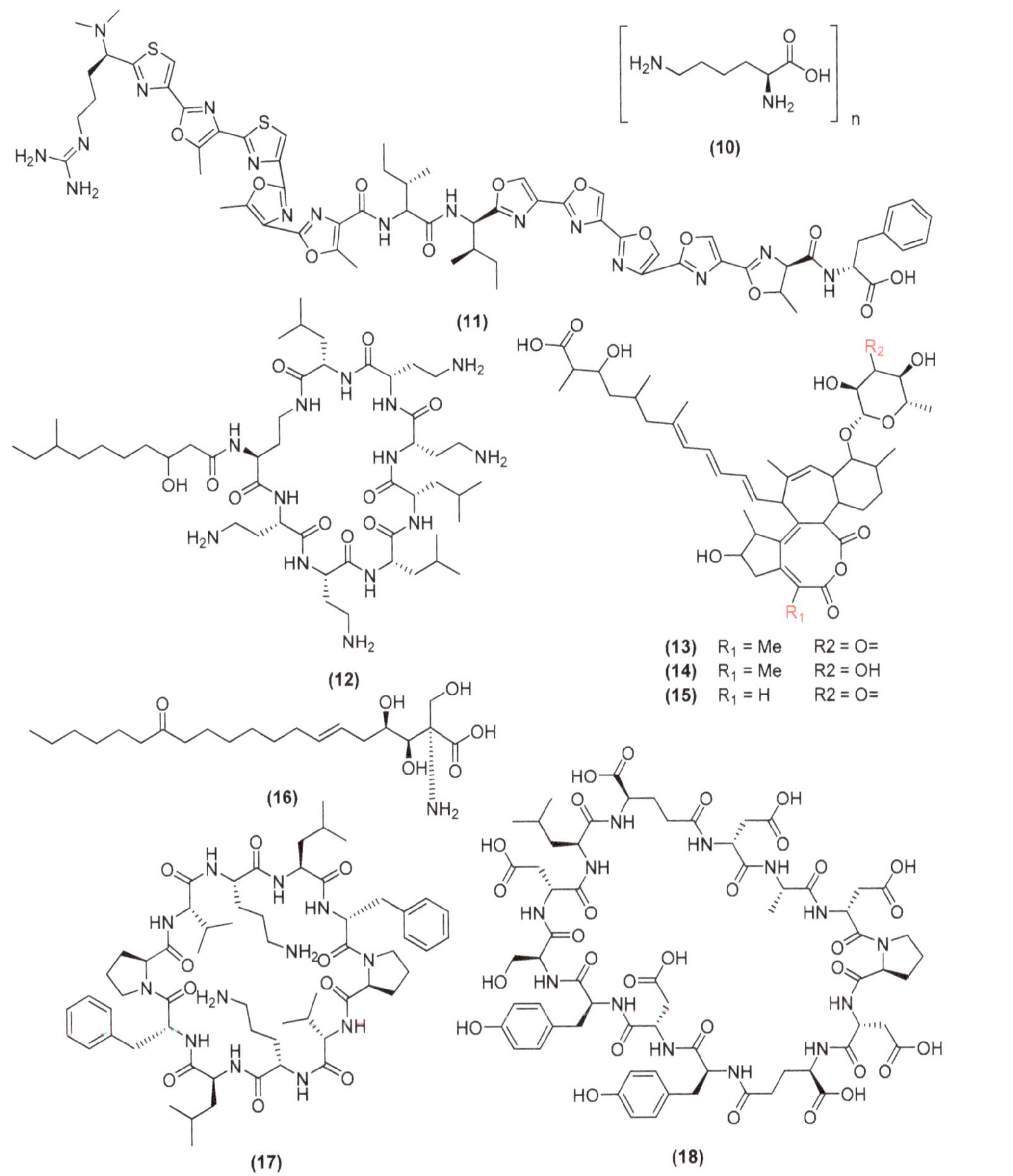

Figure 5. Chemical structures of ε-poly-L-Lysine (**10**), plantazolicin (**11**), octapeptin B (**12**), aurantinin B (**13**), aurantinin C (**14**), aurantinin (D) (**15**), myriocin (**16**), gramicidin A (**17**), and gramicidin S (**18**).

Pore-formation metabolites act in a concentration-dependent manner, by forming ion-like channels that release vital ions from the cell, leading to cell death. At low concentrations, these metabolites form unilamellar vesicles on the outer lipid membrane, distorting the shape of the cell, and eventually, lead to apoptosis [59–61]. At higher concentrations, these metabolites aggregate to form pores at the plasma membrane, causing the leakage of nucleic acids, essential ions, and ATP from the cell to cause necrosis [59,62–64]. *Bacillus* metabolites that typically utilize this mechanism includes the class of compounds known as lipopeptides. Lipopeptides are composed of a cyclic oligopeptide, attached to a flexible lipid tail, and consist of several groups including the surfactins, fengycins, and iturins [65]. Surfactins were first isolated from a culture broth of *B. subtilis* and include the compounds surfactin A (19), B (20), C (21), and lichenysin (22) (Figure 6) [66]. Surfactins exert their antibacterial activities by acting on the plasma membrane through the pore-forming mechanism [67]. Additionally, surfactins (21-22) can breakdown bacterial biofilms by decreasing the percentage of alkali-soluble polysaccharides and downregulating the expression of genes involved in biofilm formation such as icaA and icaD [68]. Lastly, surfactins can also induce the grapevine immune system in response to infection [69]. Fengycins (23-26) (Figure 6) are antifungal lipopeptides first isolated from *B. subtilis* F-29-3 (e.g., *Rolani stolonifera* with a MIC = 400 μg/mL) [70]. These fengycin molecules are often reported as membrane disruptors, either by deforming membrane shape or by causing pores, leading to cell death [71]. More recent studies have additional antimicrobial mechanisms of action for fengycin A (23) and fengycin B (24). Fengycin A can alter the gene expression related to cell wall synthesis, which alters cell components and increases hydrophobicity [72]. Furthermore, fengycin B155, a mixture of fengycin A (23) and fengycin B (24), is able to disrupt multiple intracellular components of the cell [73]. These processes include the inhibition of the mitochondria membrane potential, the condensation of chromatin involved in replication, the cleavage of DNA repair protein (poly (ARP-ribose) polymerase), and the accumulation of ROS [73]. Lastly, fengycins have been shown to inhibit quorum sensing, due to their structural similarity to *S. aureus* accessory gene regulator (Agr) [74]. Agr is a virulence factor that mediates the cell-to-cell communication between cells, and its inhibition prevents the aggregation and biofilm formation needed to promote survival [75]. Plipastatin A (26) is a lipopeptide commonly associated with the fengycin family due to its structural similarity and antifungal properties (*Fusarium oxysporum* with a MIC = 16 μg/mL) [76]. TEM analysis demonstrated that plipastatins disrupt the cell wall, membrane, and cytoskeleton of *Fusarium oxysporum*, causing intracellular leakage and eventual cell death.

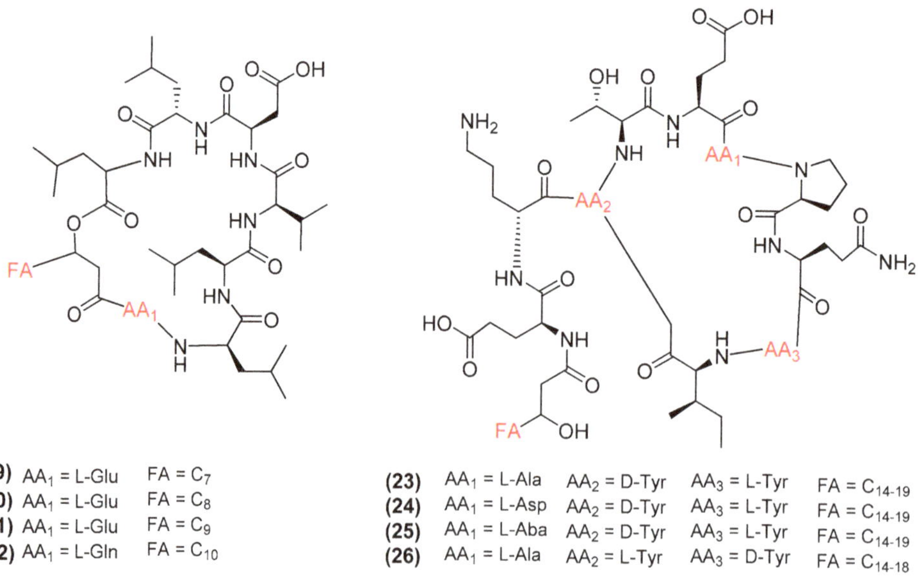

Figure 6. Chemical structures of surfactin A–C (**19–21**), lichenysin (**22**), and fengycin A–D (**23–26**).

Iturins (27–30) (Figure 7) are cyclic lipopeptides that includes iturin A (27), bacillomycin D (28), bacillomycin L (29), and mycosubtilin (30) [77]. These peptides primarily inhibit fungi by binding to the cell membrane with its fatty acid tail to form ion-conducting or phospholipid–lipopeptide sterol complexes [78]. Optical and fluorescence microscopy studies have revealed that iturin A (27) severely damages the plasma membranes of *Fusarium graminearum* at a MIC = 5 μg/mL by forming a large pore and inhibiting hyphae growth [79]. Iturin A can stimulate oxidative stress, leading to mitochondria damage and the eventual destruction of the cell [80]. Lastly, iturin A increases the transcription of immune defense genes in several plants [81]. Bacillomycin D (28) exerts antifungal properties against *Colletotrichum gloeosporioides* with an MIC of 2.2 μg/mL [82,83]. SEM and TEM analysis confirmed bacillomycin D's ability to target both cell wall and plasma membrane, leading to the leakage of intracellular organelles [82]. Bacillomycin D can disrupt the cell membrane by upregulating the expression of genes involved in ergosterol synthesis and oxidative stress [84]. These sterols adjoin to the membrane, distorting its shape and eventually releasing vital intracellular components to the environment [84]. Additionally, bacillomycin D can increase the expression of specific genes to produce ROS molecules and cellular antioxidant enzymes including deoxyivalentol, glutathione reductase, and thioredoxin [85]. Bacillomycin D has also been reported to act as a biofilm activator by binding to the matrix complex KinB-Spo0A-SinI-SinR, which triggers the production of biofilm [86]. Lastly, bacillomycin D stimulates the expression of genes involved in mediated defense responses and enzymatic proteins that can be released to target competing growth [86]. *B. amyloliquefaciens* K103 produces the potent antifungal metabolite bacillomycin L (29) (*Saccaromyces cerevisiae* with a MIC = 30 μg/mL) [78,87]. Like other iturins, bacillomycin L primarily acts on the plasma membrane, forming pores that releases its intracellular components outside the cell [88]. Studies have shown that bacillomycin L binds to sterols on the membrane, destroying the membrane and killing the cell [89]. Bacillomycin L can also alter the expression of 39 different genes in *Rhizoctonia solani* related to cellular stress, such as calcium homeostasis, energy metabolism, protein degradation, RNA processing, and carbohydrate metabolism [90]. Mycosubtilin (30), an antibiotic from the iturin group, inhibits the growth of fungal *Saccharomyces cerevisiae* with a MIC of 10 μg/mL [78]. Increased concentrations of mycosubtilin causes the lysis of the phospholipid layer, either by the aggregation of lipopeptides or clustering of mycosubtilin [91]. This binding increases membrane permeability, leading to metabolite release and the eventual lysis of the cell [92]. Mycosubtilin can also activate the salicylic acid and jasmonic acid signaling pathways involved in the immune response to pathogenic microbes [69].

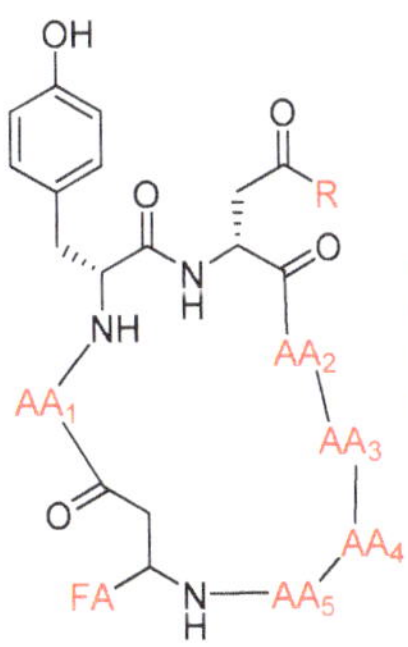

(27)	AA$_1$ = Asn	AA$_2$ = Gln	AA$_3$ = Pro	AA$_4$ = Asn	AA$_5$ = Ser	R = NH_2	FA = $C_{14\text{-}17}$
(28)	AA$_1$ = Asn	AA$_2$ = Pro	AA$_3$ = Glu	AA$_4$ = Ser	AA$_5$ = Thr	R = NH_2	FA = $C_{14\text{-}17}$
(29)	AA$_1$ = Asp	AA$_2$ = Ser	AA$_3$ = Gln	AA$_4$ = Ser	AA$_5$ = Thr	R = OH	FA = $C_{14\text{-}17}$
(30)	AA$_1$ = Asn	AA$_2$ = Gln	AA$_3$ = Pro	AA$_4$ = Ser	AA$_5$ = Asn	R = NH_2	FA = $C_{14\text{-}17}$

Figure 7. Chemical structures of iturin A (**27**), bacillomycin D (**28**), bacillomycin L (**29**), and mycosubtilin (**30**).

Mycobacillin (31) (Figure 8), an antifungal polypeptide sourced from *B. subtilis* B3, is active against *Aspergillus niger* at 20 μg/mL [93,94]. Mycobacillin has been reported to bind to ATP transporter on the plasma membrane, leading to the excessive release of ATP and the subsequent starvation of the cells [94,95]. Subtilosin A (32) (Figure 8) is a sactipeptide produced by *B. subtilis* 168 that processes antibacterial activity against both gram-positive and gram-negative pathogens (i.e., *Gardnerella vaginalis* MIC = 7.2 μg/mL) [96,97]. Its specific mechanism of action involves subtilosin A anchoring to a membrane receptor, whilst electrostatically binding to the plasma membrane [98]. This electrostatic binding dissipates the transmembrane pH gradient, causing an efflux of intracellular ATP that starves the cell and eventually leads to its death. Subtilosin A has also been shown to inhibit biofilm formation, presumably by blocking quorum sensing between cells [99].

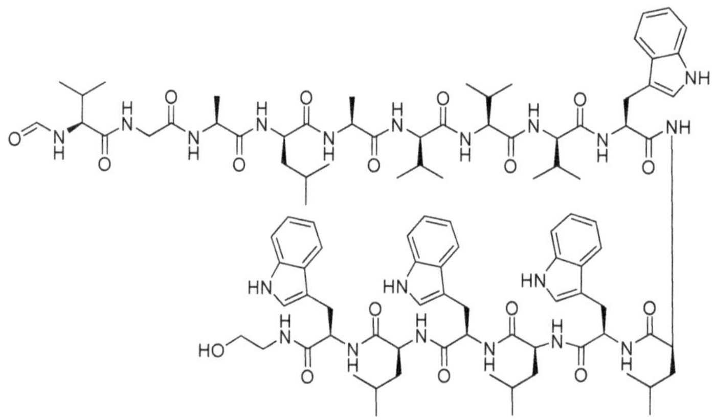

(31)

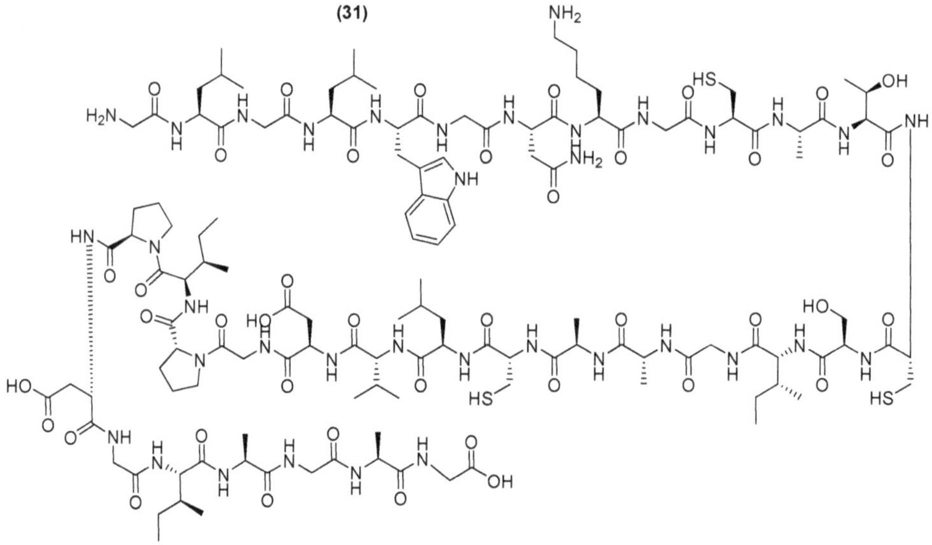

(32)

Figure 8. Chemical structures of mycobacillin (**31**) and subtilosin A (**32**).

3.3. Metabolites Targeting Intracellular Processes

Bacillus metabolites may cross the plasma membrane and bind to several intracellular targets essential for cell survival. These intracellular processes include DNA transcription, RNA translation, and protein metabolism needed for energy production. Transcription is the first step in gene expression, in which information from a gene is used to construct a functional product such as a protein. For a protein-coding gene, the RNA copy, or transcript, carries the information needed to build a protein. From the 47 compounds reviewed in this paper, 11 compounds primarily target the intracellular processes.

Zwittermicin A (33) (Figure 9), an aminopolyl antibiotic produced by *B. cereus* UW85, inhibits gram-positive and gram-negative bacteria, as well as fungi (i.e., *Erwinia herbicola* L S005 with a MIC = 60 μg/mL) [100]. Zwittermicin A disrupts cellular growth by targeting either DNA transcription and replication via inhibition of two enzymes, gyrase and topoisomerase [101]. Difficidin (34) (Figure 9), a highly unsaturated macrolide phosphate first isolated from *B. subtilis* ATCC 39320, can inhibit both gram-positive and negative strains such as *Rolani solanacearum* with a MIC value of 12.62 μg/mL of [102,103]. Microscopy analysis has revealed that difficidin downregulates the genes related to cell wall synthesis, protein production, and DNA replication [104]. Sublancin (35) (Figure 9), a glycosylated peptide produced by *B. subtilis* 168, displays antibacterial activities (i.e., methicillin-resistant *Staphylococcus aureus* ATCC43300 with a MIC = 15 μM) [105]. Mechanism investigations suggest that sublancin enters the cytoplasm and reduces DNA transcription and translation [106].

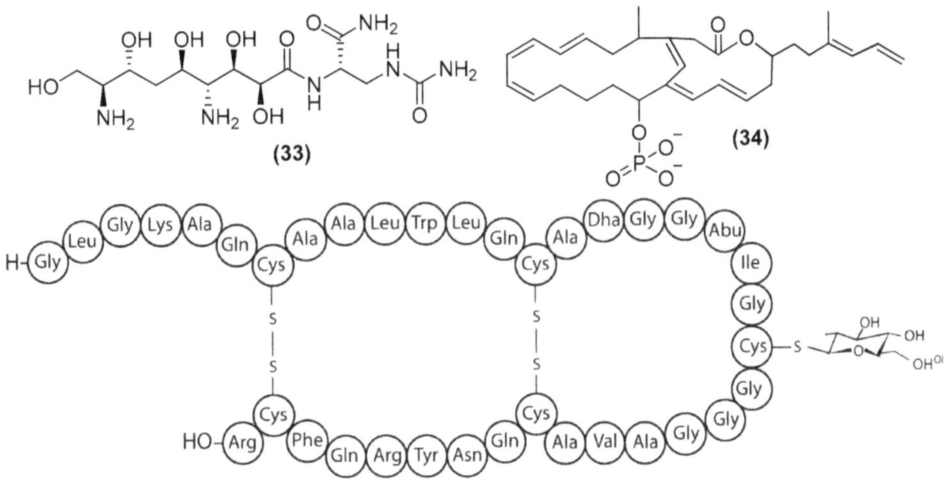

Figure 9. Chemical structures of zwittermicin A (**33**), difficidin (**34**), and sublancin (**35**).

The amicoumacins are a class of dihydroisocoumarin compounds, produced by *B. pumilus*, that exert antibacterial, antifungal, and anti-inflammatory properties. In particular, amicoumacin A (36) (Figure 10), produced by *B. pumilus* BN-103, inhibits *B. subtilis* 1779 with an MIC = 20.0 μg/mL. Further studies have shown that amicoumacin A inhibits the protein synthesis of methicillin-resistant *Staphylococcus aureus* by stabilizing the mRNA at the terminal E site on the ribosome during protein synthesis [107]. This disruption results in the perturbation of the membrane, leading to energy dissipation and eventual cell death [107,108]. Prumycin (37) (Figure 10), isolated from a culture broth of *B. amyloliquefaciens* SD-32, exerts bactericidal and fungicidal effects, such as on *S. sclerotiorum*, with an MIC value of 1.56 μg/mL [109–111]. Prumycin inhibits the protein synthesis of *Sacrina lutea*, preventing the activation of amino acids needed for protein synthesis and the transfer of amino acids to RNA [110].

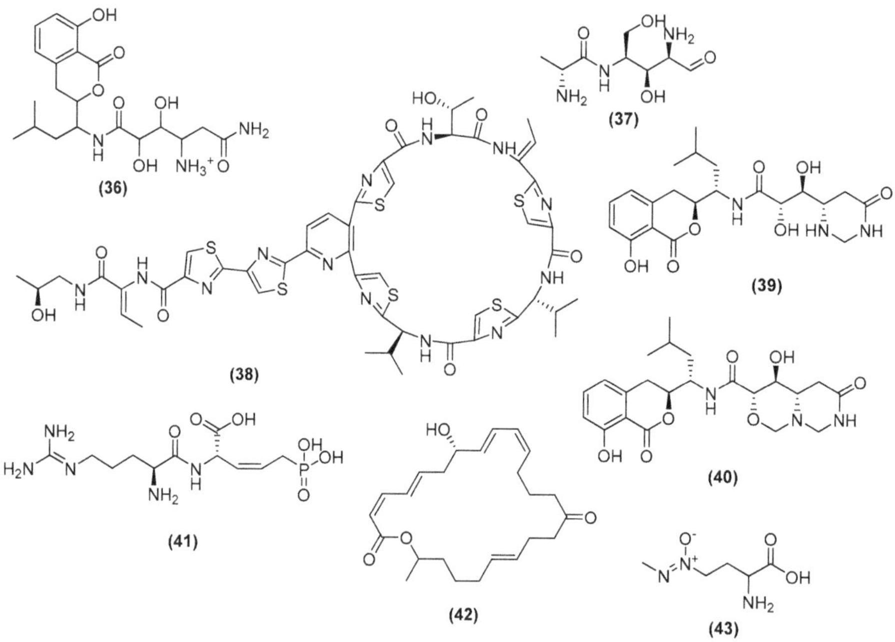

Figure 10. Chemical structures of amicoumacin A (**36**), prumycin (**37**), thiocillin (**38**), hetiamacin E (**39**), hetiamacin F (**40**), rhizocticin A (**41**), macrolactin N (**42**), and azoxybacilin (**43**).

Thiocillin (38) (Figure 10), produced by *B. cereus* ATCC 14579, has been previously reported to only target gram-positive bacteria but has recently been shown to also target gram-negative bacteria [112]. Its mechanism on gram-positive bacteria works by targeting the 50S ribosome and inhibiting its role in protein synthesis [113]. In contrast, thiocillin targets the gram-negative bacterium *Pseudomonas aeruginosa* by binding to ferrioxamine receptor FoxA, which disrupts the proton motive force to inhibit translation [113]. Hetiamacin E and F (39–40) (Figure 10) produced from *B. subtilis* PJS display antibacterial activity against methicillin-resistant *Staphylococcus aureus,* with MIC values of 8–16 μg/mL and 32 μg/mL, respectively [114]. Hetiamacin E and F inhibit protein biosynthesis, resulting in the disruption of mRNA translation, leading to cell death [114]. Rhizocticin A (41) (Figure 10) is a potent antifungal first produced from *B. subtilis* 6633. Its bioactivity data shows that it is active against a range of budding and filamentous fungi (bioactivity not avaliable) [115]. Mutant analysis suggests that rhizocticin utilizes the peptide transport system to enter the cytoplasm, where it forms the fungitoxic L-2-amino-5-phosphono-3-cis-pentenioc acid (L-APPA). L-APPA interferes with threonine metabolism, which inhibits cell growth [116].

Macrolactin N (42) (Figure 10), a novel macrolactin produced by *B. subtilis* A29, is shown to inhibit *Staphylococcus aureus* peptide deformylase (PDF), with an MIC of 100 μM [117]. PDFs are essential bacterial specific metalloenzymes, which removes formyl groups during polypeptide elongation [117]. The inhibition of these PDFs leave bacteria unable to hydrolyze these polypeptides and hinder its ability to synthesize proteins [117]. Azoxybacilin (43) (Figure 10), first isolated from *B. cereus* NR2991 and *B. cereus* Frankland, is active against a broad spectrum of mycelial fungi, such as *Candida albicans* (IC_{50} = 1.2 mg/mL) [118,119]. Its mechanism involves the interruption of the sulfur fixation pathway, an essential support system for microbial growth, by decreasing the expression of sulfate assimilation genes including MET10 and MET4 [118]. MET10 regulates the expression of sulfite reductase, and MET4 is the transactivator of MET10. The reduction of the gene expression in the sulfur-fixation pathway disrupts this support system and eventually leads to cell growth inhibition.

3.4. Metabolites Interacting with Other Emerging Targets

Quorum sensing, also known as cell-to-cell communication, is the regulation of a microbial gene expression in response to its cell density [120]. This mechanism relies on small chemical indicators and has been linked to pathogen virulence, due to its effect on cell reproduction, mobility, and biofilm formation [121]. Biofilms are extracellular adhesive structures produced by various strains of bacteria that assist in their tolerance to UV, acidity conditions, and vulnerability to antimicrobial metabolites [122]. Several key groups of *Bacillus* metabolites have been shown to interfere with this process [123]. Nonetheless, *Bacillus* metabolites such as stigmatellin Y (44) (Figure 11) have been identified as a biofilm inhibitor [124]. Stigmatellin Y is shown to inhibit *Pseudomonas aeruginosa* biofilm formation, presumably by acting as a competitive inhibitor to the quorum sensing mediator PqsR [124]. Bacillaene (45) (Figure 11) has been identified as a biofilm inhibitor produced by numerous *B. subtilis* strains [125]. Analysis of mutant strains revealed that bacillaene inhibits the biofilm of *Campylobacter jejuni*, preventing the formation of microcolonies and eventually disrupting their microbial growth.

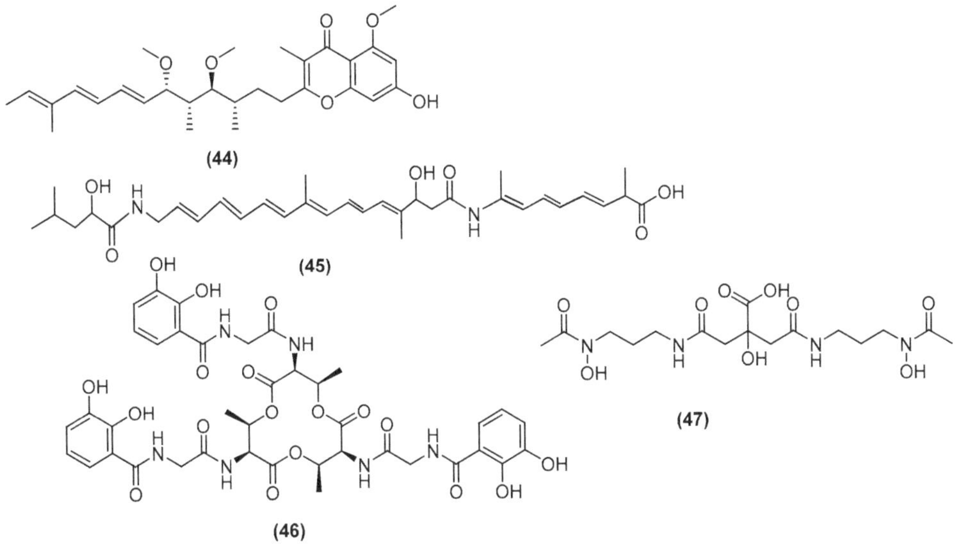

Figure 11. Chemical structures of stigmatellin Y (**44**), bacillaene (**45**), bacillibactin (**46**), and schizokinen (**47**).

Siderophores are small molecules secreted by microorganisms that are involved in iron (Fe^{2+}) uptake from the environment [126]. Iron is an essential metabolite for microbial growth and strategies have been developed to starve pathogenic microorganisms using these siderophores. Siderophores produced by *Bacillus* strains include bacillibactin (46) and schizokinen (47) (Figure 11), which were first isolated from *B. subtilis* and *B. megaterium* ATCC 19213, respectively [127,128]. These metabolites facilitate the uptake of ferric ions (Fe^{3+}) from the environment to the bacterial cell using specific membrane receptors to enter the host cell [129]. Once inside, these ions are reduced to ferrous (Fe^{2+}) ions for use in microbial growth [130].

4. Conclusions Remarks and Future Directions

This paper reviews the current literature on antimicrobial compounds from *Bacillus* sp. and their mechanism of action. Further analysis on the source of antimicrobial compounds and their mechanism of action revealed some interesting trends. In terms of number of strains that produce antimicrobial metabolites, the most prolific is *subtilis* (n = 73), followed by *amyloliquefaciens* (n = 52) and *velezensis* (n = 22) (Figure 12a). *B. subtilis* is a common bacterium in soil and one of the most-studied *Bacillus* sp. Research has shown that these species are strongly related to each other, with several papers suggesting that *amyloliquefaciens* be renamed as *velezensis* due to its similarity in conserved genomic sequence [131,132]. The least reported of these *Bacillus* sp. is *B. thuringiensis*, with only two strains producing antimicrobial compounds in the literature. This highlights the lack of studies for this species and may warrant further investigation.

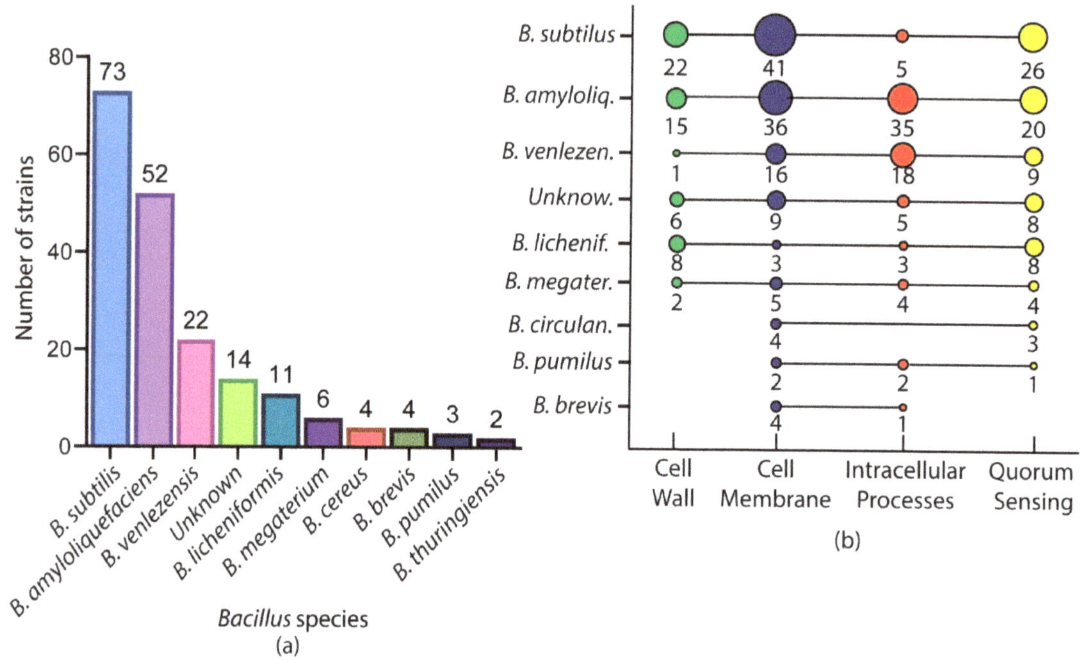

Figure 12. The analysis of (**a**) the number of strains in each species and the (**b**) mechanism of action targeted by each strain.

Further analysis on mechanism of action (Figure 12b) reveals that the cell membrane is the most popular target of different species of *Bacillus* and their metabolites (n = 122), followed by quorum sensing (n = 79), intracellular processes (n = 73), and the cell wall (n = 57). Quorum sensing is an interesting emerging target, as more species and metabolites (n = 79) hinder the process and hence, inhibit cell-to-cell communication. Further analysis also notes that many *Bacillus* species and their metabolites exert their antimicrobial activity through not only one but multiple mechanisms.

Several publications noted the geographic location of *Bacillus,* as well as the source of the bacteria. Further analysis based the information provided in the literature reveals that the majority of identified strains are from Asia (n = 37), followed by South America (n = 8) and the Middle East (n = 4). This observation may indicate that these strains share genomic similarities or properties, however, it may also stem from the research laboratories located in these sites and could be a byproduct of a focus on probiotic research at these locations. Additionally, the top three sources that these strains were isolated are from soil, local produce, and waterways. These findings reinforce the use of soil-based screening as a rich source of microorganisms. It also highlights the recent trend in investigating food produce as a source of *Bacillus* isolates. This is either guided by historical evidence of their antimicrobial properties or the anecdotal knowledge of their safe use and consumption.

The advancements of omics technologies are essential for the rapid screening of future probiotics. The characterization of the genome and biochemical properties allows the selection of particular strains with properties suitable for industrial use. A number of omics techniques have been developed to provide valuable information on the characteristics, optimization, and metabolic pathways behind antimicrobial activity [133]. One example uses omics to a rapid screen of selected *Bacillus* strains for specific gene markers known for antimicrobial activity [133]. For example, the genomic screening of *B. velezensis* CC09 revealed the loci for iturin A previously not identified in its initial screening [134].

In-depth analysis of these pathways and the precursors may reveal optimal conditions needed to produce these metabolites [135]. Wiegand utilized metabolomics and genome mining to provide insight into the expression of DNA under various fermentation conditions. These conditions includes pH levels, temperatures, and oxygen levels, which result in the discovery of optimal conditions needed to express the antimicrobial gene of interest and maximizing their yield [136]. This technique, alongside computational modelling systems, may reveal other conditions unexplored such as the ratio of carbon to nitrogen in fermentation media and the presence of small metabolites and co-culturing in order to further maximize the production of antimicrobial metabolites. As production is required, especially when optimizing for commercial purposes, these techniques can open up the field in the use of bacteria as a source of antimicrobial compounds to tackle the declining rate of antimicrobial compounds being discovered.

Supplementary Materials: The following are available online at https://www.mdpi.com/article/10.3390/antibiotics11010088/s1, Table S1: 47 antimicrobial metabolites from *Bacillus*. References [137–251] are cited in the supplementary materials.

Author Contributions: Conceptualization, X.C., I.E.C. and Y.F.; data curation, C.T.; writing—original draft preparation, C.T.; writing—review and editing, X.C., I.E.C. and Y.F.; supervision, X.C., I.E.C. and Y.F. All authors have read and agreed to the published version of the manuscript.

Funding: This research received no external funding.

Acknowledgments: This work was supported, in part, by Bioproton Pty. Ltd. (Ph.D. scholarship for C.T.).

Conflicts of Interest: The authors declare no conflict of interest.

References

1. Hill, C.; Guarner, F.; Reid, G.; Gibson, G.R.; Merenstein, D.J.; Pot, B.; Morelli, L.; Canani, R.B.; Flint, H.J.; Salminen, S.; et al. The International Scientific Association for Probiotics and Prebiotics consensus statement on the scope and appropriate use of the term probiotic. *Nat. Rev. Gastroenterol. Hepatol.* **2014**, *11*, 506–514. [CrossRef]
2. Liu, P.; Zhao, J.; Guo, P.; Lu, W.; Geng, Z.; Levesque, C.L.; Johnston, L.J.; Wang, C.; Liu, L.; Zhang, J.; et al. Dietary Corn Bran Fermented by *Bacillus subtilis* MA139 Decreased Gut Cellulolytic Bacteria and Microbiota Diversity in Finishing Pigs. *Front. Cell. Infect. Microbiol.* **2017**, *7*, 526. [CrossRef]
3. Kahn, L.H. Antimicrobial resistance: A One Health perspective. *Trans. R. Soc. Trop. Med. Hyg.* **2017**, *111*, 255–260. [CrossRef] [PubMed]
4. Lim, J.M.; Duong, M.C.; Hsu, L.Y.; Tam, C.C. Determinants influencing antibiotic use in Singapore's small-scale aquaculture sectors: A qualitative study. *PLoS ONE* **2020**, *15*, e0228701. [CrossRef]
5. Carlet, J. The gut is the epicentre of antibiotic resistance. *Antimicrob. Resist. Infect. Control* **2012**, *1*, 39. [CrossRef] [PubMed]
6. Zhang, W.; Xin, H.; Jiang, N.; Lv, Z.; Shu, J.; Shi, H. Bacillus Amyloliquefaciens-9 as an Alternative Approach to Cure Diarrhea in Saanen Kids. *Animals* **2021**, *11*, 592. [CrossRef]
7. Anisimova, E.A.; Yarullina, D.R. Antibiotic Resistance of *LACTOBACILLUS* Strains. *Curr. Microbiol.* **2019**, *76*, 1407–1416. [CrossRef]

8. Abel-Santos, E. Chapter 9—Endospores, Sporulation and Germination. In *Molecular Medical Microbiology*, 2nd ed.; Tang, Y.-W., Sussman, M., Liu, D., Poxton, I., Schwartzman, J., Eds.; Academic Press: Boston, MA, USA, 2015; pp. 163–178. [CrossRef]
9. Adeniji, A.A.L.D.; Babalola, O.O. Bacillus velezensis: Phylogeny, useful applications, and avenues for exploitation. *Appl. Microbiol. Biotechnol.* **2019**, *103*, 3669–3682. [CrossRef]
10. Caulier, S.; Nannan, C.; Gillis, A.; Licciardi, F.; Bragard, C.; Mahillon, J. Overview of the Antimicrobial Compounds Produced by Members of the *Bacillus subtilis* Group. *Front. Microbiol.* **2019**, *10*, 302. [CrossRef]
11. Ortiz, A.; Sansinenea, E. Chemical Compounds Produced by *Bacillus* sp. Factories and Their Role in Nature. *Mini Rev. Med. Chem.* **2019**, *19*, 373–380. [CrossRef] [PubMed]
12. Zhao, X.; Kuipers, O.P. Identification and classification of known and putative antimicrobial compounds produced by a wide variety of Bacillales species. *BMC Genom.* **2016**, *17*, 882. [CrossRef] [PubMed]
13. Nishino, K.; Hsu, F.-F.; Turk, J.; Cromie, M.J.; Wösten, M.M.S.M.; Groisman, E.A. Identification of the lipopolysaccharide modifications controlled by the Salmonella PmrA/PmrB system mediating resistance to Fe(III) and Al(III). *Mol. Microbiol.* **2006**, *61*, 645–654. [CrossRef]
14. Vollmer, W.; Blanot, D.; De Pedro, M.A. Peptidoglycan structure and architecture. *FEMS Microbiol. Rev.* **2008**, *32*, 149–167. [CrossRef] [PubMed]
15. Johnson, B.A.; Anker, H.; Meleney, F.L. Bacitracin: A new antibiotic produced by a member of the *B. subtilis* group. *Science* **1945**, *102*, 376–377. [CrossRef] [PubMed]
16. Zaidi, S.; Singh, S.L.; Khan, A.U. Exploring antibiofilm potential of bacitracin against streptococcus mutans. *Microb. Pathog.* **2020**, *149*, 104279. [CrossRef]
17. Siewert, G.; Strominger, J.L. Bacitracin: An inhibitor of the dephosphorylation of lipid pyrophosphate, an intermediate in the biosynthesis of the peptidoglycan of bacterial cell walls. *Proc. Natl. Acad. Sci. USA* **1967**, *57*, 767–773. [CrossRef] [PubMed]
18. Mahlstedt, S.A.; Walsh, C.T. Investigation of Anticapsin Biosynthesis Reveals a Four-Enzyme Pathway to Tetrahydrotyrosine in *Bacillus subtilis*. *Biochemistry* **2010**, *49*, 912–923. [CrossRef]
19. Newton, G.G. Antibiotics from a strain of *B. subtilis*; bacilipin A and B and bacilysin. *Br. J. Exp. Pathol.* **1949**, *30*, 306–319.
20. Wang, T.; Wu, M.B.; Chen, Z.J.; Lin, J.P.; Yang, L.R. Separation, determination and antifungal activity test of the products from a new *Bacillus amyloliquefaciens*. *Nat. Prod. Res.* **2016**, *30*, 1215–1218. [CrossRef]
21. Kenig, M.; Abraham, E.P. Antimicrobial activities and antagonists of bacilysin and anticapsin. *J. Gen. Microbiol.* **1976**, *94*, 37–45. [CrossRef]
22. Kenig, M.; Vandamme, E.; Abraham, E.P. The mode of action of bacilysin and anticapsin and biochemical properties of bacilysin-resistant mutants. *J. Gen. Microbiol.* **1976**, *94*, 46–54. [CrossRef]
23. Milner, J.L.; Silo-Suh, L.; Lee, J.C.; He, H.; Clardy, J.; Handelsman, J. Production of kanosamine by *Bacillus cereus* UW85. *Appl. Env. Microbiol.* **1996**, *62*, 3061–3065. [CrossRef]
24. Janiak, A.M.; Milewski, S. Mechanism of antifungal action of kanosamine. *Med. Mycol.* **2001**, *39*, 401–408. [CrossRef] [PubMed]
25. van Heijenoort, J. Lipid intermediates in the biosynthesis of bacterial peptidoglycan. *Microbiol. Mol. Biol. Rev.* **2007**, *71*, 620–635. [CrossRef] [PubMed]
26. Derouaux, A.; Turk, S.; Olrichs, N.K.; Gobec, S.; Breukink, E.; Amoroso, A.; Offant, J.; Bostock, J.; Mariner, K.; Chopra, I.; et al. Small molecule inhibitors of peptidoglycan synthesis targeting the lipid II precursor. *Biochem. Pharm.* **2011**, *81*, 1098–1105. [CrossRef] [PubMed]
27. Chugunov, A.; Pyrkova, D.; Nolde, D.; Polyansky, A.; Pentkovsky, V.; Efremov, R. Lipid-II forms potential "landing terrain" for lantibiotics in simulated bacterial membrane. *Sci. Rep.* **2013**, *3*, 1678. [CrossRef]
28. Ross, A.C.; Vederas, J.C. Fundamental functionality: Recent developments in understanding the structure–activity relationships of lantibiotic peptides. *J. Antibiot.* **2011**, *64*, 27–34. [CrossRef]
29. Chan, W.C.; Bycroft, B.W.; Leyland, M.L.; Lian, L.Y.; Roberts, G.C. A novel post-translational modification of the peptide antibiotic subtilin: Isolation and characterization of a natural variant from *Bacillus subtilis* A.T.C.C. 6633. *Biochem. J.* **1993**, *291 Pt 1*, 23–27. [CrossRef]
30. Parisot, J.; Carey, S.; Breukink, E.; Chan, W.C.; Narbad, A.; Bonev, B. Molecular mechanism of target recognition by subtilin, a class I lanthionine antibiotic. *Antimicrob. Agents Chemother.* **2008**, *52*, 612–618. [CrossRef]
31. Bouhss, A.; Al-Dabbagh, B.; Vincent, M.; Odaert, B.; Aumont-Nicaise, M.; Bressolier, P.; Desmadril, M.; Mengin-Lecreulx, D.; Urdaci, M.C.; Gallay, J. Specific interactions of clausin, a new lantibiotic, with lipid precursors of the bacterial cell wall. *Biophys. J.* **2009**, *97*, 1390–1397. [CrossRef]
32. Ahire, J.J.; Kashikar, M.S.; Lakshmi, S.G.; Madempudi, R. Identification and characterization of antimicrobial peptide produced by indigenously isolated *Bacillus paralicheniformis* UBBLi30 strain. *3 Biotech* **2020**, *10*, 112. [CrossRef] [PubMed]
33. Brötz, H.; Bierbaum, G.; Leopold, K.; Reynolds, P.E.; Sahl, H.G. The lantibiotic mersacidin inhibits peptidoglycan synthesis by targeting lipid II. *Antimicrob. Agents Chemother.* **1998**, *42*, 154–160. [CrossRef] [PubMed]
34. Sass, P.; Jansen, A.; Szekat, C.; Sass, V.; Sahl, H.G.; Bierbaum, G. The lantibiotic mersacidin is a strong inducer of the cell wall stress response of *Staphylococcus aureus*. *BMC Microbiol.* **2008**, *8*, 186. [CrossRef] [PubMed]
35. Brötz, H.; Bierbaum, G.; Reynolds, P.E.; Sahl, H.G. The lantibiotic mersacidin inhibits peptidoglycan biosynthesis at the level of transglycosylation. *Eur. J. Biochem.* **1997**, *246*, 193–199. [CrossRef]

36. Arguelles Arias, A.; Joris, B.; Fickers, P. Dual mode of action of amylolysin: A type-B lantibiotic produced by *Bacillus amyloliquefaciens* GA1. *Protein Pept. Lett.* **2014**, *21*, 336–340. [CrossRef]
37. Arguelles Arias, A.; Ongena, M.; Devreese, B.; Terrak, M.; Joris, B.; Fickers, P. Characterization of amylolysin, a novel lantibiotic from *Bacillus amyloliquefaciens* GA1. *PLoS ONE* **2013**, *8*, e83037. [CrossRef]
38. Oman, T.J.; van der Donk, W.A. Insights into the mode of action of the two-peptide lantibiotic haloduracin. *ACS Chem. Biol.* **2009**, *4*, 865–874. [CrossRef] [PubMed]
39. Oman, T.J.; Lupoli, T.J.; Wang, T.S.; Kahne, D.; Walker, S.; van der Donk, W.A. Haloduracin α binds the peptidoglycan precursor lipid II with 2:1 stoichiometry. *J. Am. Chem. Soc.* **2011**, *133*, 17544–17547. [CrossRef]
40. Ferrante, T.; Viola, F.; Balliano, G.; Oliaro-Bosso, S. Difference in the late ergosterol biosynthesis between yeast spheroplasts and intact cells. *Acta Biochim. Pol.* **2016**, *63*, 371–375. [CrossRef]
41. Casares, D.; Escribá, P.V.; Rosselló, C.A. Membrane Lipid Composition: Effect on Membrane and Organelle Structure, Function and Compartmentalization and Therapeutic Avenues. *Int. J. Mol. Sci.* **2019**, *20*, 2167. [CrossRef]
42. Rodrigues, B.; Morais, T.P.; Zaini, P.A.; Campos, C.S.; Almeida-Souza, H.O.; Dandekar, A.M.; Nascimento, R.; Goulart, L.R. Antimicrobial activity of Epsilon-Poly-l-lysine against phytopathogenic bacteria. *Sci. Rep.* **2020**, *10*, 11324. [CrossRef]
43. El-Sersy, N.A.; Abdelwahab, A.E.; Abouelkhiir, S.S.; Abou-Zeid, D.M.; Sabry, S.A. Antibacterial and anticancer activity of ε-poly-L-lysine (ε-PL) produced by a marine *Bacillus subtilis* sp. *J. Basic Microbiol.* **2012**, *52*, 513–522. [CrossRef]
44. Hyldgaard, M.; Mygind, T.; Vad, B.S.; Stenvang, M.; Otzen, D.E.; Meyer, R.L. The antimicrobial mechanism of action of epsilon-poly-l-lysine. *Appl. Environ. Microbiol.* **2014**, *80*, 7758–7770. [CrossRef] [PubMed]
45. Molohon, K.J.; Saint-Vincent, P.M.B.; Park, S.; Doroghazi, J.R.; Maxson, T.; Hershfield, J.R.; Flatt, K.M.; Schroeder, N.E.; Ha, T.; Mitchell, D.A. Plantazolicin is an ultra-narrow spectrum antibiotic that targets the *Bacillus anthracis* membrane. *ACS Infect. Dis.* **2016**, *2*, 207–220. [CrossRef]
46. Velkov, T.; Gallardo-Godoy, A.; Swarbrick, J.D.; Blaskovich, M.A.T.; Elliott, A.G.; Han, M.; Thompson, P.E.; Roberts, K.D.; Huang, J.X.; Becker, B.; et al. Structure, Function, and Biosynthetic Origin of Octapeptin Antibiotics Active against Extensively Drug-Resistant Gram-Negative Bacteria. *Cell Chem. Biol.* **2018**, *25*, 380–391.e385. [CrossRef]
47. Rosenthal, K.S.; Ferguson, R.A.; Storm, D.R. Mechanism of action of EM 49, membrane-active peptide antibiotic. *Antimicrob. Agents Chemother.* **1977**, *12*, 665–672. [CrossRef] [PubMed]
48. Yang, J.; Zhu, X.; Cao, M.; Wang, C.; Zhang, C.; Lu, Z.; Lu, F. Genomics-Inspired Discovery of Three Antibacterial Active Metabolites, Aurantinins B, C, and D from Compost-Associated *Bacillus subtilis* fmb60. *J. Agric. Food Chem.* **2016**, *64*, 8811–8820. [CrossRef]
49. Rollin-Pinheiro, R.; Bayona-Pacheco, B.; Domingos, L.T.S.; da Rocha Curvelo, J.A.; de Castro, G.M.M.; Barreto-Bergter, E.; Ferreira-Pereira, A. Sphingolipid Inhibitors as an Alternative to Treat Candidiasis Caused by Fluconazole-Resistant Strains. *Pathogens* **2021**, *10*, 856. [CrossRef] [PubMed]
50. Wadsworth, J.M.; Clarke, D.J.; McMahon, S.A.; Lowther, J.P.; Beattie, A.E.; Langridge-Smith, P.R.; Broughton, H.B.; Dunn, T.M.; Naismith, J.H.; Campopiano, D.J. The chemical basis of serine palmitoyltransferase inhibition by myriocin. *J. Am. Chem. Soc.* **2013**, *135*, 14276–14285. [CrossRef]
51. Wang, H.; Wang, Z.; Xu, W.; Wang, K. Comprehensive transcriptomic and proteomic analyses identify intracellular targets for myriocin to induce *Fusarium oxysporum* f. sp. *niveum cell death. Microb. Cell Fact.* **2021**, *20*, 69. [CrossRef]
52. Takada, Y.; Itoh, H.; Paudel, A.; Panthee, S.; Hamamoto, H.; Sekimizu, K.; Inoue, M. Discovery of gramicidin A analogues with altered activities by multidimensional screening of a one-bead-one-compound library. *Nat. Commun.* **2020**, *11*, 4935. [CrossRef]
53. Hladky, S.B.; Haydon, D.A. Ion transfer across lipid membranes in the presence of gramicidin A. I. Studies of the unit conductance channel. *Biochim. Biophys. Acta* **1972**, *274*, 294–312. [CrossRef]
54. Liou, J.W.; Hung, Y.J.; Yang, C.H.; Chen, Y.C. The antimicrobial activity of gramicidin A is associated with hydroxyl radical formation. *PLoS ONE* **2015**, *10*, e0117065. [CrossRef]
55. Swierstra, J.; Kapoerchan, V.; Knijnenburg, A.; van Belkum, A.; Overhand, M. Structure, toxicity and antibiotic activity of gramicidin S and derivatives. *Eur. J. Clin. Microbiol. Infect. Dis.* **2016**, *35*, 763–769. [CrossRef] [PubMed]
56. Afonin, S.; Dürr, U.H.; Wadhwani, P.; Salgado, J.; Ulrich, A.S. Solid State NMR Structure Analysis of the Antimicrobial Peptide Gramicidin S in Lipid Membranes: Concentration-Dependent Re-alignment and Self-Assembly as a β-Barrel. *Top Curr. Chem.* **2008**, *273*, 139–154. [CrossRef]
57. Kaprel'iants, A.S.; Nikiforov, V.V.; Miroshnikov, A.I.; Snezhkova, L.G.; Eremin, V.A.; Ostrovskiĭ, D.N. Membranes of bacteria and mechanism of action of the antibiotic gramicidin S. *Biokhimiia* **1977**, *42*, 329–337. [PubMed]
58. Frangou-Lazaridis, M.; Seddon, B. Effect of gramicidin S on the transcription system of the producer Bacillus brevis Nagano. *J. Gen. Microbiol.* **1985**, *131*, 437–449. [CrossRef] [PubMed]
59. Grau, A.; Ortiz, A.; de Godos, A.; Gómez-Fernández, J.C. A biophysical study of the interaction of the lipopeptide antibiotic iturin A with aqueous phospholipid bilayers. *Arch. Biochem. Biophys.* **2000**, *377*, 315–323. [CrossRef]
60. Inès, M.; Dhouha, G. Lipopeptide surfactants: Production, recovery and pore forming capacity. *Peptides* **2015**, *71*, 100–112. [CrossRef]
61. Qi, G.; Zhu, F.; Du, P.; Yang, X.; Qiu, D.; Yu, Z.; Chen, J.; Zhao, X. Lipopeptide induces apoptosis in fungal cells by a mitochondria-dependent pathway. *Peptides* **2010**, *31*, 1978–1986. [CrossRef]

62. Ambroggio, E.E.; Separovic, F.; Bowie, J.H.; Fidelio, G.D.; Bagatolli, L.A. Direct visualization of membrane leakage induced by the antibiotic peptides: Maculatin, citropin, and aurein. *Biophys. J.* **2005**, *89*, 1874–1881. [CrossRef]
63. Lei, S.; Zhao, H.; Pang, B.; Qu, R.; Lian, Z.; Jiang, C.; Shao, D.; Huang, Q.; Jin, M.; Shi, J. Capability of iturin from Bacillus subtilis to inhibit Candida albicans in vitro and in vivo. *Appl. Microbiol. Biotechnol.* **2019**, *103*, 4377–4392. [CrossRef] [PubMed]
64. Peypoux, F.; Besson, F.; Michel, G.; Delcambe, L. Preparation and antibacterial activity upon Micrococcus luteus of derivatives of iturin A, mycosubtilin and bacillomycin L, antibiotics from *Bacillus subtilis*. *J. Antibiot.* **1979**, *32*, 136–140. [CrossRef] [PubMed]
65. Romero, D.; de Vicente, A.; Rakotoaly, R.H.; Dufour, S.E.; Veening, J.W.; Arrebola, E.; Cazorla, F.M.; Kuipers, O.P.; Paquot, M.; Pérez-García, A. The iturin and fengycin families of lipopeptides are key factors in antagonism of Bacillus subtilis toward Podosphaera fusca. *Mol. Plant Microbe Interact.* **2007**, *20*, 430–440. [CrossRef]
66. Arima, K.; Kakinuma, A.; Tamura, G. Surfactin, a crystalline peptidelipid surfactant produced by Bacillus subtilis: Isolation, characterization and its inhibition of fibrin clot formation. *Biochem. Biophys. Res. Commun.* **1968**, *31*, 488–494. [CrossRef]
67. Sheppard, J.D.; Jumarie, C.; Cooper, D.G.; Laprade, R. Ionic channels induced by surfactin in planar lipid bilayer membranes. *Biochim. Biophys. Acta* **1991**, *1064*, 13–23. [CrossRef]
68. Liu, J.; Li, W.; Zhu, X.; Zhao, H.; Lu, Y.; Zhang, C.; Lu, Z. Surfactin effectively inhibits Staphylococcus aureus adhesion and biofilm formation on surfaces. *Appl. Microbiol. Biotechnol.* **2019**, *103*, 4565–4574. [CrossRef]
69. Farace, G.; Fernandez, O.; Jacquens, L.; Coutte, F.; Krier, F.; Jacques, P.; Clément, C.; Barka, E.A.; Jacquard, C.; Dorey, S. Cyclic lipopeptides from Bacillus subtilis activate distinct patterns of defence responses in grapevine. *Mol. Plant Pathol.* **2015**, *16*, 177–187. [CrossRef]
70. Tao, Y.; Bie, X.M.; Lv, F.X.; Zhao, H.Z.; Lu, Z.X. Antifungal activity and mechanism of fengycin in the presence and absence of commercial surfactin against Rhizopus stolonifer. *J. Microbiol.* **2011**, *49*, 146–150. [CrossRef]
71. Liu, Y.; Zhang, J.; Wang, S.; Guo, Y.; He, T.; Zhou, R. A Novel Adjuvant "Sublancin" Enhances Immune Response in Specific Pathogen-Free Broiler Chickens Inoculated with Newcastle Disease Vaccine. *J. Immunol. Res.* **2019**, *2019*, 1016567. [CrossRef]
72. Zhang, L.; Sun, C. Fengycins, Cyclic Lipopeptides from Marine Bacillus subtilis Strains, Kill the Plant-Pathogenic Fungus Magnaporthe grisea by Inducing Reactive Oxygen Species Production and Chromatin Condensation. *Appl. Environ. Microbiol.* **2018**, *84*, e00445-18. [CrossRef]
73. Johnson, V.L.; Ko, S.C.; Holmstrom, T.H.; Eriksson, J.E.; Chow, S.C. Effector caspases are dispensable for the early nuclear morphological changes during chemical-induced apoptosis. *J. Cell Sci.* **2000**, *113 Pt 17*, 2941–2953. [CrossRef] [PubMed]
74. Piewngam, P.; Zheng, Y.; Nguyen, T.H.; Dickey, S.W.; Joo, H.S.; Villaruz, A.E.; Glose, K.A.; Fisher, E.L.; Hunt, R.L.; Li, B.; et al. Pathogen elimination by probiotic Bacillus via signalling interference. *Nature* **2018**, *562*, 532–537. [CrossRef]
75. Chung, L.K.; Raffatellu, M. Probiotic fengycins dis(Agr)ee with Staphylococcus aureus colonization. *Cell Res.* **2019**, *29*, 93–94. [CrossRef] [PubMed]
76. Gao, L.; Han, J.; Liu, H.; Qu, X.; Lu, Z.; Bie, X. Plipastatin and surfactin coproduction by *Bacillus subtilis* pB2-L and their effects on microorganisms. *Antonie Van Leeuwenhoek* **2017**, *110*, 1007–1018. [CrossRef] [PubMed]
77. Cawoy, H.; Debois, D.; Franzil, L.; De Pauw, E.; Thonart, P.; Ongena, M. Lipopeptides as main ingredients for inhibition of fungal phytopathogens by *Bacillus subtilis/amyloliquefaciens*. *Microb. Biotechnol.* **2015**, *8*, 281–295. [CrossRef]
78. Besson, F.; Peypoux, F.; Michel, G.; Delcambe, L. Antifungal activity upon *Saccharomyces cerevisiae* of iturin A, mycosubtilin, bacillomycin L and of their derivatives; inhibition of this antifungal activity by lipid antagonists. *J. Antibiot.* **1979**, *32*, 828–833. [CrossRef] [PubMed]
79. Gong, A.D.; Li, H.P.; Yuan, Q.S.; Song, X.S.; Yao, W.; He, W.J.; Zhang, J.B.; Liao, Y.C. Antagonistic mechanism of iturin A and plipastatin A from *Bacillus amyloliquefaciens* S76-3 from wheat spikes against *Fusarium graminearum*. *PLoS ONE* **2015**, *10*, e0116871. [CrossRef]
80. Wang, Y.; Zhang, C.; Liang, J.; Wu, L.; Gao, W.; Jiang, J. Iturin A Extracted from *Bacillus subtilis* WL-2 Affects Phytophthora infestans via Cell Structure Disruption, Oxidative Stress, and Energy Supply Dysfunction. *Front. Microbiol.* **2020**, *11*, 2205. [CrossRef]
81. Tunsagool, P.; Leelasuphakul, W.; Jaresitthikunchai, J.; Phaonakrop, N.; Roytrakul, S.; Jutidamrongphan, W. Targeted transcriptional and proteomic studies explicate specific roles of *Bacillus subtilis* iturin A, fengycin, and surfactin on elicitation of defensive systems in mandarin fruit during stress. *PLoS ONE* **2019**, *14*, e0217202. [CrossRef]
82. Jin, P.; Wang, H.; Tan, Z.; Xuan, Z.; Dahar, G.Y.; Li, Q.X.; Miao, W.; Liu, W. Antifungal mechanism of bacillomycin D from *Bacillus velezensis* HN-2 against *Colletotrichum gloeosporioides* Penz. *Pestic. Biochem. Physiol.* **2020**, *163*, 102–107. [CrossRef]
83. Peypoux, F.; Besson, F.; Michel, G.; Lenzen, C.; Dierickx, L.; Delcambe, L. Characterization of a new antibiotic of iturin group: Bacillomycin D. *J. Antibiot.* **1980**, *33*, 1146–1149. [CrossRef]
84. Wu, T.; Chen, M.; Zhou, L.; Lu, F.; Bie, X.; Lu, Z. Bacillomycin D effectively controls growth of *Malassezia globosa* by disrupting the cell membrane. *Appl. Microbiol. Biotechnol.* **2020**, *104*, 3529–3540. [CrossRef]
85. Gu, Q.; Yang, Y.; Yuan, Q.; Shi, G.; Wu, L.; Lou, Z.; Huo, R.; Wu, H.; Borriss, R.; Gao, X. Bacillomycin D Produced by *Bacillus amyloliquefaciens* Is Involved in the Antagonistic Interaction with the Plant-Pathogenic Fungus *Fusarium graminearum*. *Appl. Environ. Microbiol.* **2017**, *83*, e01075-17. [CrossRef] [PubMed]
86. Xu, Z.; Mandic-Mulec, I.; Zhang, H.; Liu, Y.; Sun, X.; Feng, H.; Xun, W.; Zhang, N.; Shen, Q.; Zhang, R. Antibiotic Bacillomycin D Affects Iron Acquisition and Biofilm Formation in *Bacillus velezensis* through a Btr-Mediated FeuABC-Dependent Pathway. *Cell Rep.* **2019**, *29*, 1192–1202. [CrossRef] [PubMed]

87. Zhang, B.; Dong, C.; Shang, Q.; Cong, Y.; Kong, W.; Li, P. Purification and partial characterization of bacillomycin L produced by *Bacillus amyloliquefaciens* K103 from lemon. *Appl. Biochem. Biotechnol.* **2013**, *171*, 2262–2272. [CrossRef]
88. Quentin, M.J.; Besson, F.; Peypoux, F.; Michel, G. Action of peptidolipidic antibiotics of the iturin group on erythrocytes. Effect of some lipids on hemolysis. *Biochim. Biophys. Acta* **1982**, *684*, 207–211. [CrossRef]
89. Besson, F.; Michel, G. Action of the antibiotics of the iturin group on artificial membranes. *J. Antibiot.* **1984**, *37*, 646–651. [CrossRef] [PubMed]
90. Zhang, B.; Qin, Y.; Han, Y.; Dong, C.; Li, P.; Shang, Q. Comparative proteomic analysis reveals intracellular targets for bacillomycin L to induce *Rhizoctonia solani* Kühn hyphal cell death. *Biochim. Biophys. Acta* **2016**, *1864*, 1152–1159. [CrossRef]
91. Loison, C.; Nasir, M.N.; Benichou, E.; Besson, F.; Brevet, P.-F. Multi-scale modeling of mycosubtilin lipopeptides at the air/water interface: Structure and optical second harmonic generation. *Phys. Chem. Chem. Phys.* **2014**, *16*, 2136–2148. [CrossRef]
92. Besson, F.; Michel, G. Action of mycosubtilin, an antifungal antibiotic of *Bacillus subtilis*, on the cell membrane of *Saccharomyces cerevisiae*. *Microbios* **1989**, *59*, 113–121.
93. Bhattacharyya, P.; Bose, S.K. Effect of mycobacillin, an antifungal polypeptide antibiotic, on the producer *Bacillus subtilis* B3. *Indian J. Med. Res.* **1967**, *55*, 1025–1029. [PubMed]
94. Chowdhury, B.; Das, S.K.; Bose, S.K. Use of resistant mutants to characterize the target of mycobacillin in *Aspergillus niger* membranes. *Microbiology* **1998**, *144*, 1123–1130. [CrossRef]
95. Das, S.K.; Mukherjee, S.; Majumdar, S.; Basu, S.; Bose, S.K. Physico-chemical interaction of mycobacillin with *Aspergillus niger* protoplast membrane, the site of its action. *J. Antibiot.* **1987**, *40*, 1036–1043. [CrossRef]
96. Babasaki, K.; Takao, T.; Shimonishi, Y.; Kurahashi, K. Subtilosin A, a new antibiotic peptide produced by *Bacillus subtilis* 168: Isolation, structural analysis, and biogenesis. *J. Biochem.* **1985**, *98*, 585–603. [CrossRef]
97. Turovskiy, Y.; Cheryian, T.; Algburi, A.; Wirawan, R.E.; Takhistov, P.; Sinko, P.J.; Chikindas, M.L. Susceptibility of *Gardnerella vaginalis* biofilms to natural antimicrobials subtilosin, ε-poly-L-lysine, and lauramide arginine ethyl ester. *Infect. Dis. Obs. Gynecol.* **2012**, *2012*, 284762. [CrossRef]
98. Noll, K.S.; Sinko, P.J.; Chikindas, M.L. Elucidation of the Molecular Mechanisms of Action of the Natural Antimicrobial Peptide Subtilosin against the Bacterial Vaginosis-associated Pathogen *Gardnerella vaginalis*. *Probiotics Antimicrob. Proteins* **2011**, *3*, 41–47. [CrossRef]
99. Algburi, A.; Zehm, S.; Netrebov, V.; Bren, A.B.; Chistyakov, V.; Chikindas, M.L. Subtilosin Prevents Biofilm Formation by Inhibiting Bacterial Quorum Sensing. *Probiotics Antimicrob. Proteins* **2017**, *9*, 81–90. [CrossRef] [PubMed]
100. Hao, Z.; Yan, L.; Liu, J.; Song, F.; Zhang, J.; Li, X. Extraction of antibiotic zwittermicin A from *Bacillus thuringiensis* by macroporous resin and silica gel column chromatography. *Biotechnol. Appl. Biochem.* **2015**, *62*, 369–374. [CrossRef] [PubMed]
101. Stabb, E.V.; Handelsman, J. Genetic analysis of zwittermicin A resistance in *Escherichia coli*: Effects on membrane potential and RNA polymerase. *Mol. Microbiol.* **1998**, *27*, 311–322. [CrossRef]
102. Im, S.M.; Yu, N.H.; Joen, H.W.; Kim, S.O.; Park, H.W.; Park, A.R.; Kim, J.-C. Biological control of tomato bacterial wilt by oxydifficidin and difficidin-producing *Bacillus methylotrophicus* DR-08. *Pestic. Biochem. Physiol.* **2020**, *163*, 130–137. [CrossRef]
103. Zweerink, M.M.; Edison, A. Difficidin and oxydifficidin: Novel broad spectrum antibacterial antibiotics produced by *Bacillus subtilis*. III. Mode of action of difficidin. *J. Antibiot.* **1987**, *40*, 1692–1697. [CrossRef]
104. Wu, L.; Wu, H.; Chen, L.; Yu, X.; Borriss, R.; Gao, X. Difficidin and bacilysin from *Bacillus amyloliquefaciens* FZB42 have antibacterial activity against *Xanthomonas oryzae* rice pathogens. *Sci. Rep.* **2015**, *5*, 12975. [CrossRef]
105. Wang, S.; Wang, Q.; Zeng, X.; Ye, Q.; Huang, S.; Yu, H.; Yang, T.; Qiao, S. Use of the Antimicrobial Peptide Sublancin with Combined Antibacterial and Immunomodulatory Activities To Protect against Methicillin-Resistant *Staphylococcus aureus* Infection in Mice. *J. Agric. Food Chem.* **2017**, *65*, 8595–8605. [CrossRef] [PubMed]
106. Wu, C.; Biswas, S.; Garcia De Gonzalo, C.V.; van der Donk, W.A. Investigations into the Mechanism of Action of Sublancin. *ACS Infect. Dis* **2019**, *5*, 454–459. [CrossRef]
107. Maksimova, E.M.; Vinogradova, D.S.; Osterman, I.A.; Kasatsky, P.S.; Nikonov, O.S.; Milón, P.; Dontsova, O.A.; Sergiev, P.V.; Paleskava, A.; Konevega, A.L. Multifaceted Mechanism of Amicoumacin A Inhibition of Bacterial Translation. *Front. Microbiol.* **2021**, *12*, 172. [CrossRef] [PubMed]
108. Lama, A.; Pané-Farré, J.; Chon, T.; Wiersma, A.M.; Sit, C.S.; Vederas, J.C.; Hecker, M.; Nakano, M.M. Response of methicillin-resistant *Staphylococcus aureus* to amicoumacin A. *PLoS ONE* **2012**, *7*, e34037. [CrossRef]
109. Tanaka, K.; Fukuda, M.; Amaki, Y.; Sakaguchi, T.; Inai, K.; Ishihara, A.; Nakajima, H. Importance of prumycin produced by Bacillus amyloliquefaciens SD-32 in biocontrol against cucumber powdery mildew disease. *Pest. Manag. Sci.* **2017**, *73*, 2419–2428. [CrossRef] [PubMed]
110. Schwartz, J.L.; Katagiri, M.; Omura, S.; Tishler, M. The mechanism of prumycin action. *J. Antibiot.* **1974**, *27*, 379–385. [CrossRef]
111. Hata, T.; Omura, S.; Katagiri, M.; Atsumi, K.; Awaya, J. A new antifungal antibiotic, prumycin. *J. Antibiot.* **1971**, *24*, 900–901. [CrossRef]
112. Chan, D.C.K.; Burrows, L.L. Thiocillin and micrococcin exploit the ferrioxamine receptor of *Pseudomonas aeruginosa* for uptake. *J. Antimicrob. Chemother.* **2021**, *76*, 2029–2039. [CrossRef]
113. Bleich, R.; Watrous, J.D.; Dorrestein, P.C.; Bowers, A.A.; Shank, E.A. Thiopeptide antibiotics stimulate biofilm formation in *Bacillus subtilis*. *Proc. Natl. Acad. Sci. USA* **2015**, *112*, 3086–3091. [CrossRef]

114. Wang, T.; Lu, Q.; Sun, C.; Lukianov, D.; Osterman, I.A.; Sergiev, P.V.; Dontsova, O.A.; Hu, X.; You, X.; Liu, S.; et al. Hetiamacin E and F, New Amicoumacin Antibiotics from *Bacillus subtilis* PJS Using MS/MS-Based Molecular Networking. *Molecules* **2020**, *25*, 4446. [CrossRef]
115. Gahungu, M.; Arguelles-Arias, A.; Fickers, P.; Zervosen, A.; Joris, B.; Damblon, C.; Luxen, A. Synthesis and biological evaluation of potential threonine synthase inhibitors: Rhizocticin A and Plumbemycin A. *Bioorg. Med. Chem.* **2013**, *21*, 4958–4967. [CrossRef] [PubMed]
116. Kugler, M.; Loeffler, W.; Rapp, C.; Kern, A.; Jung, G. Rhizocticin A, an antifungal phosphono-oligopeptide of *Bacillus subtilis* ATCC 6633: Biological properties. *Arch. Microbiol.* **1990**, *153*, 276–281. [CrossRef] [PubMed]
117. Yoo, J.S.; Zheng, C.J.; Lee, S.; Kwak, J.H.; Kim, W.G. Macrolactin N, a new peptide deformylase inhibitor produced by *Bacillus subtilis*. *Bioorg. Med. Chem. Lett.* **2006**, *16*, 4889–4892. [CrossRef]
118. Aoki, Y.; Yamamoto, M.; Hosseini-Mazinani, S.M.; Koshikawa, N.; Sugimoto, K.; Arisawa, M. Antifungal azoxybacilin exhibits activity by inhibiting gene expression of sulfite reductase. *Antimicrob. Agents Chemother.* **1996**, *40*, 127–132. [CrossRef]
119. Fujiu, M.; Sawairi, S.; Shimada, H.; Takaya, H.; Aoki, Y.; Okuda, T.; Yokose, K. Azoxybacilin, a novel antifungal agent produced by *Bacillus cereus* NR2991. Production, isolation and structure elucidation. *J. Antibiot.* **1994**, *47*, 833–835. [CrossRef] [PubMed]
120. Zeriouh, H.; de Vicente, A.; Pérez-García, A.; Romero, D. Surfactin triggers biofilm formation of *Bacillus subtilis* in melon phylloplane and contributes to the biocontrol activity. *Environ. Microbiol.* **2014**, *16*, 2196–2211. [CrossRef]
121. Caro-Astorga, J.; Frenzel, E.; Perkins, J.R.; Álvarez-Mena, A.; de Vicente, A.; Ranea, J.A.G.; Kuipers, O.P.; Romero, D. Biofilm formation displays intrinsic offensive and defensive features of *Bacillus cereus*. *NPJ Biofilms Microbiomes* **2020**, *6*, 3. [CrossRef]
122. Karunakaran, E.; Biggs, C.A. Mechanisms of Bacillus cereus biofilm formation: An investigation of the physicochemical characteristics of cell surfaces and extracellular proteins. *Appl. Microbiol. Biotechnol.* **2011**, *89*, 1161–1175. [CrossRef] [PubMed]
123. Singh, A.A.; Singh, A.K.; Nerurkar, A. Disrupting the quorum sensing mediated virulence in soft rot causing *Pectobacterium carotovorum* by marine sponge associated *Bacillus* sp. OA10. *World J. Microbiol. Biotechnol.* **2021**, *37*, 5. [CrossRef]
124. Boopathi, S.; Vashisth, R.; Manoharan, P.; Kandasamy, R.; Sivakumar, N. Stigmatellin Y—An anti-biofilm compound from *Bacillus subtilis* BR4 possibly interferes in PQS-PqsR mediated quorum sensing system in *Pseudomonas aeruginosa*. *Bioorg. Med. Chem. Lett.* **2017**, *27*, 2113–2118. [CrossRef] [PubMed]
125. Li, H.; Han, X.; Dong, Y.; Xu, S.; Chen, C.; Feng, Y.; Cui, Q.; Li, W. Bacillaenes: Decomposition Trigger Point and Biofilm Enhancement in *Bacillus*. *ACS Omega* **2021**, *6*, 1093–1098. [CrossRef] [PubMed]
126. Tonziello, G.; Caraffa, E.; Pinchera, B.; Granata, G.; Petrosillo, N. Present and future of siderophore-based therapeutic and diagnostic approaches in infectious diseases. *Infect. Dis. Rep.* **2019**, *11*, 8208. [CrossRef]
127. Dertz, E.A.; Xu, J.; Stintzi, A.; Raymond, K.N. Bacillibactin-mediated iron transport in *Bacillus subtilis*. *J. Am. Chem. Soc.* **2006**, *128*, 22–23. [CrossRef]
128. Hu, X.; Boyer, G.L. Siderophore-Mediated Aluminum Uptake by *Bacillus megaterium* ATCC 19213. *Appl. Environ. Microbiol.* **1996**, *62*, 4044–4048. [CrossRef]
129. Segond, D.; Abi Khalil, E.; Buisson, C.; Daou, N.; Kallassy, M.; Lereclus, D.; Arosio, P.; Bou-Abdallah, F.; Nielsen Le Roux, C. Iron acquisition in *Bacillus cereus*: The roles of IlsA and bacillibactin in exogenous ferritin iron mobilization. *PLoS Pathog.* **2014**, *10*, e1003935. [CrossRef]
130. Chen, X.H.; Vater, J.; Piel, J.; Franke, P.; Scholz, R.; Schneider, K.; Koumoutsi, A.; Hitzeroth, G.; Grammel, N.; Strittmatter, A.W.; et al. Structural and functional characterization of three polyketide synthase gene clusters in *Bacillus amyloliquefaciens* FZB 42. *J. Bacteriol.* **2006**, *188*, 4024–4036. [CrossRef]
131. Dunlap, C.A.; Kim, S.J.; Kwon, S.W.; Rooney, A.P. Bacillus velezensis is not a later heterotypic synonym of *Bacillus amyloliquefaciens; Bacillus methylotrophicus, Bacillus amyloliquefaciens* subsp. *plantarum* and *'Bacillus oryzicola'* are later heterotypic synonyms of *Bacillus velezensis* based on phylogenomics. *Int. J. Syst. Evol. Microbiol.* **2016**, *66*, 1212–1217. [CrossRef]
132. Fan, B.; Blom, J.; Klenk, H.P.; Borriss, R. *Bacillus amyloliquefaciens, Bacillus velezensis*, and *Bacillus siamensis* Form an "Operational Group *B. amyloliquefaciens*" within the *B. subtilis* Species Complex. *Front. Microbiol.* **2017**, *8*, 22. [CrossRef]
133. Cohen, A.; Bont, L.; Engelhard, D.; Moore, E.; Fernández, D.; Kreisberg-Greenblatt, R.; Oved, K.; Eden, E.; Hays, J.P. A multifaceted 'omics' approach for addressing the challenge of antimicrobial resistance. *Future Microbiol.* **2015**, *10*, 365–376. [CrossRef]
134. Kang, X.; Zhang, W.; Cai, X.; Zhu, T.; Xue, Y.; Liu, C. Bacillus velezensis CC09: A Potential 'Vaccine' for Controlling Wheat Diseases. *Mol. Plant-Microbe Interact.* **2018**, *31*, 623–632. [CrossRef]
135. Alkema, W.; Boekhorst, J.; Wels, M.; van Hijum, S.A. Microbial bioinformatics for food safety and production. *Brief. Bioinform.* **2016**, *17*, 283–292. [CrossRef]
136. Wiegand, S.; Voigt, B.; Albrecht, D.; Bongaerts, J.; Evers, S.; Hecker, M.; Daniel, R.; Liesegang, H. Fermentation stage-dependent adaptations of *Bacillus licheniformis* during enzyme production. *Microb. Cell Factories* **2013**, *12*, 120. [CrossRef] [PubMed]
137. Amin, A.; Khan, M.A.; Ehsanullah, M.; Haroon, U.; Azam, S.M.; Hameed, A. Production of peptide antibiotics by *Bacillus* sp. GU 057 indigenously isolated from saline soil. *Braz. J. Microbiol.* **2012**, *43*, 1340–1346. [CrossRef]
138. Zhu, J.; Li, L.; Wu, F.; Wu, Y.; Wang, Z.; Chen, X.; Li, J.; Cai, D.; Chen, S. Metabolic Engineering of Aspartic Acid Supply Modules for Enhanced Production of Bacitracin in *Bacillus licheniformis*. *ACS Synth. Biol.* **2021**, *10*, 2243–2251. [CrossRef]
139. Haavik, H.I.; Vessia, B. Bacitracin production by the high-yielding mutant bacillus licheniformis strain al: Stimulatory effect of l-leucine. *Acta Pathol. Microbiol. Scand. Sect. B Microbiol.* **1978**, *86B*, 67–70. [CrossRef]
140. Ganchev, K.; Kozhukharova, L. Bacitracin biosynthesis by *Bacillus licheniformis* 16. *Acta Microbiol. Bulg.* **1984**, *15*, 38–42. [PubMed]

141. Egorov, N.S.; Loriia, Z.K.; Vybornykh, S.N.; Khamrun, R. Effect of bacitracin on the sporulation of *Bacillus licheniformis* 28 KA. *Nauchnye Dokl. vysshei shkoly. Biol. Nauk.* **1985**, *6*, 89–91.
142. Vitković, L.; Sadoff, H.L. In vitro production of bacitracin by proteolysis of vegetative *Bacillus licheniformis* cell protein. *J. Bacteriol.* **1977**, *131*, 897–905. [CrossRef] [PubMed]
143. Jamil, B.; Hasan, F.; Hameed, A.; Ahmed, S. Isolation of *Bacillus subtilis* MH-4 from soil and its potential of polypeptidic antibiotic production. *Pak. J. Pharm. Sci.* **2007**, *20*, 26–31. [PubMed]
144. Amin, A.; Khan, M.A.; Ahmad, T. Optimized antimicrobial peptide (Bacitracin) production by immobilized and free cells and of *Bacillus* Spp. GU215 using Wood chips and silicon polymer beads. *Pak. J. Pharm. Sci.* **2013**, *26*, 1077–1082. [PubMed]
145. Lu, J.Y.; Zhou, K.; Huang, W.T.; Zhou, P.; Yang, S.; Zhao, X.; Xie, J.; Xia, L.; Ding, X. A comprehensive genomic and growth proteomic analysis of antitumor lipopeptide bacillomycin Lb biosynthesis in *Bacillus amyloliquefaciens* X030. *Appl. Microbiol. Biotechnol.* **2019**, *103*, 7647–7662. [CrossRef]
146. Rogers, H.J.; Newton, G.G.F.; Abraham, E.P. Production and purification of bacilysin. *Biochem. J.* **1965**, *97*, 573–578. [CrossRef]
147. Ozcengiz, G.; Alaeddinoglu, N.G.; Demain, A.L. Regulation of biosynthesis of bacilysin by *Bacillus subtilis*. *J. Ind. Microbiol. Biotechnol.* **1990**, *6*, 91–100. [CrossRef]
148. Chung, S.; Kong, H.; Buyer, J.S.; Lakshman, D.K.; Lydon, J.; Kim, S.-D.; Roberts, D.P. Isolation and partial characterization of *Bacillus subtilis* ME488 for suppression of soilborne pathogens of cucumber and pepper. *Appl. Microbiol. Biotechnol.* **2008**, *80*, 115–123. [CrossRef]
149. Phister, T.G.; O'Sullivan, D.J.; McKay, L.L. Identification of bacilysin, chlorotetaine, and iturin a produced by *Bacillus* sp. strain CS93 isolated from pozol, a Mexican fermented maize dough. *Appl. Environ. Microbiol.* **2004**, *70*, 631–634. [CrossRef]
150. Vairagkar, U.; Mirza, Y. Antagonistic Activity of Antimicrobial Metabolites Produced from Seaweed-Associated *Bacillus amyloliquefaciens* MTCC 10456 Against *Malassezia* spp. *Probiotics Antimicrob. Proteins* **2021**, *13*, 1228–1237. [CrossRef]
151. Panneerselvam, P.; Senapati, A.; Kumar, U.; Sharma, L.; Lepcha, P.; Prabhukarthikeyan, S.R.; Jahan, A.; Parameshwaran, C.; Govindharaj, G.P.P.; Lenka, S.; et al. Antagonistic and plant-growth promoting novel Bacillus species from long-term organic farming soils from Sikkim, India. *3 Biotech* **2019**, *9*, 416. [CrossRef]
152. AlGburi, A.; Alazzawi, S.A.; Al-Ezzy, A.I.A.; Weeks, R.; Chistyakov, V.; Chikindas, M.L. Potential Probiotics *Bacillus subtilis* KATMIRA1933 and *Bacillus amyloliquefaciens* B-1895 Co-Aggregate with Clinical Isolates of Proteus mirabilis and Prevent Biofilm Formation. *Probiotics Antimicrob. Proteins* **2020**, *12*, 1471–1483. [CrossRef] [PubMed]
153. Ahire, J.J.; Kashikar, M.S.; Madempudi, R.S. Survival and Germination of *Bacillus clausii* UBBC07 Spores in in vitro Human Gastrointestinal Tract Simulation Model and Evaluation of Clausin Production. *Front. Microbiol.* **2020**, *11*, 1010. [CrossRef]
154. Brötz, H.; Bierbaum, G.; Markus, A.; Molitor, E.; Sahl, H.G. Mode of action of the lantibiotic mersacidin: Inhibition of peptidoglycan biosynthesis via a novel mechanism? *Antimicrob. Agents Chemother.* **1995**, *39*, 714–719. [CrossRef]
155. Appleyard, A.N.; Choi, S.; Read, D.M.; Lightfoot, A.; Boakes, S.; Hoffmann, A.; Chopra, I.; Bierbaum, G.; Rudd, B.A.; Dawson, M.J.; et al. Dissecting Structural and Functional Diversity of the Lantibiotic Mersacidin. *Chem. Biol.* **2009**, *16*, 490–498. [CrossRef]
156. He, P.; Hao, K.; Blom, J.; Rückert, C.; Vater, J.; Mao, Z.; Wu, Y.; Hou, M.; He, P.; He, Y.; et al. Genome sequence of the plant growth promoting strain *Bacillus amyloliquefaciens* subsp. plantarum B9601-Y2 and expression of mersacidin and other secondary metabolites. *J. Biotechnol.* **2013**, *164*, 281–291. [CrossRef] [PubMed]
157. Molinatto, G.; Puopolo, G.; Sonego, P.; Moretto, M.; Engelen, K.; Viti, C.; Ongena, M.; Pertot, I. Complete genome sequence of *Bacillus amyloliquefaciens* subsp. plantarum S499, a rhizobacterium that triggers plant defences and inhibits fungal phytopathogens. *J. Biotechnol.* **2016**, *238*, 56–59. [CrossRef] [PubMed]
158. Scholz, R.; Molohon, K.J.; Nachtigall, J.; Vater, J.; Markley, A.L.; Süssmuth, R.D.; Mitchell, D.A.; Borriss, R. Plantazolicin, a novel microcin B17/streptolysin S-like natural product from *Bacillus amyloliquefaciens* FZB42. *J. Bacteriol.* **2011**, *193*, 215–224. [CrossRef] [PubMed]
159. Chen, L.; Heng, J.; Qin, S.; Bian, K. A comprehensive understanding of the biocontrol potential of *Bacillus velezensis* LM2303 against Fusarium head blight. *PLoS ONE* **2018**, *13*, e0198560. [CrossRef] [PubMed]
160. Meyers, E.; Brown, W.E.; Principe, P.A.; Rathnum, M.L.; Parker, W.L. EM49, a new peptide antibiotic. I. Fermentation, isolation, and preliminary characterization. *J. Antibiot.* **1973**, *26*, 444–448. [CrossRef] [PubMed]
161. Wang, H.; Wang, Z.; Liu, Z.; Wang, K.; Xu, W. Membrane disruption of *Fusarium oxysporum f.* sp. niveum induced by myriocin from *Bacillus amyloliquefaciens* LZN01. *Microb. Biotechnol.* **2021**, *14*, 517–534. [CrossRef]
162. Marahiel, M.A.; Lurz, R.; Kleinkauf, H. Characterization of chromosomal and membrane associated plasmid in Bacillus brevis ATCC 9999. *J. Antibiot.* **1981**, *34*, 323–330. [CrossRef]
163. Udalova, T.P.; Gus'Kova, T.M.; Silaev, A.B. Growth of Bacillus brevis var. G.B. and formation of gramicidin S in relation to the intensity of aeration. *Mikrobiologiia* **1972**, *41*, 280–286.
164. Jiang, J.; Gao, L.; Bie, X.; Lu, Z.; Liu, H.; Zhang, C.; Lu, F.; Zhao, H. Identification of novel surfactin derivatives from NRPS modification of *Bacillus subtilis* and its antifungal activity against *Fusarium moniliforme*. *BMC Microbiol.* **2016**, *16*, 31. [CrossRef]
165. Kim, P.I.; Ryu, J.; Kim, Y.H.; Chi, Y.-T. Production of Biosurfactant Lipopeptides Iturin A, Fengycin and Surfactin A from *Bacillus subtilis* CMB32 for Control of *Colletotrichum gloeosporioides*. *J. Microbiol. Biotechnol.* **2010**, *20*, 138–145. [CrossRef]
166. Chen, M.; Wang, J.; Zhu, Y.; Liu, B.; Yang, W.; Ruan, C. Antibacterial activity against *Ralstonia solanacearum* of the lipopeptides secreted from the *Bacillus amyloliquefaciens* strain FJAT-2349. *J. Appl. Microbiol.* **2019**, *126*, 1519–1529. [CrossRef]

167. Perez, K.J.; Viana, J.D.S.; Lopes, F.C.; Pereira, J.Q.; Dos Santos, D.M.; Oliveira, J.S.; Velho, R.V.; Crispim, S.M.; Nicoli, J.R.; Brandelli, A.; et al. *Bacillus* spp. Isolated from Puba as a Source of Biosurfactants and Antimicrobial Lipopeptides. *Front. Microbiol. (Orig. Res.)* **2017**, *8*, 61. [CrossRef] [PubMed]
168. Chen, L.; Wu, Y.; Chong, X.; Xin, Q.; Wang, D.; Bian, K. Seed-borne endophytic *Bacillus velezensis* LHSB1 mediate the biocontrol of peanut stem rot caused by Sclerotium rolfsii. *J. Appl. Microbiol.* **2020**, *128*, 803–813. [CrossRef] [PubMed]
169. Cao, L.; Pan, L.; Gong, L.; Yang, Y.; He, H.; Li, Y.; Peng, Y.; Li, N.; Yan, L.; Ding, X.; et al. Interaction of a novel *Bacillus velezensis* (BvL03) against Aeromonas hydrophila in vitro and in vivo in grass carp. *Appl. Microbiol. Biotechnol.* **2019**, *103*, 8987–8999. [CrossRef]
170. Zidour, M.; Belguesmia, Y.; Cudennec, B.; Grard, T.; Flahaut, C.; Souissi, S.; Drider, D. Genome Sequencing and Analysis of *Bacillus pumilus* ICVB403 Isolated from Acartia tonsa Copepod Eggs Revealed Surfactin and Bacteriocin Production: Insights on Anti-Staphylococcus Activity. *Probiotics Antimicrob. Proteins* **2019**, *11*, 990–998. [CrossRef] [PubMed]
171. Sarwar, A.; Hassan, M.N.; Imran, M.; Iqbal, M.; Majeed, S.; Brader, G.; Sessitsch, A.; Hafeez, F.Y. Biocontrol activity of surfactin A purified from Bacillus NH-100 and NH-217 against rice bakanae disease. *Microbiol. Res.* **2018**, *209*, 1–13. [CrossRef]
172. Zhang, B.; Li, Y.; Zhang, Y.; Qiao, H.; He, J.; Yuan, Q.; Chen, X.; Fan, J. High-cell-density culture enhances the antimicrobial and freshness effects of *Bacillus subtilis* S1702 on table grapes (Vitis vinifera cv. Kyoho). *Food Chem.* **2019**, *286*, 541–549. [CrossRef] [PubMed]
173. Huang, J.; Wei, Z.; Tan, S.; Mei, X.; Shen, Q.; Xu, Y. Suppression of Bacterial Wilt of Tomato by Bioorganic Fertilizer Made from the Antibacterial Compound Producing Strain *Bacillus amyloliquefaciens* HR62. *J. Agric. Food Chem.* **2014**, *62*, 10708–10716. [CrossRef] [PubMed]
174. Aleti, G.; Lehner, S.; Bacher, M.; Compant, S.; Nikolic, B.; Plesko, M.; Schuhmacher, R.; Sessitsch, A.; Brader, G. Surfactin variants mediate species-specific biofilm formation and root colonization in Bacillus. *Environ. Microbiol.* **2016**, *18*, 2634–2645. [CrossRef] [PubMed]
175. Pueyo, M.T.; Bloch, C.; Carmona-Ribeiro, A.M.; di Mascio, P. Lipopeptides Produced by a Soil *Bacillus Megaterium* Strain. *Microb. Ecol.* **2009**, *57*, 367–378. [CrossRef] [PubMed]
176. Zhao, P.; Quan, C.; Wang, Y.; Wang, J.; Fan, S. *Bacillus amyloliquefaciens* Q-426 as a potential biocontrol agent against *Fusarium oxysporum f.* sp. spinaciae. *J. Basic Microbiol.* **2014**, *54*, 448–456. [CrossRef]
177. Malfanova, N.; Kamilova, F.; Validov, S.; Shcherbakov, A.; Chebotar, V.; Tikhonovich, I.; Lugtenberg, B. Characterization of *Bacillus subtilis* HC8, a novel plant-beneficial endophytic strain from giant hogweed. *Microb. Biotechnol.* **2011**, *4*, 523–532. [CrossRef]
178. Vanittanakom, N.; Loeffler, W.; Koch, U.; Jung, G. Fengycin—A novel antifungal lipopeptide antibiotic produced by *Bacillus subtilis* F-29-3. *J. Antibiot.* **1986**, *39*, 888–901. [CrossRef] [PubMed]
179. Ben Abdallah, D.; Frikha-Gargouri, O.; Tounsi, S. *Bacillus amyloliquefaciens* strain 32a as a source of lipopeptides for biocontrol of Agrobacterium tumefaciens strains. *J. Appl. Microbiol.* **2015**, *119*, 196–207. [CrossRef]
180. Pecci, Y.; Rivardo, F.; Martinotti, M.G.; Allegrone, G. LC/ESI-MS/MS characterisation of lipopeptide biosurfactants produced by the Bacillus licheniformis V9T14 strain. *Biol. Mass Spectrom.* **2010**, *45*, 772–778. [CrossRef]
181. Li, X.-Y.; Mao, Z.-C.; Wang, Y.-H.; Wu, Y.-X.; He, Y.-Q.; Long, C.-L. ESI LC-MS and MS/MS Characterization of Antifungal Cyclic Lipopeptides Produced by *Bacillus subtilis* XF-1. *J. Mol. Microbiol. Biotechnol.* **2012**, *22*, 83–93. [CrossRef]
182. Ma, Z.; Wang, N.; Hu, J.; Wang, S. Isolation and characterization of a new iturinic lipopeptide, mojavensin A produced by a marine-derived bacterium *Bacillus mojavensis* B0621A. *J. Antibiot.* **2012**, *65*, 317–322. [CrossRef] [PubMed]
183. Ding, L.; Guo, W.; Chen, X. Exogenous addition of alkanoic acids enhanced production of antifungal lipopeptides in *Bacillus amyloliquefaciens* Pc3. *Appl. Microbiol. Biotechnol.* **2019**, *103*, 5367–5377. [CrossRef] [PubMed]
184. Nastro, R.A.; Arguelles-Arias, A.; Ongena, M.; Di Costanzo, A.; Trifuoggi, M.; Guida, M.; Fickers, P. Antimicrobial Activity of *Bacillus amyloliquefaciens* ANT1 Toward Pathogenic Bacteria and Mold: Effects on Biofilm Formation. *Probiotics Antimicrob. Proteins* **2013**, *5*, 252–258. [CrossRef]
185. Malfanova, N.; Franzil, L.; Lugtenberg, B.; Chebotar, V.; Ongena, M. Cyclic lipopeptide profile of the plant-beneficial endophytic bacterium *Bacillus subtilis* HC8. *Arch. Microbiol.* **2012**, *194*, 893–899. [CrossRef]
186. Hernández-Morales, A.; Martínez-Peniche, R.A.; Arvizu-Gómez, J.L.; Arvizu-Medrano, S.M.; Rodríguez-Ontiveros, A.; Ramos-López, M.A.; Pacheco-Aguilar, J.R. Production of a Mixture of Fengycins with Surfactant and Antifungal Activities by *Bacillus* sp. MA04, a Versatile PGPR. *Indian J. Microbiol.* **2018**, *58*, 208–213. [CrossRef]
187. Escobar, V.V.; Ceballos, I.; Mira, J.J.; Argel, L.E.; Peralta, S.O.; Romero-Tabarez, M. Fengycin C Produced by *Bacillus subtilis* EA-CB0015. *J. Nat. Prod.* **2013**, *76*, 503–509. [CrossRef]
188. Klich, M.A.; Lax, A.R.; Bland, J. Inhibition of some mycotoxigenic fungi by iturin A, a peptidolipid produced by *Bacillus subtilis*. *Mycopathologia* **1991**, *116*, 77–80. [CrossRef] [PubMed]
189. Dang, Y.; Zhao, F.; Liu, X.; Fan, X.; Huang, R.; Gao, W.; Wang, S.; Yang, C. Enhanced production of antifungal lipopeptide iturin A by *Bacillus amyloliquefaciens* LL3 through metabolic engineering and culture conditions optimization. *Microb. Cell Factories* **2019**, *18*, 68. [CrossRef]
190. Cho, S.J.; Lee, S.K.; Cha, B.J.; Kim, Y.H.; Shin, K.S. Detection and characterization of the Gloeosporium gloeosporioides growth inhibitory compound iturin A from *Bacillus subtilis* strain KS03. *FEMS Microbiol. Lett.* **2003**, *223*, 47–51. [CrossRef]
191. Mizumoto, S.; Hirai, M.; Shoda, M. Production of lipopeptide antibiotic iturin A using soybean curd residue cultivated with *Bacillus subtilis* in solid-state fermentation. *Appl. Microbiol. Biotechnol.* **2006**, *72*, 869–875. [CrossRef]

192. Chen, H.; Yuan, C.; Cai, K.; Zheng, Z.; Yu, Z. Purification and identification of iturin A from *Bacillus subtilis* JA by electrospray ionization mass spectrometry. *Acta Microbiol. Sin.* **2008**, *48*, 116–120.
193. Thasana, N.; Prapagdee, B.; Rangkadilok, N.; Sallabhan, R.; Aye, S.L.; Ruchirawat, S.; Loprasert, S. *Bacillus subtilis* SSE4 produces subtulene A, a new lipopeptide antibiotic possessing an unusual C15 unsaturated β-amino acid. *FEBS Lett.* **2010**, *584*, 3209–3214. [CrossRef] [PubMed]
194. Arrebola, E.; Jacobs, R.; Korsten, L. Iturin A is the principal inhibitor in the biocontrol activity of *Bacillus amyloliquefaciens* PPCB004 against postharvest fungal pathogens. *J. Appl. Microbiol.* **2010**, *108*, 386–395. [CrossRef]
195. Jin, H.; Zhang, X.; Li, K.; Niu, Y.; Guo, M.; Hu, C.; Wan, X.; Gong, Y.; Huang, F. Direct Bio-Utilization of Untreated Rapeseed Meal for Effective Iturin A Production by *Bacillus subtilis* in Submerged Fermentation. *PLoS ONE* **2014**, *9*, e111171. [CrossRef]
196. Yamamoto, S.; Shiraishi, S.; Suzuki, S. Are cyclic lipopeptides produced by *Bacillus amyloliquefaciens* S13-3 responsible for the plant defence response in strawberry against Colletotrichum gloeosporioides? *Lett. Appl. Microbiol.* **2015**, *60*, 379–386. [CrossRef] [PubMed]
197. Chen, D.; Liu, X.; Li, C.; Tian, W.; Shen, Q.; Shen, B. Isolation of *Bacillus amyloliquefaciens* S20 and its application in control of eggplant bacterial wilt. *J. Environ. Manag.* **2014**, *137*, 120–127. [CrossRef]
198. Saechow, S.; Thammasittirong, A.; Kittakoop, P.; Prachya, S.; Thammasittirong, S.N.-R. Antagonistic Activity against Dirty Panicle Rice Fungal Pathogens and Plant Growth-Promoting Activity of *Bacillus amyloliquefaciens* BAS23. *J. Microbiol. Biotechnol.* **2018**, *28*, 1527–1535. [CrossRef]
199. Han, Y.; Zhang, B.; Shen, Q.; You, C.; Yu, Y.; Li, P.; Shang, Q. Purification and Identification of Two Antifungal Cyclic Peptides Produced by *Bacillus amyloliquefaciens* L-H15. *Appl. Biochem. Biotechnol.* **2015**, *176*, 2202–2212. [CrossRef] [PubMed]
200. Calvo, H.; Mendiara, I.; Arias, E.; Blanco, D.; Venturini, M. The role of iturin A from B. amyloliquefaciens BUZ-14 in the inhibition of the most common postharvest fruit rots. *Food Microbiol.* **2019**, *82*, 62–69. [CrossRef]
201. Zhang, Z.; Ding, Z.; Zhong, J.; Zhou, J.; Shu, D.; Luo, D.; Yang, J.; Tan, H. Improvement of iturin A production in *Bacillus subtilis* ZK0 by overexpression of the comA and sigA genes. *Lett. Appl. Microbiol.* **2017**, *64*, 452–458. [CrossRef]
202. Athukorala, S.N.; Fernando, W.G.; Rashid, K.Y. Identification of antifungal antibiotics of *Bacillus species* isolated from different microhabitats using polymerase chain reaction and MALDI-TOF mass spectrometry. *Can. J. Microbiol.* **2009**, *55*, 1021–1032. [CrossRef]
203. Shi, J.; Zhu, X.; Lu, Y.; Zhao, H.; Lu, F.; Lu, Z. Improving Iturin A Production of *Bacillus amyloliquefaciens* by Genome Shuffling and Its Inhibition Against Saccharomyces cerevisiae in Orange Juice. *Front. Microbiol.* **2018**, *9*, 2683. [CrossRef]
204. Kaushal, M.; Kumar, A.; Kaushal, R. *Bacillus pumilus* strain YSPMK11 as plant growth promoter and bicontrol agent against *Sclerotinia sclerotiorum*. *3 Biotech* **2017**, *7*, 90. [CrossRef]
205. Moryl, M.; Spętana, M.; Dziubek, K.; Paraszkiewicz, K.; Różalska, S.; Płaza, G.A.; Rozalski, A. Antimicrobial, antiadhesive and antibiofilm potential of lipopeptides synthesised by *Bacillus subtilis*, on uropathogenic bacteria. *Acta Biochim. Pol.* **2015**, *62*, 725–732. [CrossRef] [PubMed]
206. Habe, H.; Taira, T.; Imura, T. Screening of a *Bacillus subtilis* Strain Producing Multiple Types of Cyclic Lipopeptides and Evaluation of Their Surface-tension-lowering Activities. *J. Oleo Sci.* **2017**, *66*, 785–790. [CrossRef] [PubMed]
207. Xu, B.-H.; Ye, Z.-W.; Zheng, Q.-W.; Wei, T.; Lin, J.-F.; Guo, L.-Q. Isolation and characterization of cyclic lipopeptides with broad-spectrum antimicrobial activity from Bacillus siamensis JFL15. *3 Biotech* **2018**, *8*, 444. [CrossRef]
208. Ambrico, A.; Trupo, M.; Magarelli, R.A. Influence of Phenotypic Dissociation in *Bacillus subtilis* Strain ET-1 on Iturin A Production. *Curr. Microbiol.* **2019**, *76*, 1487–1494. [CrossRef] [PubMed]
209. Lee, T.; Park, D.; Kim, K.; Lim, S.M.; Yu, N.H.; Kim, S.; Kim, H.-Y.; Jung, K.S.; Jang, J.Y.; Park, J.-C.; et al. Characterization of *Bacillus amyloliquefaciens* DA12 Showing Potent Antifungal Activity against *Mycotoxigenic Fusarium Species*. *Plant Pathol. J.* **2017**, *33*, 499–507. [CrossRef] [PubMed]
210. Chen, W.; Li, X.; Ma, X.; Chen, S.; Kang, Y.; Yang, M.; Huang, F.; Wan, X. Simultaneous hydrolysis with lipase and fermentation of rapeseed cake for iturin A production by *Bacillus amyloliquefaciens* CX-20. *BMC Biotechnol.* **2019**, *19*, 98. [CrossRef]
211. Lin, C.; Tsai, C.-H.; Chen, P.-Y.; Wu, C.-Y.; Chang, Y.-L.; Yang, Y.-L.; Chen, Y.-L. Biological control of potato common scab by *Bacillus amyloliquefaciens* Ba01. *PLoS ONE* **2018**, *13*, e0196520. [CrossRef]
212. Waewthongrak, W.; Leelasuphakul, W.; Mccollum, G. Cyclic Lipopeptides from *Bacillus subtilis* ABS–S14 Elicit Defense-Related Gene Expression in Citrus Fruit. *PLoS ONE* **2014**, *9*, e109386. [CrossRef]
213. Wu, G.; Liu, Y.; Xu, Y.; Zhang, G.; Shen, Q.-R.; Zhang, R. Exploring Elicitors of the Beneficial Rhizobacterium *Bacillus amyloliquefaciens* SQR9 to Induce Plant Systemic Resistance and Their Interactions with Plant Signaling Pathways. *Mol. Plant-Microbe Interactions* **2018**, *31*, 560–567. [CrossRef] [PubMed]
214. Zhao, Z.; Wang, Q.; Wang, K.; Brian, K.; Liu, C.; Gu, Y. Study of the antifungal activity of Bacillus vallismortis ZZ185 in vitro and identification of its antifungal components. *Bioresour. Technol.* **2010**, *101*, 292–297. [CrossRef]
215. Mácha, H.; Marešová, H.; Juříková, T.; Švecová, M.; Benada, O.; Škríba, A.; Baránek, M.; Novotný, Č.; Palyzová, A. Killing Effect of *Bacillus velezensis* FZB42 on a *Xanthomonas campestris* pv. *Campestris* (Xcc) Strain Newly Isolated from Cabbage *Brassica oleracea* Convar. *Capitata* (L.): A Metabolomic Study. *Microorganisms* **2021**, *9*, 1410. [CrossRef]
216. Elkahoui, S.; Djébali, N.; Karkouch, I.; Ibrahim, A.H.; Kalai, L.; Bachkovel, S.; Tabbene, O.; Limam, F. Mass spectrometry identification of antifungal lipopeptides from Bacillus sp. BCLRB2 against Rhizoctonia solani and *Sclerotinia* sclerotiorum. *Microb. Pathog.* **2014**, *50*, 184–188. [CrossRef]

217. Nam, J.; Alam, S.T.; Kang, K.; Choi, J.; Seo, M.-H. Anti-staphylococcal activity of a cyclic lipopeptide, C15-bacillomycin D, produced by *Bacillus velezensis* NST. *J. Appl. Microbiol.* **2021**, *131*, 93–104. [CrossRef] [PubMed]
218. Soussi, S.; Essid, R.; Hardouin, J.; Gharbi, D.; Elkahoui, S.; Tabbene, O.; Cosette, P.; Jouenne, T.; Limam, F. Utilization of Grape Seed Flour for Antimicrobial Lipopeptide Production by *Bacillus amyloliquefaciens* C5 Strain. *Appl. Biochem. Biotechnol.* **2019**, *187*, 1460–1474. [CrossRef] [PubMed]
219. Radovanovic, N.; Milutinović, M.; Mihajlovski, K.; Jović, J.; Nastasijević, B.; Rajilic-Stojanovic, M.; Dimitrijević-Branković, S. Biocontrol and plant stimulating potential of novel strain *Bacillus* sp. PPM3 isolated from marine sediment. *Microb. Pathog.* **2018**, *120*, 71–78. [CrossRef] [PubMed]
220. Walton, R.B.; Woodruff, H.B. A crystalline antifungal agent, mycosubtilin, isolated from subtilin broth. *J. Clin. Investig.* **1949**, *28*, 924–926. [CrossRef]
221. Fickers, P.; Guez, J.-S.; Damblon, C.; Leclère, V.; Béchet, M.; Jacques, P.; Joris, B. High-Level Biosynthesis of the Anteiso-C 17 Isoform of the Antibiotic Mycosubtilin in *Bacillus subtilis* and Characterization of Its Candidacidal Activity. *Appl. Environ. Microbiol.* **2009**, *75*, 4636–4640. [CrossRef]
222. Chevanet, C.; Besson, F.; Michel, G. Effect of various growth conditions on spore formation and *Bacillomycin* L production in *Bacillus subtilis*. *Can. J. Microbiol.* **1986**, *32*, 254–258. [CrossRef] [PubMed]
223. Li, X.; Zhang, Y.; Wei, Z.; Guan, Z.; Cai, Y.; Liao, X. Antifungal Activity of Isolated *Bacillus amyloliquefaciens* SYBC H47 for the Biocontrol of Peach Gummosis. *PLoS ONE* **2016**, *11*, e0162125. [CrossRef]
224. Sengupta, S.; Bose, S. Properties and localisation of mycobacillin-synthesising enzyme system in *Bacillus subtilis* B3. *Biochim. et Biophys. Acta (BBA)-Gen. Subj.* **1971**, *237*, 120–122. [CrossRef]
225. Velho, R.V.; Caldas, D.G.G.; Medina, L.F.C.; Tsai, S.M.; Brandelli, A. Real-time PCR investigation on the expression of sboA and ituD genes in *Bacillus* spp. *Lett. Appl. Microbiol.* **2011**, *52*, 660–666. [CrossRef] [PubMed]
226. Liu, X.; Lee, J.Y.; Jeong, S.-J.; Cho, K.M.; Kim, G.M.; Shin, J.-H.; Kim, J.-S.; Kim, J.H. Properties of a Bacteriocin Produced by *Bacillus subtilis* EMD4 Isolated from Ganjang (Soy Sauce). *J. Microbiol. Biotechnol.* **2015**, *25*, 1493–1501. [CrossRef]
227. Sutyak, K.; Wirawan, R.; Aroutcheva, A.; Chikindas, M. Isolation of the *Bacillus subtilis* antimicrobial peptide subtilosin from the dairy product-derived *Bacillus amyloliquefaciens*. *J. Appl. Microbiol.* **2008**, *104*, 1067–1074. [CrossRef] [PubMed]
228. Liu, Q.; Gao, G.; Xu, H.; Qiao, M. Identification of the bacteriocin subtilosin A and loss of purL results in its high-level production in *Bacillus amyloliquefaciens*. *Res. Microbiol.* **2012**, *163*, 470–478. [CrossRef]
229. Parveen Rani, R.; Anandharaj, M.; Hema, S.; Deepika, R.; David Ravindran, A. Purification of Antilisterial Peptide (*Subtilosin* A) from Novel Bacillus tequilensis FR9 and Demonstrate Their Pathogen Invasion Protection Ability Using Human Carcinoma Cell Line. *Front Microbiol.* **2016**, *7*, 1910. [CrossRef]
230. Liu, H.; Yin, S.; An, L.; Zhang, G.; Cheng, H.; Xi, Y.; Cui, G.; Zhang, F.; Zhang, L. Complete genome sequence of *Bacillus subtilis* BSD-2, a microbial germicide isolated from cultivated cotton. *J. Biotechnol.* **2016**, *230*, 26–27. [CrossRef]
231. Li, Q.; Liao, S.; Wei, J.; Xing, D.; Xiao, Y.; Yang, Q. Isolation of *Bacillus subtilis* strain SEM-2 from silkworm excrement and characterisation of its antagonistic effect against *Fusarium* spp. *Can. J. Microbiol.* **2020**, *66*, 401–412. [CrossRef]
232. Chen, L.; Shi, H.; Heng, J.; Wang, D.; Bian, K. Antimicrobial, plant growth-promoting and genomic properties of the peanut endophyte Bacillus velezensis LDO2. *Microbiol. Res.* **2019**, *218*, 41–48. [CrossRef] [PubMed]
233. Gao, X.-Y.; Liu, Y.; Miao, L.-L.; Li, E.-W.; Sun, G.-X.; Liu, Z.-P. Characterization and mechanism of anti-*Aeromonas salmonicida* activity of a marine probiotic strain, *Bacillus velezensis* V4. *Appl. Microbiol. Biotechnol.* **2017**, *101*, 3759–3768. [CrossRef]
234. Sharma, D.; Singh, S.S.; Baindara, P.; Sharma, S.; Khatri, N.; Grover, V.; Patil, P.B.; Korpole, S. Surfactin Like Broad Spectrum Antimicrobial Lipopeptide Co-produced With Sublancin From *Bacillus subtilis* Strain A52: Dual Reservoir of Bioactives. *Front. Microbiol.* **2020**, *11*, 1167. [CrossRef] [PubMed]
235. Li, Y.; Xu, Y.; Liu, L.; Han, Z.; Lai, P.Y.; Guo, X.; Zhang, X.; Lin, W.; Qian, P.-Y. Five New Amicoumacins Isolated from a Marine-Derived Bacterium *Bacillus subtilis*. *Mar. Drugs* **2012**, *10*, 319–328. [CrossRef]
236. Park, H.B.; Perez, C.E.; Perry, E.; Crawford, J.M. Activating and Attenuating the Amicoumacin Antibiotics. *Molecules* **2016**, *21*, 824. [CrossRef] [PubMed]
237. Itoh, J.; Shomura, T.; Omoto, S.; Miyado, S.; Yuda, Y.; Shibata, U.; Inouye, S. Isolation, Physicochemical Properties and Biological Activities of Amicoumacins Produced by *Bacillus pumilus*. *Agric. Biol. Chem.* **1982**, *46*, 1255–1259. [CrossRef]
238. Wang, D.; Li, J.; Zhu, G.; Zhao, K.; Jiang, W.; Li, H.; Wang, W.; Kumar, V.; Dong, S.; Zhu, W.; et al. Mechanism of the Potential Therapeutic Candidate *Bacillus subtilis* BSXE-1601 Against Shrimp Pathogenic Vibrios and Multifunctional Metabolites Biosynthetic Capability of the Strain as Predicted by Genome Analysis. *Front. Microbiol.* **2020**, *11*, 581802. [CrossRef] [PubMed]
239. Liao, R.; Duan, L.; Lei, C.; Pan, H.; Ding, Y.; Zhang, Q.; Chen, D.; Shen, B.; Yu, Y.; Liu, W. Thiopeptide Biosynthesis Featuring Ribosomally Synthesized Precursor Peptides and Conserved Posttranslational Modifications. *Chem. Biol.* **2009**, *16*, 141–147. [CrossRef]
240. Akasapu, S.; Hinds, A.B.; Powell, W.C.; Walczak, M.A. Total synthesis of micrococcin P1 and thiocillin I enabled by Mo(vi) catalyst. *Chem. Sci.* **2019**, *10*, 1971–1975. [CrossRef]
241. Villa-Rodriguez, E.; Moreno-Ulloa, A.; Castro-Longoria, E.; Parra-Cota, F.I.; Santos-Villalobos, S.D.L. Integrated omics approaches for deciphering antifungal metabolites produced by a novel Bacillus species, B. cabrialesii TE3T, against the spot blotch disease of wheat (*Triticum turgidum* L. subsp. durum). *Microbiol. Res.* **2021**, *251*, 126826. [CrossRef]

242. Kino, K.; Kotanaka, Y.; Arai, T.; Yagasaki, M. A novel L-amino acid ligase from *Bacillus subtilis* NBRC3134, a microorganism producing peptide-antibiotic rhizocticin. *Biosci. Biotechnol. Biochem.* **2009**, *73*, 901–907. [CrossRef]
243. Patel, P.S.; Huang, S.; Fisher, S.; Pirnik, D.; Aklonis, C.; Dean, L.; Meyers, E.; Fernandes, P.; Mayerl, F. Bacillaene, a Novel Inhibitor of Procaryotic Protein Synthesis Produced by *Bacillus subtilis*: Production, Taxonomy, Isolation, Physico-chemical Characterization and Biological Activity. *J. Antibiot.* **1995**, *48*, 997–1003. [CrossRef] [PubMed]
244. Lv, J.; Da, R.; Cheng, Y.; Tuo, X.; Wei, J.; Jiang, K.; Monisayo, A.O.; Han, B. Mechanism of Antibacterial Activity of *Bacillus amyloliquefaciens* C-1 Lipopeptide toward Anaerobic *Clostridium difficile*. *BioMed Res. Int.* **2020**, *2020*, 3104613. [CrossRef] [PubMed]
245. Chen, K.; Tian, Z.; Luo, Y.; Cheng, Y.; Long, C.-A. Antagonistic Activity and the Mechanism of *Bacillus amyloliquefaciens* DH-4 Against Citrus Green Mold. *Phytopathology* **2018**, *108*, 1253–1262. [CrossRef] [PubMed]
246. Nonejuie, P.; Trial, R.M.; Newton, G.L.; Lamsa, A.; Perera, V.R.; Aguilar, J.; Liu, W.-T.; Dorrestein, P.C.; Pogliano, J.; Pogliano, K. Application of bacterial cytological profiling to crude natural product extracts reveals the antibacterial arsenal of *Bacillus subtilis*. *J. Antibiot.* **2016**, *69*, 353–361. [CrossRef]
247. Daas, M.S.; Acedo, J.; Rosana, A.R.; Orata, F.; Reiz, B.; Zheng, J.; Nateche, F.; Case, R.J.; Kebbouche-Gana, S.; Vederas, J.C. *Bacillus amyloliquefaciens* ssp. plantarum F11 isolated from Algerian salty lake as a source of biosurfactants and bioactive lipopeptides. *FEMS Microbiol. Lett.* **2018**, *365*, fnx248. [CrossRef] [PubMed]
248. Butcher, R.A.; Schroeder, F.C.; Fischbach, M.A.; Straight, P.D.; Kolter, R.; Walsh, C.T.; Clardy, J. The identification of bacillaene, the product of the PksX megacomplex in *Bacillus subtilis*. *Proc. Natl. Acad. Sci. USA* **2007**, *104*, 1506–1509. [CrossRef]
249. Dimopoulou, A.; Theologidis, I.; Benaki, D.; Koukounia, M.; Zervakou, A.; Tzima, A.; Diallinas, G.; Hatzinikolaou, D.G.; Skandalis, N. Direct Antibiotic Activity of Bacillibactin Broadens the Biocontrol Range of *Bacillus amyloliquefaciens* MBI600. *mSphere* **2021**, *6*, e0037621. [CrossRef]
250. Peters, W.J.; Warren, R.A.J. Itoic Acid Synthesis in *Bacillus subtilis*. *J. Bacteriol.* **1968**, *95*, 360–366. [CrossRef]
251. Ollinger, J.; Song, K.-B.; Antelmann, H.; Hecker, M.; Helmann, J.D. Role of the Fur Regulon in Iron Transport in *Bacillus subtilis*. *J. Bacteriol.* **2006**, *188*, 3664–3673. [CrossRef]

MDPI
St. Alban-Anlage 66
4052 Basel
Switzerland
www.mdpi.com

Antibiotics Editorial Office
E-mail: antibiotics@mdpi.com
www.mdpi.com/journal/antibiotics

www.ingramcontent.com/pod-product-compliance
Lightning Source LLC
Chambersburg PA
CBHW061456210926
51850CB00008B/1731
* 9 7 8 3 0 3 6 5 9 1 0 7 0 *